Vector-Borne and Zoonotic Diseases in Dogs and Cats

Vector-Borne and Zoonotic Diseases in Dogs and Cats

Guest Editors

Antonio Ortega-Pacheco
Matilde Jimenez-Coello

MDPI • Basel • Beijing • Wuhan • Barcelona • Belgrade • Novi Sad • Cluj • Manchester

Guest Editors
Antonio Ortega-Pacheco
Department of Animal Health
and Preventive Medicine
Autonomous University of
Yucatan
Merida
Mexico

Matilde Jimenez-Coello
Laboratory of Microbiology
Autonomous University of
Yucatan
Merida
Mexico

Editorial Office
MDPI AG
Grosspeteranlage 5
4052 Basel, Switzerland

This is a reprint of the Special Issue, published open access by the journal *Animals* (ISSN 2076-2615), freely accessible at: https://www.mdpi.com/journal/animals/special_issues/6HJ8LD8518.

For citation purposes, cite each article independently as indicated on the article page online and as indicated below:

Lastname, A.A.; Lastname, B.B. Article Title. *Journal Name* **Year**, *Volume Number*, Page Range.

ISBN 978-3-7258-4611-5 (Hbk)
ISBN 978-3-7258-4612-2 (PDF)
https://doi.org/10.3390/books978-3-7258-4612-2

Contents

About the Editors

Antonio Ortega-Pacheco

Antonio Ortega-Pacheco currently works at the Department of Animal Health and Preventive Medicine at the Faculty of Veterinary Medicine, Autonomous University of Yucatan, Mexico. He is a DVM with an MVSc and PhD in Veterinary Medicine and has been teaching and performing research studies since 1994. His research interests are focused on infectious zoonotic diseases from domestic animals, particularly in dogs and cats, looking at the epidemiology and control strategies from the One Health perspective. He is the author and co-author of several research papers and book chapters (ORCID: 0000-0002-8502-8023).

Matilde Jiménez-Coello

Matilde Jiménez-Coello holds a bachelor's degree in Veterinary Medicine, as well as an MSc and PhD in Veterinary Epidemiology and Infectious Diseases. She has been awarded the L'Oréal-UNESCO prize for Women in Science and has been a director of the Research Center of the Autonomous University of Yucatan (CIR-UADY "Dr. Hideyo Noguchi"). Currently, she is working in the microbiology laboratory at the CIR-UADY. Her research interests are the epidemiology of zoonotic diseases in domestic animals under tropical conditions from a One health perspective. She has published several research papers and book chapters (ORCID: 0000-0002-8802-2049).

animals

Editorial

Vector-Borne and Zoonotic Diseases in Dogs and Cats

Antonio Ortega-Pacheco [1,*] and Matilde Jimenez-Coello [2]

[1] Department of Animal Health and Preventive Medicine, Faculty of Veterinary Medicine, Autonomous University of Yucatan, Merida 97000, Yucatan, Mexico
[2] Regional Research Center "Dr. Hideyo Noguchi", Biomedical Unit, Autonomous University of Yucatan, Merida 97000, Yucatan, Mexico; mjcoello@correo.uady.mx
* Correspondence: opacheco@correo.uady.mx

Vector-borne and zoonotic diseases (VBZDs) remain some of the most dynamic and complex challenges in veterinary medicine, public health, and environmental sciences. Companion animals, particularly dogs and cats, have long been recognized as reservoirs of pathogens of zoonotic concern. Given the rapidly changing ecological, climatic, and socioeconomic conditions, understanding the shifting patterns of these diseases is imperative. This comprehensive volume brings together 11 peer-reviewed research articles originally published in a Special Issue of *Animals*, now consolidated into a thematic book offering a robust, multidisciplinary perspective on VBZDs in dogs and cats from a One Health perspective.

As we witness an unprecedented expansion in the distribution of vector species due to climate change, urbanization, deforestation, and globalization, the threat of emerging and re-emerging zoonoses grows. The One Health paradigm, which emphasizes the interconnectedness of human, animal, and environmental health, is the most suitable framework for addressing these complex dynamics [1,2]. Companion animals are increasingly central to these discussions, particularly dogs and cats, as their proximity to humans, vulnerability to vector-borne infections, and roles in transmission cycles render them both victims and facilitators of disease spread [3].

Leishmaniasis is a vector-borne parasitic disease of major concern in small animals and humans. Control of this infectious agent is crucial, and the search for candidate vaccines opens a wide area of research. This book opens with a study focusing on an innovative multi-epitope peptide vaccine candidate delivered via PLGA nanoparticles, which demonstrated promising immunogenic and protective effects in murine models. This research not only illustrates the advances in veterinary immunoprophylaxis but also offers hope for scalable, cost-effective preventive tools that could be adapted for field use. Complementing this, clinical evaluation of meglumine antimoniate in dogs with naturally acquired *Leishmania infantum* infection, despite potential adverse reactions, can achieve significant reductions in parasitic load and symptomatology, reinforcing its role in therapeutic protocols while underscoring the need for improved, less toxic alternatives. These studies are critical in regions where leishmaniasis remains endemic and where co-infection with other pathogens complicates clinical management.

Heartworm disease caused by *Dirofilaria immitis* is a parasitic cardiopulmonary infection with implications for both canine and feline health. This agent is widely distributed worldwide but is considered hyperendemic in some areas, such as the Canary Islands [4]. Using radiographic thoracic indices demonstrates that specific imaging markers can predict pulmonary hypertension, a serious sequela of advanced heartworm disease. Meanwhile, radiographic evaluations in cats highlight subtle *D. immitis*-induced vascular changes that precede overt clinical signs, revealing opportunities for earlier diagnosis and intervention. In hyperendemic regions, it is crucial to control *D. immitis* in dogs, so this volume presents an important review on the use of long-acting moxidectin, demonstrating its efficacy and benefits when preventative medicine is a problem for owners.

Chagas disease (American trypanosomiasis) is transmitted by vectors and caused by the protozoan *Trypanosoma cruzi*, although it is commonly associated with humans. Another study confirms that dogs are capable of developing significant cardiomyopathy, detectable through echocardiographic parameters [5]. These findings are particularly relevant in Latin America, where peridomestic transmission involving dogs is key in maintaining the life cycle of *T. cruzi* [6]. Another critical zoonotic bacterial infection examined in this volume is leptospirosis, whose global burden remains underestimated [7]. A study presented here underscores the value of region-specific antigens in serological assays, which significantly improve the sensitivity and specificity of ELISA tests. This is particularly relevant given that *Leptospira* spp. are highly diverse, and local strains often differ antigenically from reference laboratory strains.

Protozoan and nematode zoonoses have notably been reported. One article revisits the epidemiology of *Cryptosporidium* spp. and *Giardia duedenalis* in dogs and cats, where human assemblage strains demonstrate their potential cause of zoonotic transmission [8]. There is compelling molecular evidence that *G. intestinalis* assemblage E is present in symptomatic children in rural settings, implicating animal-to-human transmission and challenging established epidemiological assumptions [9]. This aligns with calls for a re-evaluation of host specificity and better integration of molecular diagnostics in surveillance systems. *Toxocara* spp. are highly relevant nematodes distributed worldwide. Their main hosts are dogs (*T. cani*) and cats (*T. cati*) [10], but they also have significant zoonotic potential with severe clinical consequences [11]. The systematic review and meta-analysis of *T. cati* presented in this book [12] describes its widespread prevalence and highlights the necessity for actions to mitigate the clinical manifestations in cats and the risk of zoonosis.

Trematode infections, particularly *Opisthorchis viverrini* and *Clonorchis sinensis*, have been reviewed in light of emerging evidence from Vietnam and Southeast Asia [13,14]. These parasites are known for their carcinogenic potential, particularly in relation to cholangiocarcinoma. However, conventional copro-parasitological diagnostics often fail to distinguish between these liver flukes and minute intestinal trematodes. The reviewed data stress the necessity for robust diagnostic alternatives such as genetic markers and the expansion of molecular tools in endemic settings where dogs and cats may be involved in the epidemiology of these parasites [15].

This book serves not only as a repository of current scientific evidence but also as a tool for veterinary practitioners, epidemiologists, policymakers, and students engaged in the control of zoonotic and vector-borne diseases. Its relevance spans endemic and non-endemic regions alike, given the increasingly transboundary nature of infectious disease emergence. We are grateful to the contributing authors for their high-quality work and to the editorial team of *Animals* for supporting the consolidation of this Special Issue into a dedicated book. We hope that this compilation will inspire further crossdisciplinary collaborations and inform more effective, evidence-based interventions in the field of zoonoses and vector-borne diseases in companion animals.

Author Contributions: Conceptualization, A.O.-P. and M.J.-C.; writing—original draft preparation, A.O.-P.; writing—review and editing, A.O.-P. and M.J.-C. All authors have read and agreed to the published version of the manuscript.

Funding: This research received no external funding.

Conflicts of Interest: The authors declare no conflicts of interest.

References

1. Zinsstag, J.; Kaiser-Grolimund, A.; Heitz-Tokpa, K.; Sreedharan, R.; Lubroth, J.; Caya, F.; Stone, M.; Brown, H.; Bonfoh, B.; Dobell, E.; et al. Advancing one human–environmental–animal health for global health security: What does the evidence say? *Lancet* **2023**, *401*, 591–604. [CrossRef] [PubMed]
2. FAO; OIE; WHO. *Taking a Multisectoral, One Health Approach: A Tripartite Guide to Addressing Zoonotic Diseases in Countries*; World Health Organization: Geneva, Switzerland, 2019.
3. Rahman, M.T.; Sobur, M.A.; Islam, M.S.; Levy, S.; Hossain, M.J.; Zowalaty, M.E.E.; Rahman, A.T.; Ashour, H.M. Zoonotic diseases: Etiology, impact, and control. *Microorganisms* **2020**, *8*, 1405. [CrossRef] [PubMed]
4. Morchón, R.; Rodríguez-Escolar, I.; Lambraño, R.E.H.; Agudo, J.Á.S.; Montoya-Alonso, J.A.; Serafín-Pérez, I.; Fernández-Serafín, C.; Carretón, E. Assessment heartworm disease in the Canary Islands (Spain): Risk of transmission in a hyperendemic area by ecological niche modeling and its future projection. *Animals* **2023**, *13*, 3251. [CrossRef] [PubMed]
5. Avalos-Borges, E.E.; Acevedo-Arcique, C.M.; Segura-Correa, J.C.; Jiménez-Coello, M.; Garg, N.J.; Ortega-Pacheco, A. Echocardiographic documentation of dilated cardiomyopathy development in dogs naturally infected with *Trypanosoma cruzi*. *Animals* **2024**, *14*, 1884. [CrossRef] [PubMed]
6. Cucunubá, Z.M.; Gutiérrez-Romero, S.A.; Ramírez, J.D.; Velásquez-Ortiz, N.; Ceccarelli, S.; Parra-Henao, G.; Henao-Martínez, A.F.; Rabinovich, J.; Basáñez, M.G.; Nouvellet, P.; et al. The epidemiology of Chagas disease in the Americas. *Lancet Regi Health Am.* **2024**, *37*, 100881. [CrossRef] [PubMed]
7. Costa, F.; Hagan, J.E.; Calcagno, J.; Kane, M.; Torgerson, P.; Martinez-Silveira, M.S.; Stein, C.; Abela-Ridder, B.; Ko, A.I. Global morbidity and mortality of Leptospirosis: A systematic review. *PLoS Negl. Trop. Dis.* **2015**, *9*, e0003898. [CrossRef] [PubMed]
8. Hsu, C.H.; Liang, C.; Chi, S.C.; Lee, K.J.; Chou, C.H.; Lin, C.S.; Yang, W.Y. An Epidemiological assessment of *Cryptosporidium* and *Giardia* spp. infection in pet animals from Taiwan. *Animals* **2023**, *13*, 3373. [CrossRef] [PubMed]
9. Abdel-Moein, K.A.; Saeed, H. The zoonotic potential of *Giardia intestinalis* assemblage E in rural settings. *Parasitol. Res.* **2016**, *115*, 3197–31202. [CrossRef] [PubMed]
10. Keegan, J.D.; Airs, P.M.; Brown, C.; Dingley, A.R.; Courtney, C.; Morgan, E.R.; Celia V Holland, C.V. Park entrances, commonly contaminated with infective *Toxocara canis* eggs, present a risk of zoonotic infection and an opportunity for focused intervention. *PLoS Negl. Trop. Dis.* **2025**, *19*, e0012917. [CrossRef] [PubMed]
11. El Meski, N.; El Ayoubi, L.W.; Hassani, S.; Bidikian, N.; Zakhour, J.; Khalil, J.B.; Kanafani, Z.A.; Kanj, S.S. Clinical Spectrum of Toxocariasis: A retrospective study from a tertiary care center in Lebanon. *Future Microbiol.* **2025**, *20*, 419–427. [CrossRef] [PubMed]
12. Bonilla-Aldana, J.L.; Espinosa-Nuñez, A.C.; Bonilla-Aldana, D.K.; Rodriguez-Morales, A.J. Toxocara cati Infection in cats (*Felis catus*): A systematic review and meta-analysis. *Animals* **2024**, *14*, 1022. [CrossRef] [PubMed]
13. Doanh, P.N.; Nawa, Y. *Clonorchis sinensis* and *Opisthorchis* spp. in Vietnam: Current status and prospects. *Trans. R. Soc. Trop Med. Hyg.* **2016**, *110*, 13–20. [CrossRef] [PubMed]
14. Petney, T.N.; Andrews, R.H.; Saijuntha, W.; Wenz-Mücke, A.; Sithithaworn, P. The zoonotic, fish-borne liver flukes *Clonorchis sinensis*, *Opisthorchis felineus* and *Opisthorchis viverrini*. *Int. J. Parasitol.* **2013**, *43*, 1031–1046. [CrossRef] [PubMed]
15. Tantrawatpan, C.; Maleewong, W.; Thanchomnang, T.; Pilap, W.; Agatsuma, T.; Andrews, R.H.; Sithithaworn, P.; Saijuntha, W. Intron regions as genetic markers for population genetic investigations of *Opisthorchis viverrini* sensu lato and *Clonorchis sinensis*. *Animals* **2023**, *13*, 3200. [CrossRef] [PubMed]

animals

MDPI

Article

Radiological Evaluation of Vascular Structures in Cats Infected with Immature Worms of *Dirofilaria immitis*

Soraya Falcón-Cordón, Yaiza Falcón-Cordón, Sara Nieves García-Rodríguez, Noelia Costa-Rodríguez, Daniel Julio Vera-Rodríguez, José Alberto Montoya-Alonso * and Elena Carretón

Internal Medicine, Veterinary Medicine and Therapeutic Research Group, Faculty of Veterinary Medicine, Research Institute of Biomedical and Health Sciences (IUIBS), Universidad de Las Palmas de Gran Canaria (ULPGC), 35016 Las Palmas de Gran Canaria, Spain; soraya.falcon@ulpgc.es (S.F.-C.); yaiza.falcon@ulpgc.es (Y.F.-C.); saranieves.garcia@ulpgc.es (S.N.G.-R.); noelia.costa@ulpgc.es (N.C.-R.); daniel.vera103@alu.ulpgc.es (D.J.V.-R.); elena.carreton@ulpgc.es (E.C.)

* Correspondence: alberto.montoya@ulpgc.es

Simple Summary: This study aimed to identify thoracic radiographic abnormalities in cats infected with immature worms of *Dirofilaria immitis*. A total of 123 cats from a hyperendemic area were included and divided into healthy cats (n = 50), asymptomatic cats who were seropositive to *D. immitis* antibodies (n = 30), and seropositive cats with clinical signs (n = 43). Different radiographic measurements were assessed including the VHS and CrPA/R4, CdPA/R9, CVC/Ao, and CVC/R4 ratios. The results showed that significant differences were observed between healthy and infected cats for all except the VHS, demonstrating enlarged vasculature in cats with *D. immitis*. Moreover, 62.8% of the cats with clinical signs showed a marked bronchointerstitial pattern, while asymptomatic cats mainly (33.3%) had a mild bronchointerstitial pattern. This study highlights the importance of thoracic radiography in diagnosing and monitoring heartworm disease in cats.

Abstract: This study aimed to assess thoracic radiographic abnormalities in cats infected with immature stages of *Dirofilaria immitis* to evaluate the utility of this diagnostic technique during early infection. A total of 123 cats from a hyperendemic area were classified into three groups: asymptomatic cats seronegative to anti-*D.-immitis* antibodies (Group A), seropositive asymptomatic cats (Group B), and seropositive cats with clinical signs that were at high risk of heartworm-associated respiratory disease (HARD) (Group C). Radiographic measurements and lung parenchymal abnormalities were analyzed and compared across the groups. Significant differences in several parameters, including CrPA/R4, and CdPA/R9 ratios, were observed between healthy and seropositive cats, suggesting early arterial damage even in the absence of adult worms. Other parameters that showed differences between healthy and infected cats were CVC/Ao and CVC/R4 ratios, but not the VHS. Group C exhibited a marked bronchointerstitial pattern, indicating severe parenchymal alterations associated with clinical signs. The study demonstrated that thoracic radiography can detect early vascular and parenchymal changes in feline *D. immitis* infections, providing valuable information for diagnosing HARD. However, it also highlights the limitations of radiographic techniques, as some seropositive cats displayed no significant abnormalities. The findings underscore the importance of combining radiography with clinical and serological assessments for a more accurate diagnosis of feline heartworm disease.

Keywords: feline heartworm disease; imaging diagnosis; thoracic radiography; vector-borne disease; radiographic indexes; cardiac silhouette; vascular enlargement

1. Introduction

Dirofilaria immitis is a nematode parasite that causes heartworm disease. It has a cosmopolitan distribution and is considered endemic in the Canary Islands [1,2]. While cats can become infected, they are more resilient to infections with adult *D. immitis* worms compared to dogs [3]. The pathophysiology of feline heartworm is basically differentiated into two stages; the first stage is associated with the arrival of immature heartworms in the pulmonary vasculature, and the second stage is related to the presence and death of adult worms [4–6].

The first stage happens approximately 3–4 months post-infection, with the arrival of immature worms in the pulmonary arteries and arterioles and subsequent death of most of them, mainly due to the action of the intravascular alveolar macrophages. This reaction causes clinical signs due to an acute vascular and parenchymal inflammatory response [6,7]. These signs are mostly respiratory in nature and this symptomatic phase is referred to as heartworm-associated respiratory disease (HARD) [8,9]. Those larvae that manage to develop and reach adulthood cause the second stage of the disease. In general, cats have a low parasite burden and the longevity of the worms is relatively short [10].

Given the complicated diagnosis of feline heartworm, a combination of serological and imaging techniques is usually recommended to detect adult parasites. Feline heartworm disease is a dynamic disease, and all the diagnostic tests carried out should be performed and studied altogether to determine whether the animal is indeed infected by adult parasites or whether there is a high index of suspicion of infection [11–13].

Among the diagnostic techniques available, thoracic radiography is widely used for evaluating the pulmonary parenchyma and vascular structures in feline cardiopulmonary diseases [14]. In *D. immitis* infections, an enlargement of the main and peripheral pulmonary arteries has been described, characterized by the loss of the conical shape, tortuosity, and truncation of the caudal lobar branches. Additionally, parenchymal alterations are commonly observed, with diffuse or focal bronchointerstitial patterns detectable on radiographs [10,15–17]. However, these lesions are not patognomonic and are similar to the lesions found in other diseases, such as infections by *Toxocara cati* or *Aerostrongylus* spp., asthma, or allergic bronchitis [10,18–20]. In most cases, these alterations are accompanied by clinical signs such as coughing or dyspnea, among others [5,10,16]. In any case, a radiological study, used in combination with the clinical history, has proven essential for establishing a correct diagnosis [21,22].

Regarding studies related to HARD induced by immature *D. immitis* in cats, it has been reported that histopathological lesions were mainly focused on the main pulmonary artery and arterioles; this caused bronchointerstitial changes, such as bronchiolar lesions and partial obstruction of some primary bronchi, hyperplasia and hypertrophy of the muscular layer and medial hypertrophy of the small pulmonary arterioles, and interstitial lung disease [6,9,10,23]. Moreover, the studies reported that these alterations may result in pulmonary endarteritis, indicating that even transient infection can cause long-term lesions in cats [10,23] that may be detectable with thoracic radiography, which is the aim of the present study. However, other studies reported that the minor distinctions observed between infected and healthy cats indicated that clinical use of thoracic radiology was very limited or, in some cases, thoracic radiographs provided no evidence of infection in cats with immature infections [10,17]. Therefore, the aim of this study was to identify thoracic radiographic abnormalities in cats infected with immature stages of *D. immitis* to determine the utility of this diagnostic technique at this stage of infection.

2. Materials and Methods

2.1. Study Animals

A total of 123 rescued and client-owned cats brought to the Veterinary Teaching Hospital of the University of Las Palmas de Gran Canaria were included in the study. These cats lived in a hyperendemic area for *D. immitis* [1,2]. They were cats that participated in a feline heartworm screening campaign for cats over 7 months of age who never received heartworm chemoprophylaxis. Clinical history and data were recorded for each animal, including age, sex, and breed. All owners provided their consent for participation in this study. The study was conducted in accordance with current European legislation on anima protection.

The cats were further examined for the presence or absence of clinical signs related to feline heartworm, such as coughing, dyspnea, tachypnea, increased respiratory effort, vomiting, systolic heart murmur, anorexia and weight loss, ascites, syncope, or neurological signs.

In addition, each feline patient underwent a thorough medical history and physical examination to eliminate the possibility of other conditions that could impact the results; cats with concomitant diseases were not included in the study. Additionally, testing for feline immunodeficiency virus (FIV) and feline leukemia virus (FeLV) was also carried out and any cat that tested positive was excluded from the study.

Out of the 123 cats surveyed, 42.35% (52/123) were male while 57.65% (68/123) were female. There were 170 European Shorthair cats, 2 Sphynx cats, 1 Maine Coon cat, and 1 Turkish Angora cat included in the breed categorization.

2.2. Serology

Blood samples were taken from either the cephalic or jugular vein and spun in dry tubes to check their serological status. The serum was stored at a temperature of −20 °C until the tests were conducted. The presence of feline *D. immitis* infection was determined through serological methods to detect specific antibodies against *D. immitis* using an indirect enzyme-linked immunosorbent assay (ELISA) (in-house ELISA, Urano Vet®, Barcelona, Spain). In short, each well of the ELISA plate was covered with recombinant *D. immitis* antigens (Di33 protein, 0.5 ug/mL). The sample diluent was mixed with the serum in a 1:100 dilution ratio. Following an initial wash to eliminate unbound molecules, the TMB substrate, labeled with horseradish peroxidase, was introduced, targeting feline IgG specifically. The readings of absorbance (or optical density) were taken at 450 nm within a 5-min time frame following the introduction of the stop solution (sulfuric acid). As per the kit manufacturer's guidelines, seronegativity was defined at a cut-off of < 1, while seropositivity was defined at a cut-off of 1 or higher. Furthermore, all samples were tested for circulating *D. immitis* antigens using a commercial immunochromatographic test kit (Uranotest Dirofilaria ©, UranoVet SL, Barcelona, Spain) according to the manufacturer's instructions.

2.3. Imaging Diagnosis

Echocardiograms were conducted on every cat to confirm the presence or absence of adult worms, as well as to exclude any other concurrent diseases. The cats were positioned on their right side with the transducer positioned in the third intercostal space to check for worms in the pulmonary arteries and right-sided heart chambers. Cats remained conscious and were continuously monitored with electrocardiography throughout the entire test.

Thoracic radiographs from all cats were taken using the same radiographic equipment (Bennett HFQ-600P, Greensboro, NC, USA) during inspiration and without sedation to minimize changes in heart size [24]. Views were obtained in both right laterolateral and dorsoventral projections, and radiographic measurements were taken using adjustable calipers by an unbiased operator (SFC), who has 10 years of clinical experience in cardiorespiratory diseases in small animals and was blinded to the clinical status of the study cats (Figure 1).

In lateral recumbent radiographs, the Vertebral Heart Score (VHS) measurement was obtained from the sum of the short-axis and the long-axis measurements as previously described [25]. The cardiac long axis was obtained from the cardiac apex to the base of the heart where it meets the trachea just cranial to the carina, expressed as the number of vertebral lengths in the lateral radiograph, measured caudally from the cranial border of T4. The short axis of the heart was measured perpendicular to the long-axis measurement at the point of maximum heart width, expressed as the number of vertebral lengths in the lateral radiograph, measured caudally from the cranial border of T4 [26].

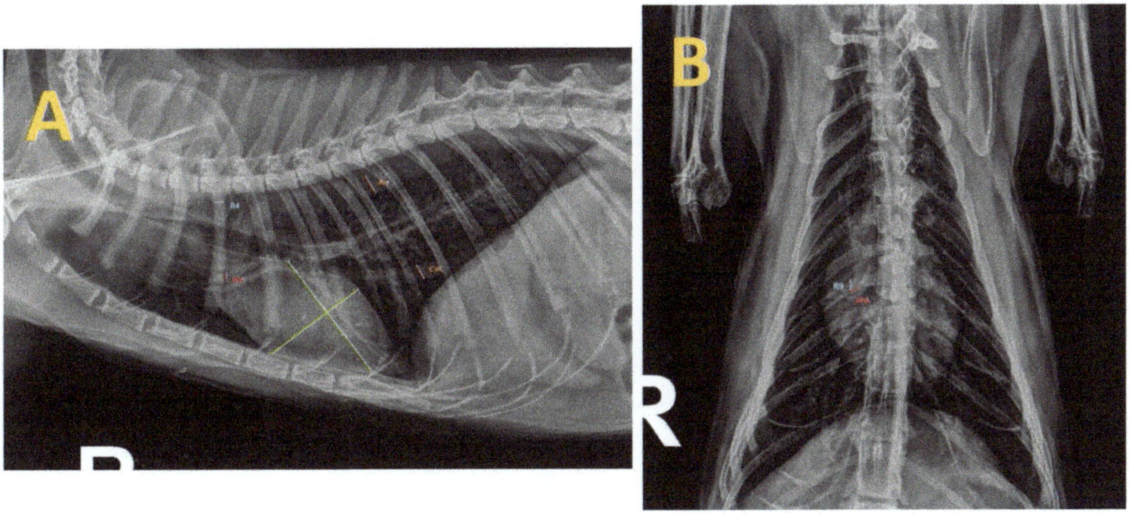

Figure 1. Thoracic radiographs of a cat seropositive for anti-Dirofilaria-immitis antibodies and presenting with clinical signs. The measurements taken during this study are shown as follows: (**A**) Right laterolateral projection illustrating the measurements of the caudal vena cava (CVC) and aorta (Ao) (in orange), the fourth rib (R4) (in blue), the right cranial pulmonary artery (CrPA) (in red) and the vertebral heart score (in green). (**B**) Dorsoventral projection displaying the measurement of the right caudal pulmonary artery (CdPA) (in red) in relation to the ninth rib (R9) (in blue).

In the laterolateral projection, the diameter of the right fourth rib (R4) just below the spine and the greatest diameter of the caudal vena cava (CVC) were also measured as described in previous studies, which included dogs with heartworm, in order to determine the mean caudal vena cava size, expressed as a ratio of the diameter of R4 [27–29]. The measurement of the diameter of the descending aorta (Ao) at the same intercostal space as the CVC was carried out as well, following previous guidelines in dogs [29]. Next, the CVC/Ao and CVC/R4 ratios were established.

Other radiological measurements were taken, including the diameter of the right cranial pulmonary artery (CrPA) passing through R4 in the laterolateral projection and the diameter of R4 at a point just distal to the spine. Moreover, in the dorsoventral projections, the distal and left sides of the summation shadow created by the right caudal pulmonary artery (CdPA) with R9 were measured. Finally, the CrPA/R4 and CdPA/R9 ratios were calculated from these measurements [30].

In addition, quantitative evaluation of the shape and tortuosity of the lung vasculature in both radiographic projections took place. Finally, the parenchyma of all radiographs was examined to determine the presence or absence of radiological abnormalities and their nature, classified as bronchial pattern, vascular pattern, interstitial pattern, alveolar pattern, or mixed patterns in all groups.

2.4. Statistical Analyses

The data were analyzed using SPSS Base 29.0 software for Windows. A Shapiro–Wilk test was performed to verify the normal distribution of the data. Additionally, a Siegel–Tukey test was performed to verify the variability of variances between groups. The chi-squared test was used to assess the association between categorical variables. A non-parametric Mann–Whitney U test was performed to determine differences between the groups for all recruited-cat measurements. In all cases, a p-value < 0.05 was determined as significant. Continuous variables were expressed as the median $\pm$ standard deviation, while qualitative variables were expressed as percentages. In all cases, a p-value < 0.05 was considered significant. In addition, Pearson's correlation coefficient was obtained to determine the relationship between variables. The strength of the correlations was categorized according to standard conventions: a correlation coefficient of $r \leq 0.30$ was considered weak, $0.31 \leq r \leq 0.50$ was classified as moderate, and $r > 0.50$ was regarded as strong.

3. Results

Based on the results, cats were divided into three groups: Group A (n = 50) consisted of cats with no clinical signs that were seronegative for anti-*D.-immitis* antibodies, Group B (n = 30) consisted of cats seropositive to *D. immitis* but who were asymptomatic, and Group C (n = 43) comprised seropositive animals with *D. immitis* exhibiting clinical signs included in the differential diagnosis of HARD. Antigen tests were negative in all cats included in the study. The list of clinical signs observed in the cats of Group C can be seen in Table 1.

Table 1. Summary of clinical signs observed in the study cats from Group C. Legend: n = number of cats showing the clinical sign described. Percentage (%) = percentage of cats in Group C showing the described clinical sign.

Clinical Sign	Number of Cats (n)	Percentage (%)
Cough	18	41.9%
Tachypnea	14	32.6%
Respiratory distress	23	53.5%
Vomiting	4	9.3%

The average age of the cats in the study was 4.60 ± 3.36 years, with Group A having an average age of 4.64 ± 3.33 years, Group B 4.27 ± 3.66 years, and Group C 4.67 ± 3.40 years. No significant statistical differences were found in age among the groups. The average weight of the cats was 3.72 ± 1.11 kg (3.51 ± 0.56 kg for cats from Group A, 3.86 ± 1.03 kg for cats from Group B, and 3.89 ± 1.57 kg cats from Group C). There were no significant differences in weight between the groups. During the echocardiographic study, adult parasites were not found in any of the cats.

No quantitative abnormalities in the shape or tortuosity of the pulmonary vasculature were found on any of the radiographs studied. The obtained radiographic measurements are shown in Table 2. A VHS value for Group A was determined as 6.43 ± 0.92 (with an upper limit of 7.3), with no significant differences between the groups in terms of the VHS value. However, 37.21% (10/43) of cats from Group C showed cardiomegaly based on established reference values [25].

For the CVC/Ao and CVC/R4 ratios, the results from Group A showed a mean value of 1.09 ± 0.12 (upper limit 1.22) and 1.62 ± 0.15 (upper limit 1.77), respectively, being significantly higher in the seropositive cats for both ratios (Table 2). Moreover, statistically significant differences were present between Groups B and C (p = 0.003 for CVC/Ao and p = 0.021 for CVC/R4).

The mean value for the CrPA/R4 ratio for Group A was 0.76 ± 0.05 (upper limit 0.81). Statistically significant differences were observed between healthy cats and the rest of the studied groups (p < 0.001); however, no significant differences were observed between Groups B and C for this parameter (Table 2).

For ventrodorsal projection, the obtained ratio for CdPA/R9 was 0.79 ± 0.05 (upper limit 0.84). Statistically significant differences were observed between Group A and the rest of the studied groups (p < 0.001). Also, a statistically significant difference was presented between Group B and Group C (p = 0.017) (Table 2).

To determine if there is a correlation between the parameters with age and weight, the Pearson correlation model was used. Initially, the results showed a low positive or negative correlation between radiographic measurements and age or weight, and only a moderate positive correlation was observed between CdPA/R9 and weight. Finally, a moderate positive correlation was determined between the CVC and VHS (Table 3).

Table 2. Results expressed by radiographic measurements and groups. Legend: VHS (Vertebral Heart Size); CrPA/R4 (right cranial pulmonary artery passing through the fourth rib in the laterolateral projection ratio); CVC/Ao (ratio of the caudal vena cava and diameter of the descending aorta in the laterolateral projection); CVC/R4 (caudal vena cava expressed as a ratio of the diameter of the fourth rib in the laterolateral projection); CdPA/R9 (right caudal pulmonary artery to the ninth rib in the dorsoventral projection ratio). Group A: seronegative asymptomatic cats; Group B: asymptomatic cats seropositive to anti-*Dirofilaria-immitis* antibodies; Group C: cats with clinical signs that were seropositive to *D. immitis*. Results are expressed as mean ± standard deviation.

Measure	Groups	Results	*p*-Value	R Effect	Interpretation
VHS	**Group A**	6.43 ± 0.92	0.27^{ns}	0.05472692	No statistically significant differences between seronegative and seropositive cats
	Group B **Group C**	5.98 ± 1.04 6.61 ± 1.62	0.41655099^{ns}	0.02500844	No statistically significant differences between Groups B and C
CrPA/R4	**Group A**	0.76 ± 0.05	3.0545×10^{-21} **	0.74419009	Statistically significant difference between seronegative and seropositive cats
	Group B **Group C**	1.12 ± 0.22 1.17 ± 0.19	0.41783262^{ns}	0.02427917	No statistically significant differences between Groups B and C
CVC/Ao	**Group A**	1.09 ± 0.12	5.02886×10^{-7} **	0.44459144	Statistically significant difference between seronegative and seropositive cats
	Group B **Group C**	1.16 ± 0.14 1.35 ± 0.25	0.00349851 **	0.31565838	Statistically significant difference between Groups B and C
CVC/R4	**Group A**	1.62 ± 0.15	1.2276×10^{-21} **	0.74896233	Statistically significant difference between seronegative and seropositive cats
	Group B **Group C**	2.41 ± 0.61 2.62 ± 0.47	0.021834542 *	0.23490465	Statistically significant difference between Groups B and C
CdPA/R9	**Group A**	0.79 ± 0.05	8.7566×10^{-22} **	0.75130749	Statistically significant difference between seronegative and seropositive cats
	Group B **Group C**	1.23 ± 0.31 1.36 ± 0.34	0.017873607 *	0.24919937	Statistically significant difference between Groups B and C

**, Correlation is significant at 0.5% ($p < 0.005$); *, correlation is significant at 5% level ($p < 0.05$); ns, correlation is not significant.

Table 3. Correlation coefficient for all studied parameters with weight and age values, as well as correlation between CVC and VHS.

Correlation	Coefficient	Interpretation
VHS–Weight	0.26410673 **	Low positive correlation
VHS–Age	0.2625266 *	Low positive correlation
CrPA/R4–Weight	0.24220781 *	Low positive correlation
CrPA/R4–Age	-0.02621345^{ns}	Low negative correlation
CVV/Ao–Weight	0.27288863 **	Low positive correlation
CVV/Ao–Age	-0.06015906^{ns}	Low negative correlation
CVV/R4–Weight	−0.0537349	Low negative correlation
CVV/R4–Age	-0.1179146^{ns}	Low negative correlation
CdPA/R9–Weight	0.41493029 **	Moderate positive correlation
CdPA/R9–Age	0.0134005^{ns}	Low positive correlation
VCC–VHS	0.4903414 **	Moderate positive correlation

**, Correlation is significant at 0.5% ($p < 0.005$); *, correlation is significant at 1% level ($p < 0.01$); ns, correlation is not significant.

Regarding the obtained results for pulmonary patterns, Group A cats did not show any lung abnormalities. The results pertaining to cats in Groups B and C showed various lung parenchymal abnormalities including bronchial, mild and marked bronchointerstitial, and alveolar patterns (Table 4). The findings indicated that the predominant pulmonary pattern in Group B was a mild bronchointerstitial pattern (33.3%; 10/30), while in Group C, a marked bronchointerstitial pattern was observed in 62.8% (27/43) of cats. On the other hand, 30% (9/30) of cats in Group B showed no radiological abnormalities at the level of the lung parenchyma. A chi-squared test was performed to confirm the correlation between the presence/absence of clinical signs and the presence/absence of lung parenchymal abnormalities in cats from Groups B and C, irrespective of the type of radiological lung abnormality and its severity. A strong correlation was found between the presence of clinical signs and the presence of lung parenchymal abnormalities, with a statistically significant difference ($p < 0.001$) (Table 4).

Table 4. Lung parenchymal abnormalities observed in the studied cats. Group B: asymptomatic cats seropositive to anti-*Dirofilaria-immitis* antibodies; Group C: cats with clinical signs that were seropositive to *D. immitis*. Results are expressed as percentage (%) as well as number of cats exhibiting each lung pattern by group.

Lung Parenchymal Abnormalities	Group B (n = 30)	Group C (n = 43)	Groups B + C (n = 73)
Bronchial pattern	20% (6/30)	4.6% (2/43)	10.9% (8/73)
Bronchointerstitial pattern (mild)	33.3% (10/33)	13.9% (6/43)	21.9% (16/73)
Bronchointerstitial pattern (marked)	16.7% (5/30)	62.8% (27/43)	43.8% (32/73)
Alveolar + interstitial pattern	0% (0/30)	18.6% (8/43)	10.9% (8/73)
Total	70% (21/30)	100% (43/43)	87.6% (64/73)

4. Discussion

The diagnosis of heartworm infection in cats is far more complex than in dogs due to the specific characteristics of the feline host, such as a low parasite load [3,10]. In the case of juvenile or pre-adult worm infections, diagnosis is virtually impossible and therefore very complicated. Thus, the objective identification of compatible radiographic changes may be useful as an indicator of a high suspicion of HARD in infected cats. In infections with adult parasites, thoracic radiography is a valuable diagnostic and monitoring tool for the diagnosis of feline heartworm disease [26,31]. However, it has never been objectively assessed in cats with high suspicion of HARD. Radiographic abnormalities may be less consistent in feline heartworm disease than in canine heartworm disease, and the absence of such abnormalities does not exclude the diagnosis of heartworm disease in cats [3,12,32]. Therefore, this study was undertaken to evaluate whether specific and objective radiographic features of the heart and pulmonary vasculature could aid in the diagnosis of HARD.

The VHS values for healthy cats were similar to those previously reported by other authors (6.7–8.1, mean 7.5) [25,33]. The results indicated no significant differences in the VHS between the different groups, in contrast to what was reported by Litster et al. [26], who found that the mean VHS for the heartworm-infected cats was significantly greater than the reference value, and Venco et al. [17], who observed a tendency for the heart silhouette to increase in size during infection and at the onset of clinical signs. These differences may be attributed to the fact that the present study focused on cats with early infections, whereas the cited studies [17,26] were performed in cats with adult parasites. Nevertheless, it should be noted that 10 cats in Group C showed cardiomegaly, suggesting a trend towards an increased cardiac silhouette in cats with clinical signs, similar to that reported by Venco et al. [17]. However, the small variations in the analyzed parameters compared to healthy cats indicate the minimal usefulness of these measures as clinical diagnostic tools [17,26]. Indeed, the heart is rarely affected in feline heartworm disease [6].

A moderate (almost strong) positive correlation between the VHS and the diameter of the CVC was identified. Quite similar results (r = 0.59) were previously obtained in cats infected with *D. immitis* [26]; these authors reported that this finding, along with the increased mean VHS, may be linked to an elevation in filling pressures during the infection. Additionally, the CVC/Ao ratios were increased in infected cats compared to the healthy group; however, while the CVC/Ao ratio may provide some insights into right-sided heart conditions, pulmonary hypertension, and right ventricular hypertrophy, these are not typically observed in cats infected with *D. immitis* [6,10,15]. Similar findings were reported by Litster et al. [26], who found that the maximum width of the CVC in heartworm-infected dogs and cats was significantly greater than that obtained in the reference group, suggesting elevated right-sided heart filling pressures in both species. The authors argued that increased cardiac size and elevated filling pressures correlated and progressed together with heartworm disease. However, while the present study has shown an increased CVC/Ao ratio in infected cats, other potential factors influencing this measurement should be considered, and it should not be solely attributed to right-sided heart disease. Moreover, the echocardiographic exam carried out in the studied cats showed no evidence of right-sided heart disease. Further research is necessary to clarify the relationship between the CVC/Ao ratio and right-sided heart disease specifically in cats, given the unique aspects of feline heartworm infection.

Regarding the results obtained for the CVC/R4 ratio, similar results were observed, with the highest mean values seen in Group C. These findings are similar to those of other studies that examined the CVC/R4 ratio in dogs with heartworm with varying degrees of cardiac enlargement and found that this ratio increased with the severity of right ventricular enlargement [27,28]. The authors stated that, assuming that the degree of right ventricular enlargement was directly related to the severity and duration of heartworm infection, the relationship between CVC and right ventricular enlargement may reflect an increase in central venous pressure due to impending *cor pulmonale* [27]. This finding would be similar to that reported in cats, as discussed earlier [26], where it was suggested that increased cardiac size and elevated filling pressures occurred and were proportional to each other as the cardiac effects of heartworm disease progressed. Although the CCV/R4 ratio is not yet widely used in feline cardiology, these canine studies, along with the aforementioned study conducted in cats with heartworm, provide a reasonable basis for hypothesizing that an increased CCV/R4 ratio may indicate cardiac stress in cats with HARD, just as other authors have seen an increased VHS in cats with *D. immitis* showing clinical signs [17]. However, as mentioned earlier, further research is needed to determine the diagnostic utility of this ratio in feline heartworm disease.

The CrPA/R4 ratios obtained for healthy cats were very similar to those previously reported, which were 0.7 ± 0.13 [30]. The results showed that this ratio was significantly increased in cats with heartworm and, regardless of the presence or absence of clinical signs, these values were elevated in a large proportion of the animals infected with *D. immitis*. This differs from that reported by other authors, who subjectively reported that cranial lobar vessels were not enlarged in cats with heartworm [32]. However, previous studies found that cats infected with immature parasites had a significant increase in wall thickness, with occlusive medial hypertrophy present in 50% of cats infected by immature worms [34]; moreover, these authors argued that it was possible that medial hypertrophy of the small pulmonary arteries in exposed cats represented a pathologic response to transient heartworm infection. The first detectable pulmonary lesions of *D. immitis* infection include arteritis, pneumonitis, and hypertrophy of smooth muscle cells in the tunica media of small pulmonary arteries, which are likely attributable to pulmonary embolization of fifth-stage larvae before the establishment of infection with adult heartworms [35,36]. Therefore, it would be expected that certain arterial abnormalities would be found in the thoracic radiographs of these cats.

Similarly, the results showed values for CdPA/R9 ratios in healthy cats within the reference ranges usually considered (<1) [37]. Other authors have established significantly higher reference values for healthy cats (1.37 ± 0.28 for CdPA/R9) [32]. The reason for these differences beyond interobserver variability is unknown. Several authors agreed that a CdPA/R9 ratio greater than 1.6 has previously been reported in association with feline heartworm disease [26,32]. The results of the present study do not show such a pronounced thickening, likely because the cats examined were in the early stages of infections with immature parasites, while other studies involved chronic infections with adult parasites and, therefore, more advanced arterial damage [9]. However, the presence of significant differences between the seropositive cats and the Control Group is indicative of vascular lesions; moreover, these ratios were higher in cats with clinical signs. As noted above, in HARD, occlusive medial hypertrophy of the small pulmonary arterioles occurs, as well as changes in the pulmonary arteries [6]. Death of *D. immitis* in the pre-cardiac stages can also lead to smooth muscle hypertrophy of pulmonary arterioles.

In cats with HARD, changes are observed in the bronchi, bronchioles, and alveoli [5,6,23,34]. In this study, the lung patterns observed were more severe in feline patients with clinical signs, as lung parenchymal abnormalities were also observed in asymptomatic patients. Cats with clinical signs showed predominantly marked bronchointerstitial patterns, while asymptomatic cats showed milder forms. This is consistent with the pathophysiology of the feline heartworm and the pathogenesis of HARD, which is caused by the death of immature worms upon reaching the lungs. Obviously, this leads to clinical signs and radiological changes. Conversely, other studies have reported no correlation between radiographic lesions, clinical signs, or antibody levels [38]; however, this was a study based on a small number of cats (n = 10), all infected with adult worms. On the other hand, it has been described that infected cats may present with apparently normal thoracic radiographs [3,10], as observed in this study. The cardiopulmonary response to heartworm infection is dynamic and radiographs may not show changes if they are produced very early or very late in the course of the disease [11]. When present, feline heartworm should be considered in cats whose clinical and epidemiological characteristics are consistent with the infection.

A cat can remain seropositive for anti-*D.-immitis* antibodies for up to a year after clearance of the infection [23], so the seropositivity of the study cats does not necessarily indicate active infections, especially in asymptomatic cats, which is a limitation of this study. Moreover, the radiographic lesions of feline heartworm infection are dynamic over time, as demonstrated in experimental cat models where the timing of infection was known with accuracy [6,10,38]; as these cats were naturally infected, the exact time of infection is unknown, which may have affected the results. Nonetheless, these were infected cats that had been exposed to larval forms of the parasite in a hyperendemic region. In a study in which cats were experimentally inoculated with 100 L3 and subsequently treated with macrocyclic lactones as early as 70 days after infection, they showed radiographic and histopathological changes consistent with HARD at necropsy [5,23]. In addition, a large study with client-owned cats showed that 28% of heartworm-infected cats were asymptomatic [39].

5. Conclusions

The radiographic changes observed in the cats of this study indicated the presence of vascular and parenchymal abnormalities in those likely infected by immature *D. immitis* parasites. This was particularly evident in cats exhibiting clinical signs consistent with HARD, suggesting early vascular damage caused by this parasite. Given the challenges in diagnosing infections by immature *D. immitis* worms in cats, the examination of these radiographic measurements could serve as a valuable diagnostic tool for veterinary clinicians when HARD is strongly suspected. The results of the present study support the diagnostic suspicion of HARD in cats with compatible clinical signs and radiographic findings; however, heartworm infection should not be entirely ruled out in cats showing normal thoracic radiographs.

Author Contributions: J.A.M.-A. and E.C. designed the study. S.F.-C., Y.F.-C. and E.C. wrote the manuscript. S.F.-C., Y.F.-C., S.N.G.-R., D.J.V.-R. and N.C.-R. performed the fieldwork, collected the data and performed the experiments. All authors participated in the discussion of the results and correction of the final manuscript. All authors have read and agreed to the published version of the manuscript.

Funding: The presented study was supported by funds from the Internal Medicine Service FULP/ULPGC (SD-240/030/0026). NCR was supported by the "Grants for the predoctoral training program for researchers" program of the Government of the Canary Islands (TESIS2021010010), financed by the Consejería de Economía, Industria, Comercio y Conocimiento, and the Fondo Social Europeo. SNRG was supported by the "Grants for the financing of predoctoral contracts" program of the Universidad de Las Palmas de Gran Canaria (PIFULPGC-2020-2-CCSALUD-2).

Institutional Review Board Statement: Ethical review and approval were not required for the animals in this study. All radiographs and echocardiographic measures were routinely collected for prescribed diagnostic purposes or official monitoring studies and subsequently made available for this study. All of the cat owners were informed about the present study and consented to participate. The study was carried out in accordance with the current Spanish and European legislation on animal protection (Spanish Royal Decree 53/2013 and 2010/63/UE Directive).

Informed Consent Statement: Not applicable.

Data Availability Statement: All data generated or analyzed during this study are included in this article. The datasets used and/or analyzed during the present study are available from the corresponding author upon reasonable request.

Acknowledgments: The authors would like to thank Uranovet for kindly providing the Urano test Dirofilaria® and the in-house ELISA Urano Vet®. Also, they would like to thank the Hospital Clínico Veterinario of the Universidad de Las Palmas de Gran Canaria (ULPGC) for their support.

Conflicts of Interest: The authors declare no conflicts of interest.

References

1. Montoya-Alonso, J.A.; Morchón, R.; García-Rodríguez, S.N.; Falcón-Cordón, Y.; Costa-Rodríguez, N.; Matos, J.I.; Rodríguez Escolar, I.; Carretón, E. Expansion of Canine Heartworm in Spain. *Animals* **2022**, *12*, 1268. [CrossRef] [PubMed]
2. Montoya-Alonso, J.A.; García-Rodríguez, S.N.; Matos, J.I.; Costa-Rodriguez, N.; Falcón-Cordón, Y.; Carretón, E.; Morchón, R. Change in the Distribution Pattern of *Dirofilaria immitis* in Gran Canaria (Hyperendemic Island) between 1994 and 2020. *Animals* **2024**, *14*, 2037. [CrossRef] [PubMed]
3. McCall, J.W.; Genchi, C.; Kramer, L.H.; Guerrero, J.; Venco, L. Heartworm disease in animals and humans. *Adv. Parasitol.* **2008**, *66*, 193–285. [PubMed]
4. European Society of Dirofilariosis and Angiostrongylosis (ESDA). Guidelines for Clinical Management of Canine Heartworm. 2017. Available online: https://www.esda.vet (accessed on 20 August 2024).
5. Dillon, A.R.; Blagburn, B.L.; Tillson, M.; Brawner, W.; Welles, B.; Johnson, C.; Cattley, R.; Rynders, P.; Barney, S. Heartworm-associated respiratory disease (HARD) induced by immature adult *Dirofilaria immitis* in cats. *Parasit. Vectors* **2017**, *10*, 514. [CrossRef]
6. American Heartworm Society. Current Canine Guidelines for the Prevention, Diagnosis and Management of Heartworm (*Dirofilaria immitis*) Infection in Dogs. 2024. Available online: https://www.heartwormsociety.org/ (accessed on 20 August 2024).
7. Dillon, A.R.; Warner, A.E.; Brawner, W.; Hudson, J.; Tillson, M. Activity of pulmonary intravascular macrophages in cats and dogs with and without adult *Dirofilaria immitis*. *Vet. Parasitol.* **2008**, *158*, 171–176. [CrossRef]
8. Simón, F.; Siles-Lucas, M.; Morchón, R.; González-Miguel, J.; Mellado, I.; Carretón, E.; Montoya-Alonso, J.A. Human and animal dirofilariasis: The emergence of a zoonotic mosaic. *Clin. Microbiol. Rev.* **2012**, *25*, 507–544. [CrossRef]
9. Dillon, A.R.; Blagburn, B.L.; Tillson, D.M.; Brawner, W.R.; Welles, B.; Johnson, C.; Spenser, J.; Kaltenboeck, B.; Rynders, P.E. Immature heartworm infection produces pulmonary parenchymal, airway, and vascular disease in cats. *J. Vet. Intern. Med.* **2007**, *21*, 608–609.
10. Venco, L.; Marchesotti, F.; Manzocchi, S. Feline heartworm disease: A 'Rubik's-cube-like' diagnostic and therapeutic challenge. *J. Vet. Cardiol.* **2015**, *17*, S190–S201. [CrossRef]
11. Brawner, W.R.; Dillon, A.R.; Robertson-Plouch, C.K.; Guerrero, J. Radiographic diagnosis of feline heartworm disease and correlation to other clinical criteria: Results of a multicenter clinical case study. *Vet. Ther.* **2000**, *1*, 81–87.
12. Lee, A.C.Y.; Atkins, C.E. Understanding Feline heartworm infection: Disease, diagnosis, and treatment. *Top. Comp. Anim. Med.* **2010**, *25*, 224–230. [CrossRef]
13. Silvestre-Ferreira, A.C.; Vieira, L.; Vilhena, H.; Cerón, J.J.; Tvarijonaviciute, A.; Montoya-Alonso, J.A.; Carretón, E.; Pastor, J. Serum acute phase proteins in *Dirofilaria immitis* and *Wolbachia* seropositive cats. *J. Feline Med. Surg.* **2017**, *19*, 693–696. [CrossRef] [PubMed]

14. Rudorf, H.; Taeymans, O.; Johnson, V. Basics of thoracic radiography and radiology. In *BSAVA Manual of Canine and Feline Thoracic Imaging*; Schwarz, T., Johnson, V., Eds.; British Small Animal Veterinary Association: Gloucester, UK, 2008; pp. 1–19.
15. Atkins, C.E.; Litster, A.L. Heartworm disease. In *Consultations in Feline Internal Medicine*; August, J.R., Ed.; Elsevier Saunders: Philadelphia, PA, USA, 2006; pp. 323–330.
16. Litster, A.L.; Atwell, R.B. Feline heartworm disease: A clinical review. *J. Feline Med. Surg.* **2008**, *10*, 137–144. [CrossRef] [PubMed]
17. Venco, L.; Genchi, C.; Genchi, M.; Grandi, G.; Kramer, L.H. Clinical evolution and radiographic findings of feline heartworm infection in asymptomatic cats. *Vet. Parasitol.* **2008**, *158*, 232–237. [CrossRef] [PubMed]
18. Dillon, A.R.; Tillson, D.M.; Wooldridge, A.; Cattley, R.; Hathcock, J.; Brawner, W.R.; Cole, R.; Welles, B.; Christopherson, P.W.; Lee-Fowler, T.; et al. Effect of pre-cardiac and adult stages of *Dirofilaria immitis* in pulmonary disease in cats: CBC, bronchial lavage cytology, serology, radiographs, CT images, bronchial reactivity, an histopathology. *Vet. Parasitol.* **2014**, *206*, 24–37. [CrossRef] [PubMed]
19. Wooldridge, A.A.; Dillon, A.R.; Tillson, D.M.; Zhong, Q.; Barney, S.R. Isometric responses of isolated intrapulmonary bronchioles from cats with and without adult heartworm infection. *Am. J. Vet. Res.* **2012**, *73*, 439–446. [CrossRef]
20. Genchi, M.; Ferrari, N.; Fonti, P.; De Francesco, I.; Piazza, C.; Viglietti, A. Relation between *Aelurostrongylus abstrusus* larvae excretion, respiratory and radiographic signs in naturally infected cats. *Vet. Parasitol.* **2014**, *206*, 182–187. [CrossRef]
21. Pinto, A.; Brunese, L. Spectrum of diagnostic errors in radiology. *World J. Radiol.* **2010**, *2*, 377–383. [CrossRef]
22. Arruda Bergamaschi, N.; Huber, L.; Ludewig, E.; Böhler, A.; Gumpenberger, M.; Hittmair, K.M.; Strohmayer, C.; Folkertsma, R.; Rowan, C. Association between clinical history in the radiographic request and diagnostic accuracy of thorax radiographs in dogs: A retrospective case-control study. *J. Vet. Intern. Med.* **2023**, *37*, 2453–2459. [CrossRef]
23. Dillon, A.R.; Blagburn, B.L.; Tillson, M.; Brawner, W.; Welles, B.; Johnson, C.; Cattley, R.; Rynders, P.; Barney, S. The progression of heartworm associated respiratory disease (HARD) in SPF cats 18 months after *Dirofilaria immitis* infection. *Parasit. Vectors* **2017**, *10*, 533. [CrossRef]
24. Silverman, S.; Suter, P.F. Influence of inspiration and expiration on canine thoracic radiographs. *J. Am. Vet. Med. Assoc.* **1975**, *166*, 502–510.
25. Litster, A.L.; Buchanan, J.W. Vertebral scale system to measure heart size in radiographs of cats. *J. Am. Vet. Med. Assoc.* **2000**, *216*, 210–214. [CrossRef] [PubMed]
26. Litster, A.; Atkins, C.; Atwell, R.; Buchanan, J. Radiographic cardiac size in cats and dogs with heartworm disease compared with reference values using the vertebral heart scale method: 53 cases. *J. Vet. Cardiol.* **2005**, *7*, 33–40. [CrossRef] [PubMed]
27. Losonsky, J.M.; Thrall, D.E.; Lewis, R.E. Thoracic radiographic abnormalities in 200 dogs with spontaneous heartworm infestation. *Vet. Radiol. Ultrasound* **1983**, *24*, 120–123.
28. Thrall, D.E.; Calvert, C.A. Radiographic evaluation of canine heartworm disease coexisting with right heart failure. *Vet. Radiol. Ultrasound* **1983**, *24*, 124–126. [CrossRef]
29. Lehmukhl, L.B.; Bonagura, J.D.; Biller, D.S.; Hartman, W.M. Radiographic evaluation of caudal vena cava size in dogs. *Vet. Radiol. Ultrasound* **1997**, *38*, 94–100. [CrossRef]
30. Hayward, N.J.; Baines, S.J.; Baines, E.A.; Herrtage, M.E. The radiographic appearance of the pulmonary vasculature in the cat. *Vet. Radiol. Ultrasound* **2004**, *45*, 501–504. [CrossRef]
31. Ackerman, N. Radiographic aspects of heartworm disease. *Semin. Vet. Med. Surg. Small Anim.* **1987**, *2*, 15–27.
32. Schafer, M.; Berry, C.R. Cardiac and pulmonary artery mensuration in feline heartworm disease. *Vet. Radiol. Ultrasound* **1995**, *36*, 462–468. [CrossRef]
33. Ghadiri, A.; Avizeh, R.; Rasekh, A.; Yadegari, A. Radiographic measurement of vertebral heart size in healthy stray cats. *J. Feline Med. Surg.* **2008**, *10*, 61–65. [CrossRef]
34. Browne, L.E.; Carter, T.D.; Levy, J.K.; Snyder, P.S.; Johnson, C.M. Pulmonary arterial disease in cats seropositive for *Dirofilaria immitis* but lacking adult heartworms in the heart and lungs. *Am. J. Vet. Res.* **2005**, *66*, 1544–1549. [CrossRef]
35. Rawlings, C.A.; Farrell, R.L.; Mahood, R.M. Morphologic changes in the lungs of cats experimentally infected with *Dirofilaria immitis*: Response to aspirin. *J. Vet. Intern. Med.* **1990**, *4*, 292–300. [CrossRef] [PubMed]
36. Holmes, R.A.; Clark, J.N.; Casey, H.W.; Henk, W.; Plue, R.E. Histopathologic and radiographic studies of the development of heartworm pulmonary vascular disease in experimentally infected cats. In Proceedings of the Heartworm Symposium '92, Austin, TX, USA, 27–29 March 1992; Soll, M.D., Ed.; American Heartworm Society: Wilmington, DE, USA, 1992; pp. 81–89.
37. Johnson, V.; Hansson, K.; Maï, W.; Dukes-McEwan, J.; Lester, N.; Schwarz, T.; Chapman, P.; Morandi, F. The heart and major vessels. In *BSAVA Manual of Canine and Feline Thoracic Imaging*; Schwarz, T., Johnson, V., Eds.; British Small Animal Veterinary Association: Gloucester, UK, 2008; pp. 86–176.
38. Dillon, A.R.; Brawner, A.R., Jr.; Robertson-Plouch, C.K.; Guerrero, J. Feline heartworm disease: Correlations of clinical signs, serology, and other diagnostics--results of a multicenter study. *Vet. Ther.* **2000**, *1*, 176–182. [PubMed]
39. Atkins, C.E.; DeFrancesco, T.C.; Coats, J.R.; Sidley, J.A.; Keene, B.W. Heartworm infection in cats: 50 cases (1985–1997). *J. Am. Vet. Med. Assoc.* **2000**, *217*, 355–358. [CrossRef] [PubMed]

animals

MDPI

Article

Assessment of Thoracic Radiographic Alterations in Dogs with Heartworm and Their Correlation with Pulmonary Hypertension, Pre- and Post-Adulticide Treatment

Soraya Falcón-Cordón, Yaiza Falcón-Cordón, Noelia Costa-Rodríguez, Jorge Isidoro Matos, José Alberto Montoya-Alonso * and Elena Carretón

Internal Medicine, Veterinary Medicine and Therapeutic Research Group, Faculty of Veterinary Medicine, Research Institute of Biomedical and Health Sciences (IUIBS), Universidad de Las Palmas de Gran Canaria (ULPGC), 35016 Las Palmas de Gran Canaria, Spain; soraya.falcon@ulpgc.es (S.F.-C.); yaiza.falcon@ulpgc.es (Y.F.-C.); noelia.costa@ulpgc.es (N.C.-R.); jorge.matos@ulpgc.es (J.I.M.); elena.carreton@ulpgc.es (E.C.)

* Correspondence: alberto.montoya@ulpgc.es

Simple Summary: Pulmonary hypertension (PH) is a common and severe complication in dogs infected with *Dirofilaria immitis*, often persisting even after treatment. This study aimed to evaluate the progression of PH in dogs by assessing changes in radiographic parameters and the Right Pulmonary Artery Distensibility (RPAD) Index before and after treatment. Parameters were measured on the day of diagnosis (Day 0), at discharge (Day 90), and six months post-discharge (Day 270). The results indicated that in non-hypertensive dogs, the RPAD Index significantly improved following treatment. In contrast, hypertensive dogs exhibited a persistently low RPAD Index throughout the study, indicating ongoing PH. Additionally, hypertensive dogs showed consistently elevated VHS, CrPA/R4, and CdPA/R9 ratios compared to non-hypertensive dogs. These findings underscore the persistence of PH despite treatment, suggesting that regular radiographic monitoring of VHS, CrPA/R4, and CdPA/R9 ratios is crucial for assessing and managing long-term outcomes in dogs with heartworm disease.

Abstract: Pulmonary hypertension (PH) is a prevalent and severe complication in dogs infected with *Dirofilaria immitis*. This study aimed to elucidate the progression of PH by analyzing radiographic parameters and the Right Pulmonary Artery Distensibility (RPAD) Index at three key time points: diagnosis (day 0), discharge (day 90), and six months post-discharge (day 270). Fifty-two heartworm-infected dogs were divided into two groups: non-hypertensive and hypertensive. Radiographic measurements, including Vertebral Heart Size (VHS), CrPA/R4 ratio, and CdPA/R9 ratio, along with the RPAD Index, were assessed on Days 0, 90, and 270. Results indicated that, in Group A, the RPAD Index improved significantly from 42% on Day 0 to 43.16% on Day 90, with no significant change by Day 270 (42%). In contrast, hypertensive dogs exhibited a persistently low RPAD Index, averaging 17% throughout this study ($p < 0.001$). Radiographic parameters in hypertensive dogs showed continuous elevation compared to non-hypertensive dogs, with significant increases in VHS, CrPA/R4, and CdPA/R9 ratios on day 270 compared to day 0 ($p < 0.05$). The results confirmed that PH persisted in dogs with *D. immitis* after adulticide treatment, highlighting the importance of regular radiographic monitoring for assessing and managing long-term outcomes in dogs with PH during and after adulticide treatment. Continuous surveillance is thus essential for the effective post-treatment management of PH in dogs.

Keywords: vector-borne disease; *Dirofilaria immitis*; pulmonary hypertension; radiographic indexes; image diagnosis; echocardiography; veterinary diagnosis; dogs

1. Introduction

Heartworm disease is caused by the nematode *Dirofilaria immitis*, with adult worms lodging in the pulmonary arteries and right ventricle and mainly affecting domestic and wild carnivores [1]. Pulmonary arterial hypertension (PH) is a frequent and generally severe condition in infected dogs; it is caused primarily by proliferative intimal changes driving irreversible structural damage to the vasculature, inflammation, loss of elasticity, and occlusion of the vascular lumen, leading to persistent PH [2,3]. In addition, embolization of dead heartworms contributes to the development and perpetuation of PH [1,4].

Previous authors agreed that severity and chronicity of pulmonary endarteritis can be assessed through the determination of PH [5–8]. To this aim, the use of echocardiography can provide valuable information for diagnosing PH based on indirect measurements, since right heart catheterization, which is the gold standard for diagnosing PH, is unacceptably invasive in routine procedures or in compromised patients [9,10].

Thoracic radiography is a valuable tool for identifying concurrent or underlying diseases in an individual patient and can also provide evidence supporting the presence of PH [9,11]. This imaging modality has the advantage that it is widely accessible for clinicians on a daily basis. Indeed, there are studies that have characterized the association of radiological and echocardiographic findings in dogs with PH [12,13]. In dogs with heartworm, thoracic radiographs may reveal right ventricular enlargement, main pulmonary artery dilation, and pulmonary artery tortuosity, although these findings are not specific [3,14]. Moreover, chest radiographs are useful in helping to estimate the severity and chronicity of heartworm disease [15]. More recently, a study in dogs with heartworm reported that evaluating certain thoracic radiographic measurements can serve as an objective preliminary screening when assessing whether complementary tests should be performed to evaluate the presence of PH [16].

Previous studies have shown that PH persisted in dogs after heartworm adulticide treatment for a minimum of 10 months following the completion of heartworm removal. The authors suggested that endarteritis may not be reversible after parasite elimination, highlighting the need for continued monitoring for PH after completion of adulticide treatment [8,17,18]. However, there is currently no research associating specific thoracic radiographic findings with the severity of PH in *D. immitis*-infected dogs and their changes following adulticide treatment. Therefore, the objective of this study was to determine if specific radiographic findings are linked to the severity of PH in naturally infected dogs before and after adulticide treatment.

2. Materials and Methods

2.1. Enrollment and Treatment of the Dogs

This research included 52 dogs owned by clients, which were taken to the Veterinary Medicine Service of the Veterinary Teaching Hospital of the University of Las Palmas de Gran Canaria (Spain). These dogs resided in an area with a high prevalence of heartworm [19–21] and were selected for this study based on a positive test result for circulating *D. immitis* antigens (Urano test Dirofilaria®, Urano Vet SL, Barcelona, Spain). Clinical history and information of each animal, such as age, sex, and breed, were recorded. A thorough medical history and examination were conducted for each dog to rule out the presence of other diseases that could impact the results. Any animal that was given medication for cardiovascular conditions was not included in this research. Likewise, dogs showing symptoms of heart disease (such as valvular heart disease, cardiomyopathy, and congenital defects) were not included in this research. Diagnostic imaging tests (thoracic radiography and echocardiography), as well as physical examination, clinical history, and anamnesis, were also used to rule out other coexisting cardiorespiratory conditions.

The infected canines underwent adulticide treatment as per the treatment protocol advised by the international Heartworm Societies, incorporating the recently published modifications [22–24]. To summarize, upon diagnosis on day 0, the dog started doxycycline administration (10 mg/kg BID) for 4 weeks for the treatment of the endosymbiont bacteria

Wolbachia pipientis, along with monthly oral tablets for heartworm prevention containing ivermectin (≥6 mcg/kg) and pyrantel pamoate (≥5 mg/kg). On days 30, 60, and 61, the dog received intramuscular melarsomine doses (2.5 mg/kg). A follow-up examination on day 90 determined discharge eligibility based on the absence of abnormalities (adult parasites in echocardiography, radiographic irregularities, or cardiorespiratory symptoms). After 6 months from discharge, on day 270, adulticidal efficacy was confirmed through an antigen detection test. It was advised to restrict exercise during the treatment period, especially from the first melarsomine dose until discharge.

2.2. Radiograph and Ultrasound Evaluations

On day 0 (diagnosis), day 90 (discharge), and 270 (6 months after discharge), thoracic radiographs were digitally captured using an RX generator (HFQ 300 P, Bennett, NC, USA) during peak inspiration, without sedation. The radiographic parameters (kVP and mAs) were adjusted individually for each dog based on thoracic thickness. Both right laterolateral and dorsoventral views were taken. Vertebral Heart Size (VHS) was determined following the guidelines previously established by Buchanan and Bücheler [25]. While potential dorsal spine alterations were not a basis for exclusion in VHS interpretation, none of the dogs in this study exhibited such alterations [26]. Furthermore, the CrPA/R4 ratio was calculated based on the measurement of the diameter of the right cranial pulmonary artery (CrPA) passing through the fourth rib (R4) in the laterolateral view, as well as the diameter of R4 just distal to the spine [16,27]. Finally, the CdPA/R9 ratio was determined in the dorsoventral view by the measurement of the diameter of the right caudal pulmonary artery (CdPA) overlapping the ninth rib (R9), in accordance with established guidelines and previous studies [16,27]. All measurements were conducted using electronic callipers on a DICOM workstation (DAIPACS, version 2.71) by a researcher who was unaware of the clinical status of the dogs involved in this study.

The canines also underwent echocardiographic evaluations on days 0, 90, and 270 utilizing an ultrasound device equipped with spectral and color Doppler as well as multifrequency probes (5.5–10 MHz) (Logic P5, General Electric, New York, NY, USA) to determine the presence or absence of PH, in accordance with the guidelines set by the American College of Veterinary Internal Medicine (ACVIM) [11]. The dogs were positioned in right laterolateral recumbence with the transducer placed in the third intercostal space. They remained conscious and were continuously monitored electrocardiographically. Each measurement involved recording six consecutive cardiac cycles, and all echocardiographic assessments were performed by the same researcher. To the aims of this study, the Right Pulmonary Artery Distensibility Index (RPAD Index) was utilized for comparison with thoracic radiographic measurements, with dogs being more likely to exhibit moderate to severe PH if their RPAD Index was <29.5%, a criterion that has been described and validated in dogs affected by heartworm disease [28–30]. The dogs were divided into 2 groups: Group A included dogs without PH, and Group B included dogs with PH. All dogs with PH showed a RPAD Index < 29.5%, and all dogs without PH showed a RPAD Index > 29.5%.

2.3. Statistical Analysis

The data were analyzed using the SPSS Base 29.0 software for Windows (SPSS Inc./IBM, Chicago, IL, USA). A Shapiro–Wilk test was performed to verify the normal distribution of the data. Additionally, the Siegel–Tukey test was performed to verify the variability of variances between groups. Continuous variables were expressed as mean ± standard deviation. Qualitative variables are expressed as percentages. The non-parametric test of Wilcoxon was used to determine the differences in the different stages (day 0, 90, and 270). In addition, a U-Mann–Whitney test was used to determine differences between dogs with and without PH. In all cases, a p-value < 0.05 was determined as significant. Furthermore, Cohen's D was employed to interpret the differential magnitude between the studied statistical groups. Considering a statistical difference with values > 0.70 for this study.

All dog owners gave their approval for their pets to take part in this research, and an informed consent was specifically signed for this purpose. This study was carried out in compliance with the existing animal welfare laws in Europe.

3. Results

Of the dogs studied, 57.7% (30/52) were males and 42.3% (22/52) were females, ranging in age from 1 to 14.5 years (mean: 5.5 ± 0.4 years). Regarding breed, 75% (39/52) were mixed-breed dogs, and 25% (13/52) were purebred. Of the latter, PH was present in the following breeds: Labrador Retriever (n = 5), American Pit Bull Terrier (n = 1), and Rat Terrier (n = 1). On the other hand, PH was absent in the following purebred dogs: Garafian Shepherd (n = 2), Canarian Mastiff (n = 1), Spanish Water Dog (n = 1), Miniature Pinscher (n = 1), and Rat Terrier (n = 1). Regarding sex, 42.3% (22/52) were females and 57.7% (30/52) were males. In addition, the mean weight of the animals included in this study was 15.33 ± 1.35 kg. On day 0, 40.4% (21/52) of the dogs had PH with a mean RPAD Index of 17%, whereas 59.6% (31/52) were normotensive with a mean RPAD Index of 42%.

Symptoms were observed in 19.35% of dogs without PH, whereas all dogs diagnosed with PH exhibited symptoms, with cough and dyspnea being the most common (Table 1).

Table 1. Distribution of symptoms and frequency by groups. Group A: dogs without pulmonary hypertension. Group B: dogs with pulmonary hypertension.

Groups	Symptoms	Percentage
Group A	Asymptomatic	80.65% (25/31)
	Cough	19.35% (6/31)
Group B	Asymptomatic	0% (0/21)
	Cough	95.24% (20/21)
	Dyspnea	57.14 (12/21)
	Weight loss	4.76% (1/21)

The Wilcoxon-signed rank test was used to compare the differences in scores between patients on days 0, 90, and 270. For the RPAD Index, only significant differences were observed between days 0 and 90 (p = 0.011). For VHS, statistically significant differences were observed between day 0 and day 90 (p = 0.001) and between day 0 and day 270 (p = 0.000). However, no statistically significant differences were observed between the measurements at the three time points for the CrPA/R4 and CdPA/R9 ratios (Table 2).

As described in the Section 2, the dogs were further divided into two groups: Group A included dogs without PH (n = 31), and Group B included dogs with PH (n = 21).

The results showed that RPAD Index values were significantly lower in the group of dogs with PH at all time points ($p = 2.1218 \times 10^{-14}$ for day 0, $p = 2.1218 \times 10^{-14}$ for day 90, and $p = 1.4216 \times 10^{-12}$ for day 270). In addition, while the RPAD Index increased significantly between days 0 and 90 in dogs from group A, the RPAD Index did not change significantly throughout the treatment in dogs with PH (Table 3). Additionally, a dog in group A experienced a decline in pulmonary arterial values, leading to the end of treatment with the presence of PH and an RPAD Index of 22%.

Table 2. Results of radiographic measures in all studied dogs in different time points. RPAD Index (Right Pulmonary Artery Distensibility Index); VHS (Vertebral Heart Size); CrPA/R4 (right cranial pulmonary artery passing through the fourth rib in the laterolateral projection ratio); CdPA/R9 (right caudal pulmonary artery to the ninth rib in the dorsoventral projection ratio). (*): $p < 0.05$.

Measure	Time Point	*p*-Value
RPAD Index	Day 0–Day 90	0.01108686 *
	Day 0–Day 270	0.37772975
	Day 90–Day 270	0.14238119
VHS	Day 0–Day 90	0.00137051 *
	Day 0–Day 270	0.00027207 *
	Day 90–Day 270	0.05733536
CrPA/R4	Day 0–Day 90	0.16833147
	Day 0–Day 270	0.2023426
	Day 90–Day 270	0.36260403
CdPA/R9	Day 0–Day 90	0.16833147
	Day 0–Day 270	0.22349953
	Day 90–Day 270	0.18534784

Table 3. Results of the Wilcoxon test for all measures by time points and groups of studied dogs. RPAD Index (Right Pulmonary Artery Distensibility Index); VHS (Vertebral Heart Size); CrPA/R4 (right cranial pulmonary artery passing through the fourth rib in the laterolateral projection ratio); CdPA/R9 (right caudal pulmonary artery to the ninth rib in the dorsoventral projection ratio). Group A: dogs without pulmonary hypertension. Group B: dogs with pulmonary hypertension. Results are displayed as mean ± standard deviation. NS: No significant differences were found between time points. (*): $p < 0.05$.

Measure	Group	Day 0	Day 90	Day 270	Time Points with Significant Differences between Measures	*p* Value
RPAD Index	Group A	42% ± 0.06	43.16% ± 0.07	42% ± 0.07	Day 0–Day 90	0.043641 *
	Group B	17% ± 0.11	17.96% ± 0.11	17.84% ± 0.11	NS	>0.05
VHS	Group A	9.85 ± 0.74	9.92 ± 0.68	9.98 ± 0.72	Day 0–Day 270	0.01313415 *
	Group B	10.29 ± 0.74	10.43 ± 0.76	10.48 ± 0.76	Day 0–Day 90 Day 0–Day 270	0.01844421 * 0.01313415 *
CrPA/R4	Group A	1.08 ± 0.16	1.06 ± 0.18	1.06 ± 0.16	NS	>0.05
	Group B	1.54 ± 0.36	1.59 ± 0.41	1.60 ± 0.47	Day 0–Day 90 Day 0–Day 270	0.01448255 * 0.04484749 *
CdPA/R9	Group A	1.14 ± 0.15	1.13 ± 0.16	1.12 ± 0.15	NS	>0.05
	Group B	1.68 ± 0.38	1.77 ± 0.39	1.79 ± 0.39	Day 0–Day 90 Day 0–Day 270	0.02499993 * 0.01999474 *

Regarding the radiographic indices, the mean VHS for each group on days 0, 90, and 270 are shown in Table 3. These results indicated that the mean VHS was significantly higher in Group B than in Group A at all time points ($p = 0.03645206$ for day 0, $p = 0.0131386$ for day 90, and $p = 0.01618378$ for day 270). In addition, a significant increase in VHS was observed between days 0 and 270 for Group A dogs and between days 0–90 and days 0–270 for Group B dogs. These results show that VHS tended to increase over time from day 0 to 6 months after treatment, particularly in the group of hypertensive dogs. When compared with the established reference values [25], the number of dogs in Group B exceeding the reference range increased towards the end of this study, whereas the number of dogs in Group A exceeding the reference value remained stable throughout this study (Table 4).

Table 4. Results for Vertebral Heart Score (VHS) results are distributed by time points in the studied dogs. Group A: dogs without pulmonary hypertension. Group B: dogs with pulmonary hypertension. Results are displayed as mean ± standard deviation.

	Groups	VHS	Dogs above Reference Value (≥10.5) [25]
Day 0	GROUP A	9.85 ± 0.74	12.90% (4/31)
	GROUP B	10.29 ± 0.74	28.57% (6/21)
Day 90	GROUP A	9.92 ± 0.68	12.90% (4/31)
	GROUP B	10.43 ± 0.76	42.86% (9/21)
Day 270	GROUP A	9.98 ± 0.72	12.90% (4/31)
	GROUP B	10.48 ± 0.76	47.62% (10/21)

The CrPA/R4 and CdPA/R9 ratios were significantly higher in Group B dogs at all time points (CrPA/R4: $p = 2.5007 \times 10^{-8}$ for day 0, $p = 2.5007 \times 10^{-8}$ for day 90, and $p = 7.5505 \times 10^{-8}$ for day 270; CdPA/R9: $p = 2.0504 \times 10^{-9}$ for day 0, $p = 5.7586 \times 10^{-11}$ for day 90, and $p = 1.4513 \times 10^{-11}$ for day 270). In addition, the CrPA/R4 and CdPA/R9 ratios remained constant without significant variation throughout the treatment in dogs from Group A; however, significant modifications were observed in dogs from Group B during adulticide treatment, which were statistically significant between days 0 and 90 and tended to increase during this study (Table 3).

Cohen's D, a standardized measure of effect size, was used to interpret the magnitude of differences between the statistical groups studied. For VHS, the calculated Cohen's D value was between 20.99 and 21.29, indicating a substantial difference between normotensive and hypertensive dogs. This extremely large effect size suggests a clinically meaningful distinction between the two groups in terms of VHS measurements. Similarly, for the CrPA/R4 ratio, the Cohen's D value was between 2.87 and 2.89, indicating a significant difference between the normotensive and hypertensive groups. Although this was slightly smaller than the effect size for VHS and the CdPA/R9 ratio, it was still a large effect size, indicating a notable difference in the CRPA/R4 ratio between the two groups. In the case of the CdPA/R9 ratio, the Cohen's D value was between 3.08 and 3.15, while indicating a smaller effect size compared to VHS, still reflected a moderate difference between the normotensive and hypertensive groups, remaining a relevant and potentially informative parameter to discriminate between these groups, underscoring its importance in the assessment of PH (Table 5).

Table 5. Cohen's D for the studied parameters. RPAD Index (Right Pulmonary Artery Distensibility Index); VHS (Vertebral Heart Size); CrPA/R4 (right cranial pulmonary artery passing through the fourth rib in the laterolateral projection ratio); CdPA/R9 (right caudal pulmonary artery to the ninth rib in the dorsoventral projection ratio).

Measure	Cohen's D	Value	Interpretation
VHS	Day 0	20.9929921	Large effect
	Day 90	21.1714818	Large effect
	Day 270	21.2935957	Large effect
CrPA/R4	Day 0	2.86616835	Large effect
	Day 90	2.88762566	Large effect
	Day 270	2.84047934	Large effect
CdPA/R9	Day 0	3.07601672	Large effect
	Day 90	3.15304145	Large effect
	Day 270	3.11923901	Large effect

Cohen's D: d = 0.2 small effect, d = 0.5 moderate effect; d = 0.8 large effect.

4. Discussion

Several studies have suggested that adult *D. immitis* parasites begin to produce lesions once they reach the pulmonary arteries, resulting in proliferative endarteritis that chronically leads to PH and heart failure [3]. PH is therefore a lesion associated with proliferative endarteritis and is consequently common in this disease.

The determination of PH can be indirect, as its direct estimation by right heart catheterization is neither clinically nor practically feasible [31]. Therefore, echocardiography is considered the best method of measurement. Furthermore, previous studies have shown that thoracic radiographs are useful in determining the presence and severity of lesions associated with PH, both in heartworm infection and in other cardiopulmonary pathologies [2,32,33], and some studies have even characterized the association between some radiological changes and echocardiographic findings in dogs with PH [12,13]. As in the present study, these authors reported the presence of cardiomegaly and enlarged pulmonary arteries, among other findings. Recently, a study showed that the assessment of VHS, CrPA/R4, and CdPA/R9 ratios may be used objectively in the preliminary evaluation of chest radiographs of dogs with PH as a screening tool when deciding whether to perform additional tests to assess the presence of PH [16]. The results of the present study confirmed the findings of the previous study, as statistically significant increases in the parameters evaluated (VHS, CrPA/R4, and CdPA/R9 ratios) were observed at all time points between dogs with and without PH.

Furthermore, other previous studies using echocardiography and specific serum biomarkers (i.e., endothelin-1 and acute phase proteins) have shown that PH persisted in dogs after the end of adulticide treatment, at least 10 months after the last dose of melarsomine, reporting that proliferative endarteritis may not be reversible and may persist even after the elimination of adult parasites, thus being a chronic problem that may affect the quality of life and life expectancy of the dog [8,17,18]. Therefore, its study and early detection should be a priority. The results of this study would add to this evidence by demonstrating the persistence of PH, in this case by means of a thoracic radiographic study.

Regarding radiographic indices, the results show an increase in VHS in dogs with PH throughout this study. These results are similar to those reported by other authors who have suggested that heartworm-infected dogs often have an enlarged cardiac silhouette on thoracic radiographs, which is a reliable parameter to discriminate severe PH from non-pulmonary hypertensive dogs [34]. When compared to the reference values [25], 28.6% of dogs with PH had a VHS above the reference value on day 0, rising to 47.6% by the end of this study. Chronic persistence of PH eventually leads to right-sided heart disease, and the observed increase in cardiac silhouette in dogs with PH, despite parasite clearance, may be due to this persistence [5,9]. On the other hand, 12.9% of dogs without PH were observed to have increased VHS on day 0, which remained constant throughout this study. These dogs did not have any other associated pathology, as this was an exclusion criterion, and the animals were examined prior to inclusion in this study. Therefore, this could be due to the use of a generic reference value established without taking into account breed-specific indices, and some dog breeds have variations in VHS reference values [35,36].

Similarly, a significant increase in CrPA/R4 and CdPA/R9 ratios was observed throughout this study, indicating a slight worsening of arterial damage. This is consistent with previous reports by other authors, indicating persistence of endarteritis in dogs after parasite clearance [8,17]. In this research, even one of the dogs went from having normal pulmonary arterial pressure to being hypertensive at the end of the treatment, which had also been observed in a previous study [8]. Furthermore, as mentioned before, a study observed a possible persistence of inflammation at the vascular level by serological detection of acute phase proteins and endothelin-1 in dogs with PH, 7 months after the end of adulticidal treatment [18]. In this context, the results of the present study are in agreement with previous reports, showing that CrPA/R4 and CdPA/R9 ratios could be used to determine the persistence of PH after the end of adulticide treatment.

Given the accessibility of radiological techniques to veterinary clinicians, the low technical requirements, and the low cost, the results of this study are of great interest as an aid in the post-treatment evaluation of dogs with heartworm. Given the high incidence of PH in dogs with *D. immitis* and the persistence of endarteritis and PH in chronically infected dogs, the objective determination of VHS, CrPA/R4, and CdPA/R9 ratios may be useful for long-term monitoring when other imaging tools are not available or in support of other imaging tools. This could have a beneficial impact on the animal, as it would allow for close monitoring, leading to better quality and life expectancy.

5. Conclusions

In conclusion, knowledge of the response and possible changes in the pulmonary vasculature after adulticide treatment by radiological assessment of VHS, CrPA/R4, and CdPA/R9 ratios could be useful in the detection and monitoring of PH in dogs during adulticide treatment and for close monitoring thereafter.

Author Contributions: J.A.M.-A. and E.C. designed this study. S.F.-C., Y.F.-C. and E.C. wrote the manuscript. S.F.-C., Y.F.-C., J.I.M. and N.C.-R. performed the fieldwork, collected the data, and performed the experiments. All authors participated in the discussion of the results, corrected, read, and approved the final manuscript. All authors have read and agreed to the published version of the manuscript.

Funding: The presented study was supported by own funds from the Internal Medicine Service FULP/ULPGC (SD-240/030/0026). N.C.-R. was supported by the "Grants for the predoctoral training program for researchers" program of the Government of the Canary Islands (TESIS 2021010010).

Institutional Review Board Statement: Ethical review and approval were not required for the animal in this study. All radiographs and echocardiographic measures were routinely collected for prescribed diagnostic purposes or official monitoring studies and subsequently made available to this study. All of the dog owners were informed about the present study and consented to participate. This study was carried out in accordance with the current Spanish and European legislation on animal protection (Spanish Royal Decree 53/2013 and 2010/63/UE Directive).

Informed Consent Statement: Informed consent was obtained from all subjects involved in the study.

Data Availability Statement: All data generated or analyzed during this study are included in this article. The datasets used and/or analyzed during the present study are available from the corresponding author upon reasonable request.

Acknowledgments: The authors would like to thank Uranovet for kindly providing the Urano test Dirofilaria®. Also, they would like to thank the Hospital Clínico Veterinario of the Universidad de Las Palmas de Gran Canaria (ULPGC) for their support.

Conflicts of Interest: The authors declare no conflicts of interest.

References

1. Simón, F.; Siles-Lucas, M.; Morchón, R.; González-Miguel, J.; Mellado, I.; Carretón, E.; Montoya-Alonso, J.A. Human and animal dirofilariasis: The emergence of a zoonotic mosaic. *Clin. Microbiol. Rev.* **2012**, *25*, 507–544. [CrossRef]
2. Johnson, L.; Boon, J.; Orton, E.C. Clinical characteristics of 53 dogs with Doppler-derived evidence of pulmonary hypertension: 1992–1996. *J. Vet. Intern. Med.* **1999**, *13*, 440–447.
3. McCall, J.W.; Genchi, C.; Kramer, L.H.; Guerrero, J.; Venco, L. Heartworm disease in animals and humans. *Adv. Parasitol.* **2008**, *66*, 193–285.
4. Bowman, D.D.; Atkins, C.E. Heartworm biology, treatment, and control. *Vet. Clin. N. Am. Small Anim. Pract.* **2009**, *39*, 1127–1158. [CrossRef] [PubMed]
5. Sasaki, Y.; Kitagawa, H.; Hirano, Y. Relationship between pulmonary arterial pressure and lesions in the pulmonary arteries and parenchyma, and cardiac valves in canine dirofilariasis. *J. Vet. Med. Sci.* **1992**, *54*, 739–744. [CrossRef]
6. Uchide, T.; Saida, K. Elevated endothelin-1 expression in dogs with heartworm disease. *J. Vet. Med. Sci.* **2005**, *67*, 1155–1161. [CrossRef] [PubMed]
7. Venco, L.; Bertazzolo, W.; Giordano, G.; Paltrinieri, S. Evaluation of C-reactive protein as a clinical biomarker in naturally heartworm-infected dogs: A field study. *Vet. Parasitol.* **2014**, *206*, 48–54. [PubMed]

8. Falcón-Cordón, Y.; Montoya-Alonso, J.A.; Caro-Vadillo, A.; Matos-Rivero, J.I.; Carretón, E. Persistence of pulmonary endarteritis in canine heartworm infection 10 months after the eradication of adult parasites of *Dirofilaria immitis*. *Vet. Parasitol.* **2019**, *273*, 1–4. [CrossRef]
9. Kellihan, H.B.; Stepien, R.L. Pulmonary hypertension in dogs: Diagnosis and therapy. *Vet. Clin. N. Am. Small Anim. Pract.* **2010**, *40*, 623–641. [CrossRef]
10. Kellihan, H.B.; Waller, K.R.; Pinkos, A.; Steinberg, H.; Bates, M.L. Acute resolution of pulmonary alveolar infiltrates in 10 dogs with pulmonary hypertension treated with sildenafil citrate: 2005–2014. *J. Vet. Cardiol.* **2015**, *17*, 182–191. [CrossRef]
11. Reinero, C.; Visser, L.C.; Kellihan, H.B.; Masseau, I.; Rozanski, E.; Clercx, C.; Williams, K.; Abbott, J.; Borgarelli, M.; Scansen, B.A. ACVIM consensus statement guidelines for the diagnosis, classification, treatment, and monitoring of pulmonary hypertension in dogs. *J. Vet. Intern. Med.* **2020**, *34*, 549–573. [CrossRef]
12. Adams, D.S.; Marolf, A.J.; Valdes-Martinez, A.; Randal, E.K.; Bachand, A.M. Associations between thoracic radiographic changes and severity of pulmonary arterial hypertension diagnosed in 60 dogs via Doppler echocardiography: A retrospective study. *Vet. Radiol. Ultrasound* **2017**, *58*, 454–462. [CrossRef]
13. Chanroon, K.; Tuntipas, T.; Sukpanich, P.; Punyarit, P.; Surachetpong, S. Novel methods to diagnose pulmonary hypertension in dogs assessed by thoracic radiography. *Thai J. Vet. Med.* **2018**, *48*, 385–391. [CrossRef]
14. Kellum, H.B.; Stepien, R.L. Sildenafil citrate therapy in 22 dogs with pulmonary hypertension. *J. Vet. Intern. Med.* **2007**, *21*, 1258–1264. [CrossRef] [PubMed]
15. Ackerman, N. Radiographic aspects of heartworm disease. *Semin. Vet. Med. Surg. (Small Anim.)* **1987**, *2*, 15–27. [PubMed]
16. Falcón-Cordón, S.; Falcón-Cordón, Y.; Caro-Vadillo, A.; Costa-Rodríguez, N.; Montoya-Alonso, J.A.; Carretón, E. Association between Thoracic Radiographic Changes and Indicators of Pulmonary Hypertension in Dogs with Heartworm Disease. *Animals* **2024**, *14*, 1900. [CrossRef] [PubMed] [PubMed Central]
17. Serrano-Parreño, B.; Carretón, E.; Caro-Vadillo, A.; Falcón-Cordón, S.; Falcón-Cordón, Y.; Montoya-Alonso, J.A. Pulmonary hypertension in dogs with heartworm before and after the adulticide protocol recommended by the American Heartworm Society. *Vet. Parasitol.* **2017**, *236*, 34–37. [CrossRef]
18. Falcón-Cordón, Y.; Tvarijonaviciute, A.; Montoya-Alonso, J.A.; Muñoz-Prieto, A.; Caro-Vadillo, A.; Carretón, E. Evaluation of acute phase proteins, adiponectin and endothelin-1 to determine vascular damage in dogs with heartworm disease (*Dirofilaria immitis*), before and after adulticide treatment. *Vet. Parasitol.* **2022**, *309*, 109759. [CrossRef] [PubMed]
19. Montoya-Alonso, J.A.; Carretón, E.; Morchón, R.; Silveira-Viera, L.; Falcón, Y.; Simón, F. The impact of the climate on the epidemiology of *Dirofilaria immitis* in the pet population of the Canary Islands. *Vet. Parasitol.* **2016**, *216*, 66–71. [CrossRef]
20. Montoya-Alonso, J.A.; Morchón, R.; García-Rodríguez, S.N.; Falcón-Cordón, Y.; Costa-Rodríguez, N.; Matos, J.I.; Rodríguez Escolar, I.; Carretón, E. Expansion of Canine Heartworm in Spain. *Animals* **2022**, *12*, 1268. [CrossRef]
21. Montoya-Alonso, J.A.; García-Rodríguez, S.N.; Matos, J.I.; Costa-Rodriguez, N.; Falcón-Cordón, Y.; Carretón, E.; Morchón, R. Change in the Distribution Pattern of *Dirofilaria immitis* in Gran Canaria (Hyperendemic Island) between 1994 and 2020. *Animals* **2024**, *14*, 2037. [CrossRef] [PubMed]
22. European Society of Dirofilariosis and Angiostrongylosis (ESDA). Guidelines for Clinical Management of Canine Heartworm. 2017. Available online: https://www.esda.vet (accessed on 25 July 2024).
23. Carretón, E.; Falcón-Cordón, Y.; Falcón-Cordón, S.; Morchón, R.; Matos, J.I.; Montoya-Alonso, J.A. Variation of the adulticide protocol for the treatment of canine heartworm infection: Can it be shorter? *Vet. Parasitol.* **2019**, *271*, 54–56. [CrossRef] [PubMed]
24. American Heartworm Society. Current Canine Guidelines for the Prevention, Diagnosis and Management of Heartworm (*Dirofilaria immitis*) Infection in Dogs. 2024. Available online: https://www.heartwormsociety.org/ (accessed on 25 July 2024).
25. Buchanan, J.W.; Bücheler, J. Vertebral scale system to measure canine heart size in radiographs. *J. Am. Vet. Med. Assoc.* **1995**, *206*, 194–199. [CrossRef] [PubMed]
26. Costanza, D.; Greco, A.; Piantedosi, D.; Bruzzese, D.; Pasolini, M.P.; Coluccia, P.; Castiello, E.; Baptista, C.S.; Meomartino, L. The heart to single vertebra ratio: A new objective method for radiographic assessment of cardiac silhouette size in dogs. *Vet. Radiol. Ultrasound* **2023**, *64*, 378–384. [CrossRef]
27. Oui, H.; Oh, J.; Keh, S.; Lee, G.; Jeon, S.; Kim, H.; Yoon, J.; Choi, J. Measurements of the pulmonary vasculature on thoracic radiographs in healthy dogs compared to dogs with mitral regurgitation. *Vet. Radiol.* **2015**, *56*, 251–256. [CrossRef]
28. Venco, L.; Genchi, C.; Vigevani Colson, P.; Kramer, L. Relative utility of echocardiography, radiography, serologic testing and microfilariae counts to predict adult worm burden in dogs naturally infected with heartworms. In *Recent Advances in Heartworm Disease. Symposium'01*; Seward, R.L., Knight, D.H., Eds.; American Heartworm Society: Batavia, IL, USA, 2003; pp. 111–124.
29. Venco, L.; Mihaylova, L.; Boon, J.A. Right pulmonary artery distensibility index (RPAD index). A field study of an echocardiographic method to detect early development of pulmonary hypertension and its severity even in the absence of regurgitant jets for doppler evaluation in heartworm-infected dogs. *Vet. Parasitol.* **2014**, *206*, 60–66. [PubMed]
30. Visser, L.C.; Im, M.K.; Johnson, L.R.; Stern, J.A. Diagnostic value of right pulmonary artery distensibility index in dogs with pulmonary hypertension: Comparison with doppler echocardiographic estimates of pulmonary arterial pressure. *J. Vet. Intern. Med.* **2016**, *30*, 543–552. [CrossRef]
31. D'Alto, M.; Dimopoulos, K.; Coghlan, J.G.; Kovacs, G.; Rosenkranz, S.; Naeije, R. Right Heart Catheterization for the Diagnosis of Pulmonary Hypertension: Controversies and Practical Issues. *Heart Fail. Clin.* **2018**, *14*, 467–477. [CrossRef]

32. Serres, F.J.; Chetboul, V.; Tissier, R.; Sampedrano, C.C.; Gouni, V.; Nicolle, A.P.; Pouchelon, J.L. Doppler echocardiography-derived evidence of pulmonary arterial hypertension in dogs withdegenerative mitral valve disease: 86 cases (2001–2005). *J. Am. Vet. Med. Assoc.* **2006**, *229*, 1772–1778. [CrossRef]
33. Stepien, R.L.; Rak, M.B.; Blume, L.M. Use of radiographic measurements to diagnose stage B2 preclinical myxomatous mitral valve disease in dogs. *J. Am. Vet. Med. Assoc.* **2020**, *256*, 1129–1136. [CrossRef] [PubMed]
34. Litster, A.; Atkins, C.; Atwell, R.; Buchanan, J. Radiographic cardiac size in cats and dogs with heartworm disease compared with reference values using the vertebral heart scale method: 53 cases. *J. Vet. Cardiol.* **2005**, *7*, 33–40. [CrossRef] [PubMed]
35. Jepsen-Grant, K.; Pollard, R.E.; Johnson, L.R. Vertebral heart scores in eight dog breeds. *Vet. Radiol. Ultrasound* **2013**, *54*, 3–8. [CrossRef] [PubMed]
36. Lamb, C.R.; Wikeley, H.; Boswood, A.; Pfeiffer, D.U. Use of breed-specific ranges for the vertebral heart scale as an aid to the radiographic diagnosis of cardiac disease in dogs. *Vet. Rec.* **2001**, *148*, 707–711. [CrossRef] [PubMed]

animals MDPI

Article

Treatment of Canine Leishmaniasis with Meglumine Antimoniate: A Clinical Study of Tolerability and Efficacy

Serena Digiaro [1], Alessandra Recchia [1], Antonella Colella [1], Sara Cucciniello [1], Beatrice Greco [1], Dora Buonfrate [2] and Paola Paradies [1,*]

1 Department DiMePre-J, Veterinary Section, University of Bari "Aldo Moro", Strada Provinciale Casamassima km3, Valenzano, 70010 Bari, Italy; serena.digiaro@uniba.it (S.D.); alessandra.recchia1@uniba.it (A.R.); antonella.colella@uniba.it (A.C.); s-cucciniello@libero.it (S.C.); beatrice.greco@uniba.it (B.G.)

2 Department of Infectious, Tropical Diseases and Microbiology, IRCCS Sacro Cuore Don Calabria Hospital, Negrar, 37024 Verona, Italy; dora.buonfrate@sacrocuore.it

* Correspondence: paola.paradies@uniba.it; Tel.: +0039-080467884

Simple Summary: This study evaluates the potential adverse effects associated with meglumine antimoniate (aNm) in the treatment of canine leishmaniasis through both a retrospective analysis and a prospective study. The retrospective study comprised records of 87 dogs treated with aNm. Adverse effects during treatment emerged in about one third of dogs, including local reactions at the injection site, injection site pain-related systemic reaction, systemic disease due to renal function worsening, acute pancreatitis, gastrointestinal self-limiting signs or severe idiosyncratic skin reactions. Half of these animals required treatment suspension. The prospective study included 16 dogs (LeishVet stages II and III) treated with the same aNm; of these, 2 dogs were excluded for severe reactions at the injection site and 4 dogs reported mild and transient adverse events. In the prospective study, treatment efficacy was also evaluated at 1-year follow-up. No animals showed clinical laboratory relapse during the study and interestingly, PCR turned negative between D0 and D60 in 78.5% of animals. Adverse severe events associated with aNm are possible, but the high rate of parasitic clearance supports its use as first line treatment, at least in dogs with normal renal function.

Abstract: Antimoniate therapy, in association with allopurinol, is one of the first-line treatments of canine leishmaniasis (CanL). This study evaluates the potential adverse effects associated with aNm in the treatment of CanL through both a retrospective analysis and a long-term prospective study also aimed to investigate its efficacy. The retrospective study reviewed records of 87 dogs with CanL with at least one follow-up available during or at the end of therapy with aNm (Glucantime®) at a dose of 50 mg/kg administered subcutaneously twice a day in association with allopurinol. In total, 29.8% of dogs showed adverse effects during treatment as local reactions at the injection site (n = 6), severe systemic reaction to pain (originating from the inoculation site) with depression and anorexia (n = 4), systemic disease due to renal function worsening (n = 4), acute pancreatitis (n = 1), diarrhea (n = 5), vomiting (n = 3) and severe idiosyncratic skin reactions (n = 3). Of these dogs, 13 (14.9%) required treatment suspension. The prospective study included 16 dogs, selected among the LeishVet stages II and III CKD IRIS stage 1 (International Renal Interest Society staging of canine Chronic Kidney Disease) and treated with the same aNm plus allopurinol protocol as in the retrospective study and observed for 360 days; 2 dogs were excluded for severe reactions at the injection site. Mild and transient adverse events were reported in the other 4 dogs. The criteria used to evaluate the efficacy of treatment with aNm were as follows: a reduction in the clinical score and improvement and/or normalization of laboratory parameters, negativization of PCR on the bone marrow samples and disease-free interval time. The proportion of reduction in the clinical score reached 91.9% at D180. No animals showed clinical laboratory relapse during the whole study duration and interestingly, the PCR results showed complete negativity between D0 and D60 in 78.5% of animals. Veterinarians must be vigilant regarding the potentially serious adverse effects associated with aNm and promptly stop drug administration if unexpected clinical manifestations occur. On the other hand, they should not discard its use for CanL treatment since it is confirmed that aNm in association with allopurinol is highly effective in controlling CanL.

Keywords: canine leishmaniasis; meglumine antimoniate; adverse effects; efficacy

1. Introduction

Canine leishmaniasis (CanL) is a protozoal disease caused by *Leishmania infantum*. The life cycle of *L. infantum* involves definitive mammal hosts and vector insects, including the sand fly [1]. CanL is endemic in the Mediterranean basin, South America, and central and southwestern Asia [1]. *L. infantum* commonly causes chronic infection with subclinical or clinical course, depending on parasite strain, host genetics and immune status [2]. Based on this, some dogs will be able to control the infection for many years without clinical signs. On the other hand, some infected dogs may present acute evolution and severe disease, or a progressive disease that could be fatal if proper management and supportive therapy are not promptly provided [2].

Several pharmacological treatment protocols have been proposed for CanL. The purpose of CanL treatment is to control clinical signs and hematobiochemical alterations, improve the dog's cellular immunity, reduce the parasitic load, prevent relapses, and decrease the transmission rate to the vector [3]. It is common opinion that treatment can lead to parasitological clearance, though this is not frequent [1,4,5]. Consequently, clinical remission may only be temporary [1,6] and clinical relapse can occur, although reinfection is always possible in endemic areas. First-choice treatments used for CanL are the combinations of N-methylglucamine antimoniate (aNm) plus allopurinol and miltefosine (MIL) plus allopurinol [1]. Although aNm is a much older and effective treatment for CanL, upon introduction in the veterinary market in the last 15 years, MIL has increasingly been used by veterinarians with some undeniable advantages, such as the once-a-day oral administration and the mild adverse effects, which are mostly self-limiting and result in dose-dependent gastrointestinal reactions [7]. In this study, we explored whether there was enough evidence to suggest treatment with MIL instead of treatment with aNm in all dogs affected by CanL. Several authors have compared the efficacy of aNm and allopurinol treatment with MIL and allopurinol, finding similar clinical efficacy, although treatment response is faster [8,9] and clinical relapses are less frequent [9–11] when the aNm protocol is used. Other studies have suggested that aNm compared to MIL may be more effective in controlling inflammation and oxidation in CanL [12–15], while few comparing data are available on the parasitological clearance after treatment.

On the other hand, the possibility of clinical adverse events associated to aNm treatment is known [1,16], but mostly based on fragmentary descriptions rather than on large-scale studies focusing on this aspect [6,9,17]. Furthermore, based on the current literature, the impact of aNm on renal function is controversial, with some authors concerned about potential impairment [8,17–20], and others suggesting the absence of renal involvement [6,9,12,21].

The purpose of this work was to report the potential clinical adverse events associated with the administration of aNm as well as the efficacy of the treatment with aNm plus allopurinol.

2. Materials and Methods

The aNm tolerability was evaluated with a retrospective and a prospective study, while the efficacy was estimated with the prospective study only. Procedures to confirm diagnosis and treatment protocol were the same both in the retrospective and prospective study (see below).

2.1. Retrospective Study

Clinical records of patients diagnosed with CanL, treated with aNm and allopurinol (standard protocol) and with at least one follow-up available during and/or immediately after aNm discontinuation were reviewed.

Dogs were included regardless of the LeishVet clinical stage. Dogs presented for consultation due to the appearance of new clinical signs during standard treatment set by a practitioner were also included. Exclusion criteria were as follows: (i) treatment with different aNm protocols; (ii) treatment with further drugs for concomitant diseases; (iii) severe concomitant diseases.

Available clinical records during the aNm 30-day treatment and/or the first post-aNm treatment record were reviewed to identify described or reported potential adverse events. Available collateral exams (i.e., laboratory and imaging examinations) were also reviewed to better understand the clinical picture in dogs presenting adverse events.

2.2. Prospective Study

Dogs with a CanL diagnosis of stage II or III (IRIS 1) according to the LeishVet classification [22] were included in the study, irrespective of breed, sex or age.

Exclusion criteria were as follows: (i) treatment with effective drugs against CanL in the previous 3 months; (ii) treatment with long-acting corticosteroids and other immunomodulatory drugs within 1 month before inclusion; (iii) concurrent disorders and potentially lethal diseases; (iv) pregnancy or lactation. Dogs were housed, fed regularly, and received treatment by owners. Dogs were tested for other vector-borne diseases (i.e., *Ehrlichia* spp.) and only negative dogs were included. Dogs were treated with the aNm plus allopurinol standard protocol.

Dogs whose pharmacological side effects required treatment interruption during follow-up evaluation were excluded. All possible side effects reported by owners were documented during the month of aNm therapy.

Dogs were observed for 360 days. Clinical scores and body weights were recorded on individual files at day 0, 30, 60, 90, 180, and 360 (D0–D360). Clinical scores were obtained by assessing the presence and severity of 26 clinical signs according to Mirò et al. [23]. Blood samples were collected at each follow-up to record the complete blood count (CBC), biochemical data (e.g., urea, creatinine, total plasma proteins, albumin concentration, globulin concentration and fractions, albumin/globulin ratio, protein electrophoresis, alanine-aminotransferase (ALT), alkaline phosphatase (ALP), plasma bilirubin), and *Leishmania* spp. IFAT. At D0 and D60, fine needle aspiration on popliteal lymph nodes was performed to microscopically identify *Leishmania* spp. amastigotes [24], and bone marrow samples were collected to perform qPCR analysis from sternum bones. Urine samples were collected under ultrasound-guide at D0, D30 and D60 to perform a complete urine analysis and to evaluate the urine protein/creatinine ratio (UPC). The study was approved by the ethics committee of the Department of Emergency and Organ Transplantation, Uniba (Prot. DETO 226/2018).

Treatment safety was assessed by the incidence of adverse events observed daily by the owner. Evaluation of urea, creatinine, and liver enzymes as well as the UPC ratio before and after treatment were also considered to assess treatment safety.

Clinical score reduction, improvement and/or normalization of laboratory parameters, serological data, PCR negativization on bone marrow, and disease-free interval time were used as criteria to assess treatment efficacy.

These parameters were evaluated during each follow-up in order to identify dogs under relapse. Numbers and times of relapses were assessed at the end of the study. Relapse was defined by the reappearance of clinical signs and/or clinical–pathological alterations indicative of leishmaniasis [3,22].

2.3. Diagnostic Procedures

In both retrospective and prospective studies, diagnosis was confirmed by cytological positivity in lymph node aspirates stained with May–Grüwndal–Giemsa [24] and observed under a microscope by an experienced operator and/or by IFAT (cut off dilution 1:80) in patients with a compatible, strong clinical suspicion. An in-house IFAT assay was performed using *L. infantum* promastigotes as the antigen, and following a previously recommended protocol [25].

Bone marrow fine needle aspirate material was examined a posteriori by q-PCR assay following the methodology used by the Italian National Reference Center for Leishmaniasis (C.Re.Na.L., Palermo, Italy). In brief, DNA from bone marrow was subjected to two consecutive PCR amplifications using the kinetoplast-specific primers R221 and R332 in the first run, and the *Leishmania*-specific primers R223 and R333 in the second run [26].

2.4. Treatment

Dogs were treated with aNm (Glucantime®, Boehringer, Milan, Italy) at a dose of 50 mg/kg/twice a day (BID) subcutaneously (SC) for 30 days and allopurinol (Zyloric®, Teofarma, Pavia, Italy) at a dosage of 10 mg/kg/BID per os (OS) for 6 months (standard protocol) in both retrospective and prospective studies. As a matter of practice, at the time of diagnosis, the owners were instructed to use the following injection pattern: left side of midline in the morning; right side in the evening moving cranio-caudally from the neck to the thorax each day to avoid proximity between injections. In the retrospective study, only dogs under the standard aNm protocol were included to avoid variability due to dosage.

3. Results

3.1. Retrospective Study

The records of 87 dogs were reviewed. Data on signalment, presenting complaints, and clinical signs at the time of diagnosis are reported in Tables 1 and 2. Diagnosis was achieved through lymph node cytology and IFAT in 80/87 dogs and with just IFAT in 7 dogs. Clinical signs and pathological conditions occurring during treatment and associated with aNm adverse events are reported in Table 3. In brief, 26 (29.8%) animals showed clinical evidence of adverse events of various types and severity. Treatment suspension was necessary in 13 animals (14.9%) (Table 3). In particular, six (6.9%) dogs showed a local reaction at the inoculation site (swelling/granuloma/abscess), which required interruption of treatment in one case; four (4.6%) dogs showed a severe systemic reaction to pain (originating from the inoculation site) and characterized by lethargy, anorexia and reluctance to move (with normal blood test results and with return to normality within a few days after treatment discontinuation); five patients developed clinical signs associated with renal failure (4.6%) and acute pancreatitis (0.87%); five (5.7%) patients manifested diarrhea and three (3.4%) vomiting during treatment. Severe cutaneous reactions were reported in three other patients that needed immediate aNm discontinuation and were reported as idiosyncratic reactions. In particular, a 3-year-old male Great Dane presented with acute onset of facial and ventral erythema, itching, and erosions (Figure 1), which appeared 3 days after treatment and worsened (consultation was carried out after 7 days from treatment onset); a 5-year-old female Yorkshire presented to consultation for a progressive serpiginous cutaneous lesion on their back with a burn-like appearance (Figure 2) that manifested during treatment; a 6-year-old medium-sized female mongrel initially presented with a deep linear skin ulcer on one limb, which, in a few days, extended to all four limbs and led to complete detachment of the foot despite treatment suspension (Figure 3).

Table 1. Signalment data of the 87 dogs included in the retrospective study.

		Breed	
		Mongrel	26 (29.9%)
Females	n: 40 (46%)	English setter	12 (13.8%)
Males	n: 47 (54%)	German Shepherd	7 (8%)
		Doberman pinscher	7 (8%)
		Great Dane	4 (4.6%)
		Beagle	4 (4.6%)
		Boxer	4 (4.6%)
Age (years)	Mean: 5.25 (Min: 0.75; Max: 14)	Breton Spaniel	3 (3.4%)
Weight (kg)	Mean: 22.47 (Min: 4.5 Max: 50)	English pointer	2 (2.3%)
		Labrador retriever	2 (2.3%)
		Pomeranian	2 (2.3%)
		Others *	14 (16%)

* One dog for each of the following breeds: American Pit Bull Terrier, Border Collie, Cocker Spaniel, Argentine mastiff, Golden Retriever, Greyhound, Husky, Italian hound, Italian Spinone, Little Italian Greyhound, Maremma Sheepdog, Pinscher, Shar Pey; Weimaraner.

Table 2. Clinical signs and/or presenting complaints at diagnosis in the 87 dogs included in the retrospective study.

Clinical Signs	Dogs n	% (tot 87)
Lymphadenomegaly	69	79.3
Dermatological signs	61	70
Weight loss	48	55.2
Pale mucous membranes	14	16
Ophthalmological signs	13	15
Lameness	9	10.3
Dysorexia	7	8.1
Splenomegaly	7	8.1
Lethargy	6	6.9
Epistaxis	5	5.7
Ophthalmological signs	5	5.7
Gastrointestinal signs	5	5.7
Joint swelling	5	5.7
Abdominal pain	4	4.6
Others *	31	35.6

* "Other" clinical signs/presenting complaint at the time of diagnosis include unilateral epistaxis (n = 3), lameness (n = 3), polyuria/polydipsia (n = 3), respiratory sounds (n = 3), unformed soft stools (n = 2), polyuria/polydipsia (n = 2), tonsillitis (n = 2), tremors (n = 2), gingivitis (n = 1), and sneezing (n = 2), oral cavity mouth ulcer (n = 1), posterior train pain (n = 1), vulvar discharge (n = 1), frontal muscle atrophy (n = 1), cough (n = 1), hyperthermia (n = 1), muscular hypotrophy (n = 1), and dysorexia (n = 1).

Table 3. Number and relative percentage of dogs that developed clinical signs and pathological conditions during treatment associated with N-methylglucamine antimoniate (aNm) adverse events; number and relative percentage of dogs in which treatment was discontinued.

Clinical Signs/Pathological Conditions	Dogs n (tot 87)	% (tot 87)	Treatment Suspension (Dogs n)	% (tot 87)
Local reaction at the site of inoculation	6	6.9	1	
Diarrhea	5	5.7	1	
Severe systemic reaction to pain originating from the inoculation site associated with lethargy, anorexia, and reluctance to move	4	4.6	3	
Severe systemic reaction due to renal failure	4	4.6	4	
Vomiting	3	3.4	0	
Severe cutaneous idiosyncratic reactions	3	3.4	3	
Severe systemic reaction due to acute pancreatitis	1	1.1	1	
Total	26	29.8	13	14.9

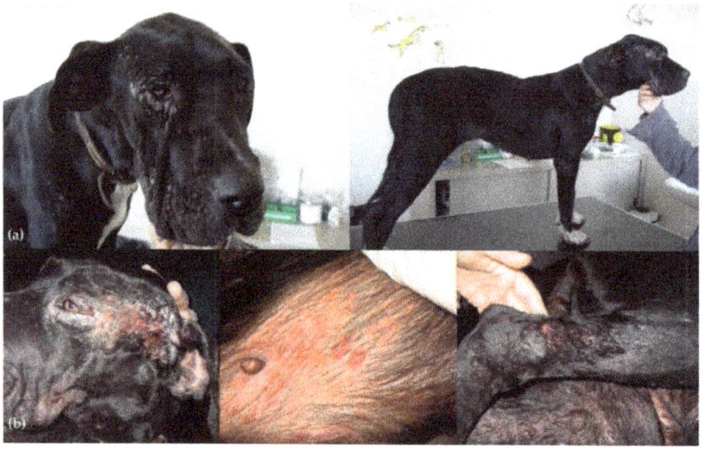

Figure 1. Meglumine antimoniate (aNm) cutaneous idiosyncratic reaction on a 3-year-old male Great Dane. (**a**) Mild clinical signs at the time of diagnosis before aNm treatment started; (**b**) acute onset of facial and ventral erythema and erosion, which appeared 3 days after aNm treatment started and rapidly progressed until aNm was stopped at day 7. Pictures (**b**) refer to the day of consultation (i.e., day 7).

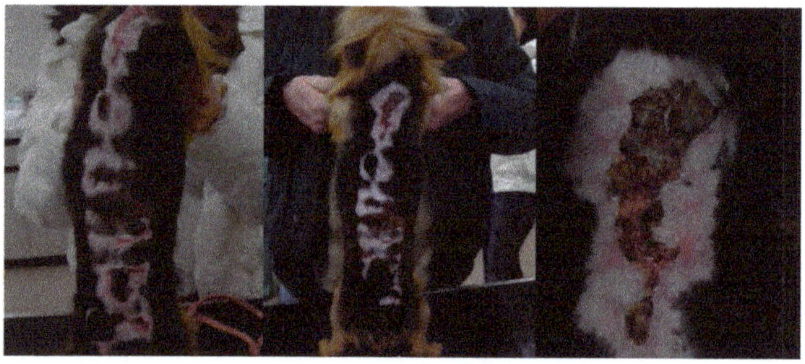

Figure 2. A 5-year-old female Yorkshire. Progressive serpiginous cutaneous lesion on the back with a burn-like appearance that manifested during aNm treatment.

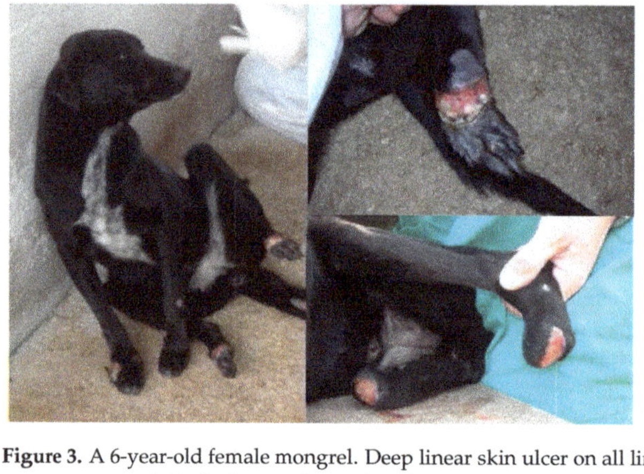

Figure 3. A 6-year-old female mongrel. Deep linear skin ulcer on all limbs that appeared after aNm treatment started. Evolution until complete detachment of the feet despite treatment suspension.

3.2. Prospective Study

Sixteen dogs met the inclusion criteria and were enrolled. The dogs' data, clinical score, laboratory parameters, and LeishVet classification at the time of inclusion are reported in Table 4. Nine dogs were classified as stage II, and seven as stage III, IRIS 1. Two dogs were excluded from the study before D30 due to a severe local reaction (Figure 4); thus, they were treated with an alternative drug (i.e., MIL) (dog 5 and 12). The remaining 14 dogs presented the following adverse events during treatment: a mild and transient subcutaneous reaction appearing in the first week of treatment and self-limiting in 2 dogs (dog 13 and 16), episodic recurrent diarrhea throughout the treatment duration (dog 6), laboratory alterations indicative of liver impairment (ALT 478 IU/L 20–75, AST 301 IU/L 18–50, bilirubin 1.25 mg/dL 0.10–0.30) in the absence of clinical signs at D30 and self-limiting (dog 3).

Table 4. Dogs' data, clinical score, laboratory parameters, and LeishVet classification at the time of inclusion in the prospective study. Dog 5 and dog 12 were excluded from the study before D30 due to severe local reaction.

Dog	Breed	Sex/Age (Years)/ Weight (kg)	Clinical Score	HCT (%)	RBC (×1000/µL)	HGB (gr/dL)	TP (gr/dL)	Alb (gr/dL)	Glob γ (%)	Glob γ (gr/dL)	A/G	IFAT Titer	Urea (mg/dL)	Crea (mg/dL)	UPC	LeishVet Clinical Stage
C1	Mongrel	M/8/32	5	42	7.34	14.5	9	2.07	38.3	3.45	0.3	320	28	0.99	0.26	II
C2	Mongrel	F/3/20	14	40.3	6.57	13.5	13	2.31	61.8	8.03	0.22	320	31	1.02	0.61	II
C3	Mongrel	M/9/30	9	42.4	6.87	14.3	9	2.67	33.5	3.02	0.42	320	34	1.06	0.28	II
C4	German Sheperd	F/6/18	11	38.3	5.97	13.2	7.5	2.66	27.2	2.04	0.55	320	58	1.01	1.1	II
C5	Mongrel	M/7/13	6	40.6	6.14	13.4	11.3	2.85	45.7	5.16	0.34	320	40	1.29	1.1	III IRIS 1
C6	Boxer	F/4/26	10	28.2	4.49	9.2	10.2	1.91	55.4	5.65	0.23	320	30	0.78	3.55	III IRIS 1
C7	Rottweiler	M/6/20	13	30.8	4.65	10.1	8.4	2.27	39.6	3.33	0.37	320	21	0.83	0.4	II
C8	Setter	M/5/17	6	44.6	6.96	15.2	7.5	2.81	36.3	2.72	0.6	320	39	1.07	1.59	III IRIS 1
C9	Mongrel	M/8/13	5	43.1	6.45	14.6	8.1	3.02	14.6	1.18	0.59	320	55	0.95	0.4	II
C10	German Sheperd	F/7/25	7	38.5	6.08	12.8	7.3	2.47	21.6	1.58	0.51	320	32	1.13	0.18	II
C11	Italian Spinone	M/5/20	9	32.9	4.93	10.6	9.2	2.60	38.8	3.57	0.39	320	46	1.07	1.62	III IRIS 1
C12	Labrador	M/12/30	10	24.8	3.80	7.2	10	18.2	53.5	5.32	0.22	320	41	0.78	0.96	III IRIS 1
C13	Mongrel	M/8/22	15	33.9	5.38	11.1	9.4	2.48	46.2	4.34	0.36	320	30	0.95	0.94	II
C14	Mongrel	F/7/18	11	32.5	5.20	10	9	1.75	49.1	4.42	0.24	320	32	1.18	1.49	III IRIS 1
C15	Mongrel	F/5/4.8	12	28.2	4.45	9.4	9.2	2.94	34.3	3.16	0.47	320	28	0.87	1.24	III IRIS 1
C16	Mongrel	F/8/12	22	27.3	4.28	8.3	8.2	1.41	42.3	3.47	0.21	320	24	0.6	1.1	III IRIS 1

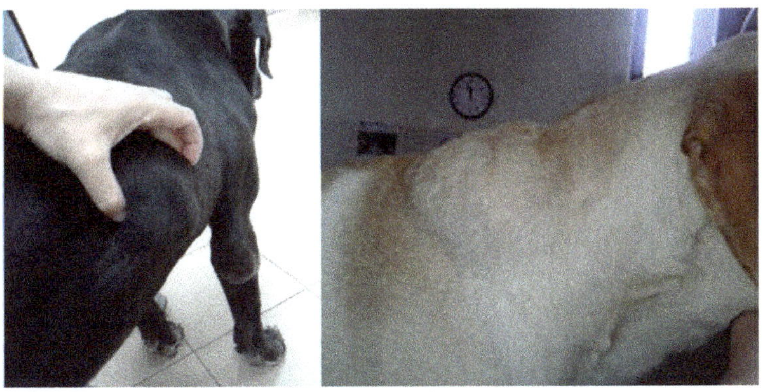

Figure 4. Severe local reaction (granuloma and abscess) after aNm treatment started in two dogs (C5 and C12) that led to their exclusion from the prospective study before D30.

Table 5 displays the parameters of renal function in dogs under aNm treatment, with urea, creatinine, and UPC values at D0, D30, and D60. No animal showed alterations in urea or creatinine parameters during or after therapy; UPC did not worsen in any of the dogs. In all the 6 dogs with proteinuria (UPC > 1) at D0 (LeishVet stage III CKD IRIS 1), a reduction in UPC was registered. The UPC trend is reported in Figure 5.

Table 5. Urine protein/creatinine ratio (UPC), urea, and creatinine parameters of the 16 dogs enrolled in the prospective study evaluated at D0, D30, and D60 to assess the renal impact of N-methylglucamine antimoniate (aNm) treatment. Dogs 5 and 12 were excluded from the study before D30 due to severe local reaction.

	UPC			UREA			CREATININE		
DOGS	**D0**	**D30**	**D60**	**D0**	**D30**	**D60**	**D0**	**D30**	**D60**
C1	0.26	0.24	0.33	28	31	32	0.99	1.04	1.08
C2	0.61	0.27	0.14	31	43	26	1.02	0.89	0.27
C3	0.28	0.97	0.23	34	16	28	1.06	0.69	0.98
C4	0.21	0.17	0.17	58	32	27	1.01	0.71	0.89
C5	1.1			40			1.29		
C6	3.55	1.66	0.39	30	21	15	0.78	0.85	0.95
C7	0.4	0.61	0.28	21	19	23	0.83	0.84	0.8
C8	1.59	0.24	0.24	39	24	34	1.07	0.96	1.08
C9	0.4	0.47	0.55	55	28	30	0.95	0.77	0.73
C10	0.18	0.23	0.27	32	28	36	1.13	1.04	1.02
C11	1.62	0.51	0.2	46	38	38	1.07	0.9	0.89
C12	0.96			41			0.78		
C13	0.94	0.33	0.47	30	32	25	0.95	0.99	1.13
C14	1.49	0.31	0.5	32	22	43	1.18	1.01	1.31
C15	1.24	0.89	0.95	28	23	45	0.87	0.76	0.9
C16	1.1	2.98	0.83	24	40	38	0.6	0.88	0.86

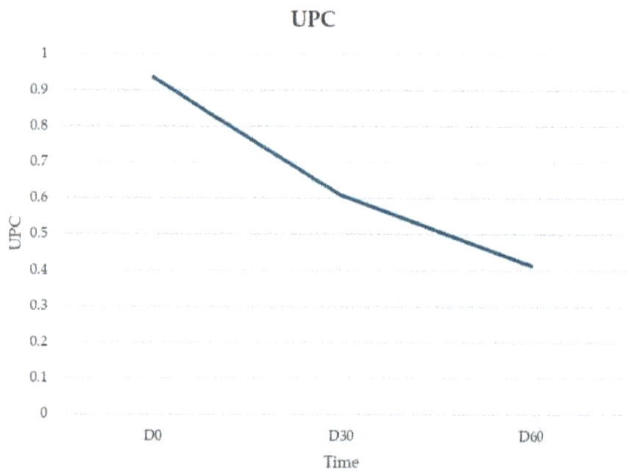

Figure 5. Urine protein/creatinine ratio (UPC) trend assessed at D0, D30 and D60.

The clinical score decreased from 10.64 (average value) to 0.86 from D0 to D180, while the average weight of the 14 dogs included in the study increased from 20.05 to 26 kg. The clinical score showed a 53.7% reduction from D0 to D30, reaching 91.9% at D180. The trend of clinical and laboratory parameters through the six time points is reported in Figure 6. None of the dogs showed clinical/laboratory relapse until the end of the study. In all dogs, IFAT titers decreased or remained stable during the study, but none of the patients showed serological negative results. Lymph node cytology, which was positive in all dogs at D0, turned negative in all dogs at D60. PCR results on bone marrow showed complete parasitological clearance between D0 and D60 in 11/14 animals (Table 6).

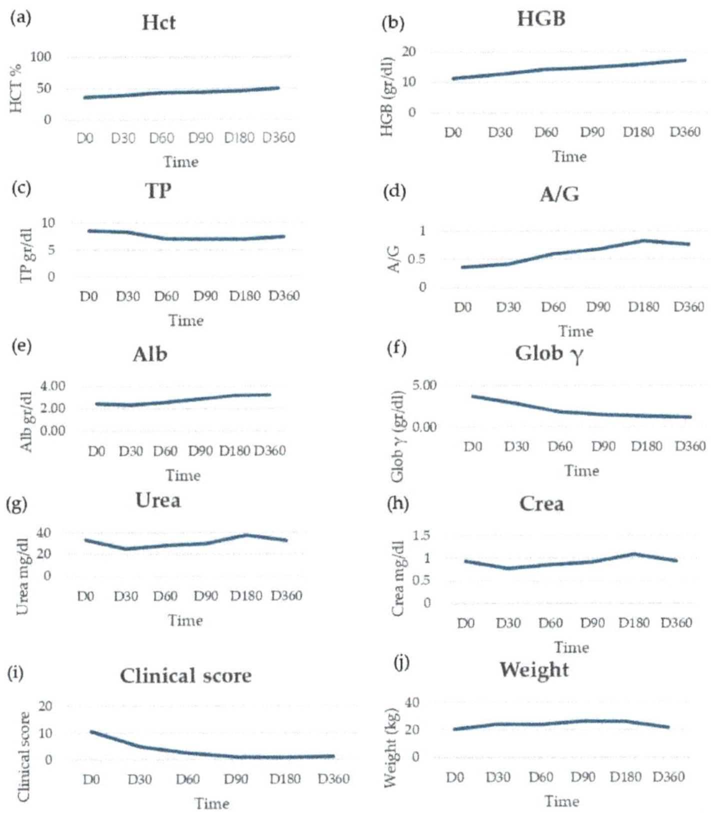

Figure 6. Selected clinical and hematobiochemical parameters trend assessed throughout the study, from D0 to D360. (**a**) Hematocrit (Hct) trend; (**b**) hemoglobin (HGB) trend; (**c**) total protein (TP) trend; (**d**) albumin/globulin ratio (A/G) trend; (**e**) albumin (Alb) trend; (**f**) gamma globulin (Glob γ) trend; (**g**) urea trend; (**h**) creatinine (Crea) trend; (**i**) clinical score trend; (**j**) weight trend.

Table 6. qPCR results on bone marrow at D0 and D60 of the 14 dogs included in the prospective study. In dogs 5 and 12, PCR was not performed since both patients were excluded from the study.

DOG	D0	D60
C1	1.1×10^3	(−)
C2	3.3×10^5	(−)
C3	1.3×10^4	(−)
C4	1.1×10^3	(−)
C6	8.7×10^4	(−)
C7	(−)	(−)
C8	3.5×10^4	1.7×10^4
C9	2.8×10^5	(−)
C10	2.3×10^5	2.4×10^2
C11	1.1×10^6	(−)
C13	9.2×10^5	(−)
C14	6.6×10^6	(−)
C15	6.7×10^5	(−)
C16	6.0×10^6	1.7×10^4

4. Discussion

Potential adverse events associated with aNm treatment were registered in nearly one-third (29.8%) of dogs in the retrospective study. This percentage may have been overestimated, since it includes seven dogs that manifested self-limiting vomiting and diarrhea where dietary indiscretions cannot be completely excluded due to the retrospective nature of the study. Furthermore, the percentage of adverse events decreases to 14.9%, if severe adverse events requiring treatment discontinuation are considered.

The most frequently reported clinical events include apathy, anorexia, vomiting, diarrhea, and pain at the inoculation site [6,9,19,27]. Increased liver transaminases (for transient hepatotoxicity in the absence of clinical signs), and decreased iron levels have also been described [1,16]. It has also been hypothesized that aNm may cause cardiotoxicity (arrhythmias, QT prolongation, and sudden death) in human patients with various forms of leishmaniasis treated with antimonials [28,29]. However, Xenoulis et al. [30] compared serum concentrations of troponin, a marker of heart disease, in dogs with leishmaniasis treated either with aNm or with another drug, showing that no dogs developed signs of cardiotoxicity.

Local reaction at the inoculation site was the most common side effect in the retrospective study, but treatment discontinuation was needed just in one dog. In the prospective study, two dogs developed a severe local reaction resulting in their exclusion from the study, with one requiring surgical curettage. Local reactions were described as warm and painful edematous areas and granulomas localized at the injection sites. As mentioned, these reactions occurred although owners were instructed on how to administer injections. The possibility of local reactions should always be considered, and owners should be instructed to monitor the sites of injections. Noteworthy are the severe systemic reactions characterized by a reluctance to move, tremors and generalized pain affecting mostly small breed dogs and resolving with treatment discontinuation.

One dog developed acute pancreatitis with severe symptoms, requiring treatment suspension and hospitalization. Acute pancreatitis had already been described in rare single cases [18,31]. However, in a study of 30 dogs, specific canine pancreatic lipase concentration (cPLI) did not show a significant increase during aNm treatment, and no signs or alterations attributable to pancreatitis were recorded [30].

Based on the available literature, there are conflicting data concerning the possibility that aNm may induce renal damage and worsen an already compromised renal situation. Some studies have documented a worse impact of aNm on kidneys, evaluated through biochemical parameters and urine examination, in comparison with MIL [8,17,18,20]. Tubular damage was identified through renal biopsy in healthy animals treated with aNm but not in those treated with MIL [19]. On the other site, some studies have reported no impact of aNm on renal function [6,9,12,21]. Furthermore, the use of aNm in patients with documented renal injury associated with CanL is a highly debated issue, but most of the authors suggest that in patients with advanced renal impairment, the use of aNm may be contraindicated [12,27] and in this circumstance, an alternative pharmacological treatment such as MIL is often preferred. Our retrospective study again calls for caution in the use of aNm in patients with renal damage, as three dogs presented worsening renal function parameters associated with clinical signs of renal failure. Moreover, one dog developed severe renal failure, despite normal renal function before treatment (LeishVet stage II). Daza Gonzáles et al. [12] already described the development of acute renal failure after treatment in a single dog and attributed it to antimonate toxicity inducing tubular necrosis or to glomerular disease related to immunocomplex deposition due to rapid parasite death. The dog in our study, a 2-year-old male Dogue de Bordeaux, showed sudden lethargy, anorexia, and weakness about 1 week after the beginning of aNm treatment. The dog's renal condition never reverted to normal and the dog died 1 month later. The authors are more prone to consider aNm-associated direct tubular damage rather than damage due to the rapid death of amastigotes with the development of an acute immune-mediated glomerulopathy because of the absence of proteinuria. Acute pancreatitis and tubular renal damage are rare but possible, probably due to individual predisposition or unknown concomitant factors.

In the prospective study, we decided to select a population of dogs with clinical leishmaniasis associated with no or mild compromised renal function (LeishVet stage II and III, CKD IRIS 1). This decision was made due to ethical reasons and based on the retrospective study results, showing that the impact of aNm on subjects with compromised renal function can lead to clinical worsening. In this selected population, none of the dogs presented worsening of renal function during or after treatment. Only one dog (14) showed temporary hyperazotemia at D180, but timing excludes the hypothesis of an association with aNm.

Regarding the interpretation of UPC, proteinuria was considered of renal–glomerular origin in the absence of urinary active sediment. In most dogs with CKD IRIS 1 (UPC > 1), treatment with aNm was associated with a reduction in proteinuria as previously reported by other authors [12,21,32]. In two dogs, a temporary increase in UPC at D30 was observed (dog 3 and 16). Transient increments in proteinuria had already been reported and attributed to the release of antigens after the parasite's death [33].

From our prospective study data, we can deduce that in dogs with normal renal function or with CKD IRIS 1, treatment with aNm had no negative impact on renal function itself. On the other hand, data from our retrospective study suggest caution in the use of aNm, particularly in subjects with advanced renal injury and/or compromised renal function and again document that acute renal damage can be a possible but rare event in single dogs with normal renal function.

Noteworthy are the severe idiosyncratic cutaneous reactions not previously reported to the author's knowledge. It is interesting that in the three dogs presented for consultation due to the rapid onset of cutaneous signs under treatment, therapy was not immediately discontinued because neither the owners nor the referral vet quickly associated them with aNm adverse events. Treatment discontinuation allowed complete recovery in two dogs. On the contrary, in the third dog, despite treatment suspension, the process progressed. In this extremely severe case, it is possible to hypothesize that immune-mediated vasculitis probably due to the rapid death of parasites could have contributed to the dramatic evolution. Cutaneous vasculitis has already been described as a post-treatment complication in CanL [9].

Due to the severity of some reported conditions, it is critical that clinicians and owners are aware of the possibility of immediately discontinuing treatment if unexpected signs appear during therapy.

Concerning aNm clinical efficacy, the results of our prospective study confirm what has already been reported by numerous authors, including the following: clinical score improvement already evident at the end of treatment, normalization of hematobiochemical alterations, and absence of clinical relapses until the end of the study.

It was not surprising that IFAT titers after treatment slowly reduced in most of the dogs without seroconversion in any dog. The relevance of monitoring IFAT titers for disease follow-up and post-treatment monitoring has long been controversial [22], but common opinion is now that it is of limited value. In fact, IFAT does not show any significant correlation with the clinical status of the animals and in some subjects, the antibody titers may remain elevated for years after treatment, particularly in endemic areas [9,34].

The most interesting finding of our study is the high number of dogs (78.6%) that tested negative with qPCR at D60, suggesting that the parasitological clearance after treatment could be more frequent than previously suggested; in fact, it is reported in the literature as a possible but rare event under leishmanicidal treatment [1]. Efficacy of the aNm tested protocol in terms of parasitic clearance was very high in our study. In a recent study by our own research group that aimed to compare two dosing regimens of MIL [7], and working with a similar study design and dog population, bone marrow qPCR was negative in 50% and 14.3% of the dogs treated with a modified or standard dose of MIL, respectively [7]. The high percentage of parasitic clearance after treatment leads the authors to suggest that the potential side effects do not justify discarding the aNm protocol, which remains an excellent option at least in the absence of renal damage.

A limitation of the retrospective study is that the adverse effects were extrapolated based solely on the presence of clinical manifestations. Any possible subclinical laboratory alterations presented during treatment were not investigated. Only in dogs for whom clinical adverse events were registered, collateral tests were evaluated for a better understanding of the condition (i.e., renal failure, pancreatitis).

A limitation of the prospective study is the limited number of dogs and the lack of a specific control group. To address the latter limitation, the authors referred to a previous work of the same research group to compare efficacy in terms of parasitic clearance.

5. Conclusions

In conclusion, treatment with aNm in dogs with CanL can be associated with mild to severe adverse events including local, systemic and idiosyncratic reactions, and can have a potential impact on renal function, especially if already compromised. Veterinarians must be vigilant regarding the potentially serious adverse effects associated with aNm and immediately stop drug administration if unexpected clinical manifestations occur. Nevertheless, it is the authors' opinion that the aNm protocol should still be proposed as the first-choice treatment for CanL, at least in dogs with preserved renal function and in CKD IRIS 1. Monitoring during treatment and owner compliance are essential to its safe use; what does not kill you makes you stronger!

Author Contributions: Conceptualization, P.P.; methodology, P.P.; software, S.C.; investigation, S.D., A.C. and B.G. data curation, S.C., S.D. and A.R.; writing—original draft preparation, P.P. and S.D.; writing—review and editing, P.P., S.D., A.R. and D.B.; supervision, P.P. All authors have read and agreed to the published version of the manuscript.

Funding: This research received no external funding.

Institutional Review Board Statement: The animal study protocol was approved by the Ethics Committee of the Department of Emergency and Organ Transplantation, (Prot. DETO 226/2018, 26 January 2018).

Informed Consent Statement: Written informed consent was obtained from all owners of the animals included in the perspective study.

Data Availability Statement: The dataset is available from the authors.

Conflicts of Interest: The authors declare no conflicts of interest.

References

1. Morales-Yuste, M.; Martín-Sánchez, J.; Corpas-Lopez, V. Canine Leishmaniasis: Update on Epidemiology, Diagnosis, Treatment, and Prevention. *Vet. Sci.* **2022**, *9*, 387. [CrossRef] [PubMed]
2. Ribeiro, P.A.F.; Dias, D.S.; Novais, M.V.M.; Lage, D.P.; Tavares, G.S.V.; Mendonça, D.V.C.; Oliveira, J.S.; Chávez-Fumagalli, M.A.; Roatt, B.M.; Duarte, M.C.; et al. A *Leishmania* Hypothetical Protein-Containing Liposome-Based Formulation Is Highly Immunogenic and Induces Protection against Visceral Leishmaniasis. *Cytokine* **2018**, *111*, 131–139. [CrossRef] [PubMed]
3. Noli, C.; Saridomichelakis, M.N. An Update on the Diagnosis and Treatment of Canine Leishmaniosis Caused by *Leishmania Infantum* (Syn. *L. Chagasi*). *Vet. J.* **2014**, *202*, 425–435. [CrossRef] [PubMed]
4. Baneth, G.; Shaw, S.E. Chemotherapy of canine leishmaniosis. *Vet. Parasitol.* **2002**, *106*, 315–324. [CrossRef] [PubMed]
5. Paradies, P.; Sasanelli, M.; de Caprariis, D.; Testini, G.; Traversa, D.; Lia, R.P.; Dantas-Torres, F.; Otranto, D. Clinical and Laboratory Monitoring of Dogs Naturally Infected by *Leishmania infantum*. *Vet. J.* **2010**, *186*, 370–373. [CrossRef] [PubMed]
6. Rambaldi, J.; Barbarossa, A.; Morbidelli, E.; Zaghini, A. Tolerability and Pharmacokinetics of Two Antimony Products after Subcutaneous Administration in Dogs. *J. Infect. Dev. Ctries* **2018**, *12*, 279–283. [CrossRef] [PubMed]
7. Iarussi, F.; Paradies, P.; Foglia Manzillo, V.; Gizzarelli, M.; Caratozzolo, M.F.; Navarro, C.; Greco, B.; Rubino, G.T.R.; Oliva, G.; Sasanelli, M. Comparison of Two Dosing Regimens of Miltefosine, Both in Combination with Allopurinol, on Clinical and Parasitological Findings of Dogs with Leishmaniosis: A Pilot Study. *Front. Vet. Sci.* **2020**, *7*, 577395. [CrossRef] [PubMed]
8. Manna, L.; Corso, R.; Galiero, G.; Cerrone, A.; Muzj, P.; Gravino, A.E. Long-Term Follow-up of Dogs with Leishmaniosis Treated with Meglumine Antimoniate plus Allopurinol versus Miltefosine plus Allopurinol. *Parasites Vectors* **2015**, *8*, 289. [CrossRef] [PubMed]

9. Torres, M.; Bardagí, M.; Roura, X.; Zanna, G.; Ravera, I.; Ferrer, L. Long Term Follow-up of Dogs Diagnosed with Leishmaniosis (Clinical Stage II) and Treated with Meglumine Antimoniate and Allopurinol. *Vet. J.* **2011**, *188*, 346–351. [CrossRef]
10. Oliva, G.; Roura, X.; Crotti, A.; Maroli, M.; Castagnaro, M.; Gradoni, L.; Lubas, G.; Paltrinieri, S.; Zatelli, A.; Zini, E. Guidelines for Treatment of Leishmaniasis in Dogs. *J. Am. Vet. Med. Assoc.* **2010**, *236*, 1192–1198. [CrossRef]
11. Paradies, P.; Sasanelli, M.; Amato, M.E.; Greco, B.; De Palo, P.; Lubas, G. Monitoring the Reverse to Normal of Clinico-Pathological Findings and the Disease Free Interval Time Using Four Different Treatment Protocols for Canine Leishmaniosis in an Endemic Area. *Res. Vet. Sci.* **2012**, *93*, 843–847. [CrossRef] [PubMed]
12. Daza González, M.A.; Miró, G.; Fermín Rodríguez, M.; Rupérez Noguer, C.; Fragío Arnold, C. Short Term Impacts of Meglumine Antimoniate Treatment on Kidney Function in Dogs with Clinical Leishmaniosis. *Res. Vet. Sci.* **2019**, *126*, 131–138. [CrossRef] [PubMed]
13. Rossi, G.; Ibba, F.; Meazzi, S.; Giordano, A.; Paltrinieri, S. Paraoxonase Activity as a Tool for Clinical Monitoring of Dogs Treated for Canine Leishmaniasis. *Vet. J.* **2014**, *199*, 143–149. [CrossRef] [PubMed]
14. Martinez-Subiela, S.; Pardo-Marín, L.; Tecles, F.; Baneth, G.; Cerón, J.J. Serum C-Reactive Protein and Ferritin Concentrations in Dogs Undergoing Leishmaniosis Treatment. *Res. Vet. Sci.* **2016**, *109*, 17–20. [CrossRef] [PubMed]
15. Santos, M.F.; Alexandre-Pires, G.; Pereira, M.A.; Marques, C.S.; Gomes, J.; Correia, J.; Duarte, A.; Gomes, L.; Rodrigues, A.V.; Basso, A.; et al. Meglumine Antimoniate and Miltefosine Combined With Allopurinol Sustain Pro-Inflammatory Immune Environments During Canine Leishmaniosis Treatment. *Front. Vet. Sci.* **2019**, *6*, 362. [CrossRef] [PubMed]
16. Ortega, V.; Radaic, A.; de Jesus, M.B.; de Paula, E.; Giorgio, S. Improved Efficacy of Meglumine Antimoniate Incorporated in Anionic Liposomes against Leishmania Infantum Infecting Canine Macrophages. *J. Pharm. Pharmacol.* **2022**, *74*, 896–904. [CrossRef] [PubMed]
17. Mateo, M.; Maynard, L.; Vischer, C.; Bianciardi, P.; Miró, G. Comparative Study on the Short Term Efficacy and Adverse Effects of Miltefosine and Meglumine Antimoniate in Dogs with Natural Leishmaniosis. *Parasitol. Res.* **2009**, *105*, 155–162. [CrossRef] [PubMed]
18. Ikeda-Garcia, F.A.; Lopes, R.S.; Ciarlini, P.C.; Marques, F.J.; Lima, V.M.F.; Perri, S.H.V.; Feitosa, M.M. Evaluation of Renal and Hepatic Functions in Dogs Naturally Infected by Visceral Leishmaniasis Submitted to Treatment with Meglumine Antimoniate. *Res. Vet. Sci.* **2007**, *83*, 105–108. [CrossRef] [PubMed]
19. Bianciardi, P.; Brovida, C.; Valente, M.; Aresu, L.; Cavicchioli, L.; Vischer, C.; Giroud, L.; Castagnaro, M. Administration of Miltefosine and Meglumine Antimoniate in Healthy Dogs: Clinicopathological Evaluation of the Impact on the Kidneys. *Toxicol. Pathol.* **2009**, *37*, 770–775. [CrossRef]
20. Proverbio, D.; Spada, E.; de Giorgi, G.B.; Perego, A.R. Proteinuria Reduction after Treatment with Miltefosine and Allopurinol in Dogs Naturally Infected with Leishmaniasis. *Vet. World.* **2016**, *9*, 904–908. [CrossRef]
21. Roura, X.; Fondati, A.; Lubas, G.; Gradoni, L.; Maroli, M.; Oliva, G.; Paltrinieri, S.; Zatelli, A.; Zini, E. Prognosis and Monitoring of Leishmaniasis in Dogs: A Working Group Report. *Vet. J.* **2013**, *198*, 43–47. [CrossRef] [PubMed]
22. Solano-Gallego, L.; Miró, G.; Koutinas, A.; Cardoso, L.; Pennisi, M.G.; Ferrer, L.; Bourdeau, P.; Oliva, G.; Baneth, G. LeishVet Guidelines for the Practical Management of Canine Leishmaniosis. *Parasites Vectors* **2011**, *4*, 86. [CrossRef] [PubMed]
23. Miró, G.; Oliva, G.; Cruz, I.; Cañavate, C.; Mortarino, M.; Vischer, C.; Bianciardi, P. Multicentric, controlled clinical study to evaluate effectiveness and safety of miltefosine and allopurinol for canine leishmaniosis. *Vet. Dermatol.* **2009**, *20*, 397–404. [CrossRef] [PubMed]
24. Saridomichelakis, M.N.; Mylonakis, M.E.; Leontides, L.S.; Koutinas, A.F.; Billinis, C.; Kontos, V.I. Evaluation of Lymph Node and Bone Marrow Cytology in the Diagnosis of Canine Leishmaniasis (*Leishmania infantum*) in Symptomatic and Asymptomatic Dogs. *Am. J. Trop. Med. Hyg.* **2005**, *73*, 82–86. [CrossRef] [PubMed]
25. Gradoni, L.; Gramiccia, M. *"Leishmaniosis" in OIE Manual of Diagnostic Tests and Vaccines for Terrestrial Animals (Mammals, Birds and Bees)*, 6th ed.; Office International des Epizooties: Paris, France, 2008; pp. 240–250.
26. van Eys, G.J.J.M.; Schoone, G.J.; Kroon, N.C.M.; Ebeling, S.B. Sequence Analysis of Small Subunit Ribosomal RNA Genes and Its Use for Detection and Identification of *Leishmania* Parasites. *Mol. Biochem. Parasitol.* **1992**, *51*, 133–142. [PubMed]
27. Pereira, M.A.; Santos, R.; Oliveira, R.; Costa, L.; Prata, A.; Gonçalves, V.; Roquette, M.; Vala, H.; Santos-Gomes, G. Prognostic Factors and Life Expectancy in Canine Leishmaniosis. *Vet. Sci.* **2020**, *7*, 128. [CrossRef] [PubMed]
28. Berhe, N.; Abraham, Y.; Hailu, A.; Ali, A.; Mengistu, G.; Tsige, K.; Abebe, Y. Electrocardiographic Findings in Ethiopians on Pentavalent Antimony Therapy for Visceral Leishmaniasis. *East. Afr. Med. J.* **2001**, *78*, 608–610. [CrossRef] [PubMed]
29. Oliveira, L.F.; Schubach, A.O.; Martins, M.M.; Passos, S.L.; Oliveira, R.V.; Marzochi, M.C.; Andrade, C.A. Systematic Review of the Adverse Effects of Cutaneous Leishmaniasis Treatment in the New World. *Acta Trop.* **2011**, *118*, 87–96. [CrossRef]
30. Xenoulis, P.G.; Saridomichelakis, M.N.; Chatzis, M.K.; Kasabalis, D.; Petanides, T.; Suchodolski, J.S.; Steiner, J.M. Prospective Evaluation of Serum Pancreatic Lipase Immunoreactivity and Troponin I Concentrations in Leishmania Infantum-Infected Dogs Treated with Meglumine Antimonate. *Vet. Parasitol.* **2014**, *203*, 326–330. [CrossRef]
31. Aste, G.; Di Tommaso, M.; Steiner, J.M.; Williams, D.A.; Boari, A. Pancreatitis Associated with N-Methyl-Glucamine Therapy in a Dog with Leishmaniasis. *Vet. Res. Commun.* **2005**, *29* (Suppl. S2), 269–272. [CrossRef]
32. Plevraki, K.; Koutinas, A.F.; Kaldrymidou, H.; Roumpies, N.; Papazoglou, L.G.; Saridomichelakis, M.N.; Savvas, I.; Leondides, L. Effects of Allopurinol Treatment on the Progression of Chronic Nephritis in Canine Leishmaniosis (Leishmania Infantum). *J. Vet. Intern. Med.* **2006**, *20*, 228–233. [CrossRef] [PubMed]

33. Paltrinieri, S.; Mangiagalli, G.; Ibba, F. Use of Urinary γ-Glutamyl Transferase (GGT) to Monitor the Pattern of Proteinuria in Dogs with Leishmaniasis Treated with N-Methylglucamine Antimoniate. *Res. Vet. Sci.* **2018**, *119*, 52–55. [CrossRef] [PubMed]
34. Olías-Molero, A.; Corral, M.J.; Jiménez-Antón, M.D.; Alunda, J.M. Early antibody response and clinical outcome in experimental canine leishmaniasis. *Sci. Rep.* **2019**, *9*, 18606. [CrossRef] [PubMed]

animals

MDPI

Article

Association between Thoracic Radiographic Changes and Indicators of Pulmonary Hypertension in Dogs with Heartworm Disease

Soraya Falcón-Cordón [1], Yaiza Falcón-Cordón [1], Alicia Caro-Vadillo [2], Noelia Costa-Rodríguez [1], José Alberto Montoya-Alonso [1,*] and Elena Carretón [1]

[1] Internal Medicine, Veterinary Medicine and Therapeutic Research Group, Faculty of Veterinary Medicine, Research Institute of Biomedical and Health Sciences (IUIBS), Universidad de Las Palmas de Gran Canaria (ULPGC), 35016 Las Palmas de Gran Canaria, Spain; soraya.falcon@ulpgc.es (S.F.-C.); yaiza.falcon@ulpgc.es (Y.F.-C.); noelia.costa@ulpgc.es (N.C.-R.); elena.carreton@ulpgc.es (E.C.)

[2] Department of Animal Medicine and Surgery, Faculty of Veterinary Medicine, Complutense University, 28040 Madrid, Spain; aliciac@vet.ucm.es

* Correspondence: alberto.montoya@ulpgc.es

Simple Summary: Pulmonary hypertension (PH) is a high-risk condition in dogs with heartworm disease (*Dirofilaria immitis*). Echocardiography is the diagnostic technique of choice to detect PH; however, it is not accessible to all routine clinicians. Therefore, given the importance of this condition during an infection with *D. immitis*, the aim of this study was to evaluate the association of the radiological findings in dogs with heartworm disease and the presence or absence of echocardiographically characterised PH. The results obtained suggest that the evaluation of certain radiographic measures may be useful in the preliminary evaluation of the thoracic radiographs of a dog as a preliminary screening when assessing whether to perform complementary tests to evaluate the presence of PH in dogs with heartworm disease.

Abstract: Pulmonary hypertension (PH) is a consequence of pulmonary endarteritis during infection with *Dirofilaria immitis* in dogs. Echocardiography is the technique of choice but is not always accessible to all clinicians. This study aimed to evaluate the association of the radiological findings in dogs with heartworm disease and the presence or absence of echocardiographically characterised PH. The study included 62 heartworm-infected dogs that underwent thoracic radiographs and echocardiography. The studied dogs showed moderate to severe PH when the Right Pulmonary Artery Distensibility (RPAD) Index was <29.5%. The RPAD Index was used for comparison with thoracic radiographs. The Vertebral Heart Size (VHS), right cranial pulmonary artery passing through the fourth rib in the laterolateral projection (CrPA/R4) ratio, and right caudal pulmonary artery to the ninth rib in the dorsoventral projection (CdPA/R9) ratio showed significant differences between dogs with/without PH ($p < 0.001$). Sensitivity (sen) and specificity (sp) cut-off values were obtained: VHS $\geq$ 9.53 (sen 93.75%, sp 63.33%); CrPA/R4 $\geq$ 1.08 (sen 87.5%, sp 70%); and CdPA/R9 $\geq$ 1.10 (sen 96.88%, sp 76.66%). The CrPA/R4 and CdPA/R9 ratios showed potential as a preliminary screening tool for PH in heartworm-infected dogs, suggesting that they may reliably indicate the presence of PH and guide the decision for further diagnostic testing.

Keywords: vector-borne disease; *Dirofilaria immitis*; pulmonary hypertension; radiographic indexes; image diagnosis; echocardiography; veterinary diagnosis

1. Introduction

Pulmonary hypertension (PH) is one of the most common findings in dogs infected by *Dirofilaria immitis* (heartworm disease) as a consequence of the chronic development of proliferative endarteritis within the pulmonary vasculature [1]. The diagnosis of PH is mainly based on transthoracic Doppler echocardiography, which provides a non-invasive and reliable method for estimating pulmonary arterial pressure since right heart catheterisation, the gold standard for diagnosing PH, is unavailable and unacceptably invasive in compromised patients [2]. However, this method has limitations as the diagnosis is often based on indirect and subjective parameters. In addition, some of these echocardiographic measurements, such as tricuspid valve regurgitation, can be difficult to achieve. On the other hand, other estimators, such as the Right Pulmonary Artery Distensibility Index (RPAD Index), have been shown to be of great help in estimating the presence of PH [3], especially in dogs with heartworm disease [4], which can be useful in cases in which tricuspid regurgitation or pulmonary regurgitation cannot be measured.

The radiographic changes that occur in canine heartworm disease also provide important information. In infected dogs, the main findings are dilatation of the main pulmonary artery and tortuosity of the pulmonary arteries; right ventricular enlargement may also be observed in chronic infestations [5,6]. Furthermore, these signs are associated with the presence of PH in dogs with heartworm disease [7,8]. Thoracic radiography can provide supportive evidence for PH and information on concomitant or causative diseases in an individual dog [2,9], and, unlike echocardiography, this imaging technique is mostly available to the everyday clinician and does not require such specific training or dedicated equipment. Therefore, it would be interesting to perform studies aimed at evaluating its usefulness in detecting the presence of PH in this disease.

In fact, there are studies that have characterised the association of radiological and echocardiographic findings in dogs with PH [10,11], but there are not as many studies performed in dogs with heartworm disease. Given that PH is a common and serious condition in this pathology and given the unique and characteristic changes that pulmonary endarteritis produces in this pathology, this research aims to evaluate the association of the radiological findings in dogs with heartworm disease and the presence or absence of echocardiographically characterised PH.

2. Materials and Methods

The study included 62 dogs owned by clients who were brought to the Veterinary Service of the University of Las Palmas of Gran Canaria. The dogs lived in a hyperendemic area for *D. immitis* [12–14]. Inclusion in the study was based on a positive result for circulating *D. immitis* antigens (Urano test Dirofilaria®, Urano Vet SL, Barcelona, Spain). Dogs were also examined for the presence or absence of microfilariae using a modified Knott test. Clinical history and data including age, sex, and breed were recorded for each animal. A complete history and examination were performed on each dog to rule out the presence of other pathologies that might influence the results, and animals with concomitant diseases were excluded from the study.

On the day of diagnosis and the start of treatment (day 0), digital thoracic radiographs were taken using a radiographic unit (RX generator; model: HFQ 300 P, Bennett, NC, USA) at the time of peak inspiration without sedation. The examination protocols (kVP and mAs) were adapted specifically for each dog according to the thoracic thickness of the dog. Views were obtained in its right laterolateral and dorsoventral projections. Vertebral Heart Size (VHS) was measured according to the guidelines of Buchanan and Bücheler 1995 [15]. Although no consideration was given to excluding dogs that might have some type of alteration in the dorsal spine when interpreting VHS, none of the dogs in the present study had such alterations [16].

In addition, the diameter of the right cranial pulmonary artery (CrPA) passing through the fourth rib (R4) in the laterolateral projection and the diameter of R4 at a point just distal to the spine, as well as the distal and left sides of the summation shadow created by the right caudal pulmonary artery (CdPA) with the ninth rib (R9) in the dorsoventral projection, were also measured according to previous guidelines [17]. CrPA/R4 and CdPA/R9 ratios were calculated from these measurements (Figure 1). Measurements were performed using electronic callipers on a DICOM workstation (DAIPACS. 2.71 version). All measurements were performed by the same technician, blinded to the clinical status of the dogs included in this study.

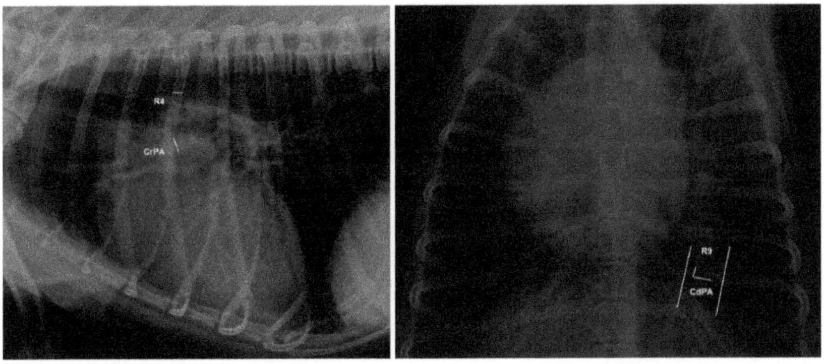

(A) **(B)**

Figure 1. Right laterolateral (**A**) and dorsoventral (**B**) thoracic radiographs illustrating the measurement methods for this study. (**A**) The diameter of the right cranial lobar artery (CrPA) at the level of the fourth rib (R4) and the fourth rib just distal to the spine were measured, and (**B**) the diameter of the caudal lobar artery (CdPA) passing through the ninth rib (R9).

Dogs underwent echocardiographic examination using an ultrasound machine with spectral and colour Doppler and multifrequency probes (5.5–10 MHz) (Logic P5, General Electric, New York, NY, USA). Dogs were placed in the right laterolateral position with the transducer in the third intercostal space. Dogs were conscious and monitored electrocardiographically throughout the study. Six continuous cardiac cycles were recorded for each measurement. All echocardiography exams were performed by the same technician. The presence or absence of PH was determined according to the American College of Veterinary Internal Medicine (ACVIM) guidelines [9]. Of all the echocardiographic indices studied, the determination of the Right Pulmonary Artery Distensibility Index (RPAD Index) was used in this study for comparison with thoracic radiographs as all dogs showed higher likelihood of moderate to severe PH when the RPAD Index was <29.5%, as previously described and validated in dogs with heartworm disease [3,4,18].

In addition, other echocardiographic findings were used to estimate worm burden [19], and a score of 1 to 4 was assigned from low to high worm burden as follows: (1) no worms visualised, (2) few worm echoes in the distal part of the right pulmonary artery, (3) worm echoes occupying the right pulmonary artery and extending to the main pulmonary artery, and (4) worm echoes occupying the entire right pulmonary artery and the main pulmonary artery to the level of the pulmonary valve. Scores of 1 and 2 corresponded to low parasite burden, and scores of 3 and 4 corresponded to high burden.

The data were analysed using the SPSS Base 29.0 software for Windows (SPSS Inc./IBM, Chicago, IL, USA). A Shapiro–Wilk test was performed to verify the normal distribution of the data. Continuous variables were expressed as median ± standard deviation. Qualitative variables were expressed as percentage. The chi-squared test or Fisher's exact test was used to assess the association between categorical variables. In all cases, a p value < 0.05 was determined as significant. The results of the statistical procedures were also graphed by scatter plot. A simple linear regression was performed between the RPAD Index values and the other variables studied (VHS, CrPA/R4, and CdPA/R9 ratios) to identify the best one-variable model, and a regression analysis of all subsets was performed with a maximum improvement of R^2 as a selection criterion. Receiver operator characteristic curve (ROC) analyses were performed to determine the optimal cut-off values for the prediction of the RPAD Index being <29.5% (moderate or severe hypertension). For all results, $p < 0.05$ was considered statistically significant. Furthermore, Cohen's D was employed to interpret the differential magnitude between the studied statistical groups considering a statistical difference with values > 0.70 for this study.

All the owners provided their consent to participate in this study, which was carried out in accordance with the current European legislation on animal protection.

3. Results

Of the studied dogs, 27 were male and 35 were female, with the ages ranging from 1.5 to 12 years (mean: 5.25 years). Based on breed, 37 were mixed-breed dogs and 25 were pure-bred dogs. PH was present in 32 dogs (51.6%), with a mean RPAD Index of 29.1%. Microfilaremia was present in 40.3% of the dogs. The parasite burden was low in 87.1% of the dogs and high in 12.9% of them.

The results showed significant differences between the body weight and worm burden ($p = 0.023$), with the mean weight being 18.3 ± 10.3 kg for the dogs with a low worm burden and 11.1 ± 7.8 kg for the dogs with a high worm burden. Additionally, the dogs with PH were significantly older ($p = 0.021$).

The VHS showed a mean value of 10.1 ± 0.8 for all the studied dogs, with significant differences observed when differentiating based on the presence or absence of PH (10.41 ± 0.81 vs. 9.72 ± 0.81, respectively) ($p < 0.001$). There were no statistically significant differences in the VHS values based on the microfilaremia, parasite load, age, or sex.

The results for the CrPA/R4 and CdPA/R9 ratios were 1.21 ± 0.39 and 1.4 ± 0.5, respectively. Significant differences were observed between the groups for the CrPA/R4 ratio (1.37 ± 0.44 vs. 1.01 ± 0.23, respectively) ($p < 0.001$) and for the CdPA/R9 ratio (1.63 ± 0.56 vs. 1.16 ± 0.27, respectively) ($p < 0.001$) (Table 1). Statistically significant differences were also found for the CdPA/R9 ratio in relation to the presence or absence of microfilaremia (1.45 ± 0.35 vs. 1.37 ± 0.58, respectively) ($p = 0.031$). No significant differences were found for the rest of the studied parameters (parasite load, age, or sex).

Table 1. Correlation coefficients for all studied radiographic parameters.

Correlation	Coefficient	Interpretation
IDAPD-VHS	−0.4776417 ***	Moderate negative correlation
IDAPD-CRPA/R4	−0.4344274 ***	Moderate negative correlation
IDAPD-CDPA/R9	−0.53781329 ***	Moderate negative correlation

*** Correlation is significant at 0.5% ($p < 0.005$).

The Pearson correlation model was used to determine whether there was a correlation between the presence of PH, based on the RPAD Index, and the studied radiographic parameters (VHS, CrPA/R4, and CdPA/R9 ratios). The correlations obtained for all three were moderately negative, indicating that, as the RPAD Index decreased, the other parameters increased. Furthermore, the results for all three correlations were statistically significant ($p < 0.005$) (Table 1).

In this study, regression analysis was performed to determine the area under the curve (AUC), coefficient of determination (R^2), and the specificity and sensitivity of the radiographic indicators (VHS, CrPA/R4, and CdPA/R9 ratios). In addition, cut-off values were established for each of these parameters.

The CdPA/R9 ratio model proved to be the most effective in explaining the variability of the dependent variable (RPAD Index), with an R^2 of 0.976, followed by the CrPA/R4 ratio and VHS. This indicated that CdPA/R9 had a superior ability to model the influence of the independent variables on the RPAD Index.

AUC values were calculated for each radiographic indicator and showed that all the values were above 0.5 but below 1. This suggested that the models provided a more accurate classification than would be achieved by chance, enabling the correct prediction of positive and negative cases. Specifically, the AUC for VHS was 0.78, indicating good discriminatory ability, with a 78% probability of distinguishing between the positive and negative cases.

Similarly, the CrPA/R4 ratio showed an AUC of 0.77, and the CdPA/R9 ratio, the highest, reached an AUC of 0.82, highlighting its superior performance in model discrimination.

For VHS, a cut-off of 9.53 or higher resulted in a sensitivity of 93.75% and a specificity of 63.33%. This indicated a high ability to detect true positive cases, but with a moderate false positive rate. For the CrPA/R4 ratio, a cut-off of 1.08 or higher yielded a sensitivity of 87.5% and a specificity of 70%, providing a reasonable balance between detecting positive cases and minimising false positives. Finally, the cut-off for the CdPA/R9 ratio was set at 1.10 or higher, providing a sensitivity of 96.88% and a specificity of 76.66%, making it the most efficient of the three in terms of correctly identifying both positive and negative cases (Table 2).

Table 2. Results of simple regression analyses for the prediction of PH using the Right Pulmonary Artery Distensibility Index (RPADI < 29.5%). R^2 (coefficient of determination); AUC (area under receiver operating characteristic curve); VHS (Vertebral Heart Size); CrPA/R4 (right cranial pulmonary artery passing through the fourth rib in the laterolateral projection ratio); CdPA/R9 (right caudal pulmonary artery to the ninth rib in the dorsoventral projection ratio).

Parameter	R^2	AUC	Cut-Off Value	Sensitivity	Specificity
VHS	0.853	0.78	≥9.53	93.75%	63.33%
CRPA/R4	0.959	0.77	≥1.08	87.5%	70%
CDPA/R9	0.976	0.82	≥1.10	96.88%	76.66%

In addition, Cohen's D, a standardised measure of effect size, was used to interpret the magnitude of the differences between the statistical groups studied. For VHS, the calculated Cohen's D value of 22.11 indicated a substantial difference between the dogs with the presence of pulmonary hypertension and those in absence. This extremely large effect size suggests a clinically meaningful distinction between the two groups in terms of the VHS measurements. Similarly, for the CRPA/R4 ratio, the Cohen's D value of 4.31 indicated a significant difference between the groups. Although this is slightly smaller than the effect size for VHS, it is still a large effect size, indicating a notable difference in the CRPA/R4 ratio between the two groups. In the case of the CDPA/R9 ratio, the Cohen's D value of 0.702, while indicating a smaller effect size compared to VHS and the CDPA/R4 ratio, still reflects a moderate difference between the dogs without PH and the group with the presence of PH. This suggests that, although the effect size was smaller, the CDPA/R9 ratio remains a relevant and potentially informative parameter to discriminate between these groups, underlining its importance in the assessment of PH (Table 3).

Table 3. Cohen's D for the three studied parameters. VHS (Vertebral Heart Size); CrPA/R4 (right cranial pulmonary artery passing through the fourth rib in the laterolateral projection ratio); CdPA/R9 (right caudal pulmonary artery to the ninth rib in the dorsoventral projection ratio).

Measure	Cohen's D	Interpretation
VHS	22.1	Large effect
CrPA/R4	4.31	Large effect
CdPA/R9	0.702	Moderate effect

Cohen's D: d = 0.2 small effect; d = 0.5 moderate effect; d = 0.8 large effect.

4. Discussion

Pulmonary hypertension is caused by the presence of adult *D. immitis* parasites as a result of the lesions they cause in the pulmonary arteries from the early stages of parasitism. These lesions lead to proliferative endarteritis, resulting in the enlargement, tortuosity, and loss of elasticity of the pulmonary vasculature [1]. PH is a common phenomenon in this pathology, severe and apparently irreversible in most cases [20,21], so its study and early detection should be a priority.

Echocardiography is the method of choice to detect this condition, but this technique is often inaccessible to clinical veterinarians, either due to a lack of knowledge or equipment, or to animal owners due to financial constraints. However, radiology is a diagnostic imaging technique widely used in all veterinary clinics that is less technically demanding and more affordable [22]. Therefore, in this study, the authors aimed to evaluate its usefulness as a first approximation imaging tool to determine the presence or absence of PH in dogs with heartworm disease.

The results showed significant differences between the VHS indices and the CrPA/R4 and CdPA/R9 ratios in animals with PH compared with those without PH in dogs with heartworm disease. Previous studies have demonstrated radiological changes in dogs with PH caused by several pathologies, mainly cardiomegaly, with other findings described as infiltration of the lung parenchyma and enlargement of the pulmonary arteries [10,23]. In addition, the ECVIM guidelines recommend that echocardiography be performed when a dog has radiological changes consistent with PH [9]. Indeed, thoracic radiography is a useful technique for detecting cardiopulmonary abnormalities in dogs with *D. immitis*, with the enlargement of the right ventricle, the main pulmonary artery, and the right lobar pulmonary artery being the most commonly reported abnormalities [24–26]. However, no studies were found on the use of radiological changes objectively as an estimator of PH, so the results of this study validate the clinical usefulness of radiographs in dogs with heartworms to estimate the presence of PH.

The results obtained showed that the VHS was increased in the group with PH compared to the group without PH. Other authors have reported increased cardiac silhouettes on thoracic radiographs in dogs with PH, especially when symptomatic [25,27]. The cut-off value of 9.53 provides high sensitivity but low specificity for the detection of PH, which is consistent as the value of 9.53 is within the normal range for healthy dogs [15]. Therefore, although statistically there are differences between VHS as a function of the presence or absence of PH, and there is a significant correlation between the RPAD Index and VHS, it could not be considered an adequate value to determine the presence of this condition with the results of this study. In dogs with heartworm disease, cardiomegaly occurs only in the final stages of the disease and in dogs with severe PH, whereas this study included dogs with PH considered to be moderate to severe.

The comparison between the pulmonary arteries and the rib is a useful and common criterion to assess the pulmonary vasculature in dogs, as pointed out by previous authors [17,28], also in canine heartworm disease [24,25]. In *D. immitis* infection, there is enlargement, dilation, and increased tortuosity of the pulmonary arteries as a result of endarteritis, leading to PH. However, whether these vascular changes can act as predictors of PH in infected dogs has never been investigated. The results showed that both ratios showed significant differences between the presence/absence of PH. Furthermore, there was a correlation between both ratios and the RPAD Index, especially for the CdPA/R9 ratio. The cut-off values for the CrPA/R4 ($\geq$1.08) and CdPA/R9 ($\geq$1.10) ratios showed good sensitivity, excellent in the case of the CdPA/R9 ratio, with good specificity, indicating that these parameters could be used in the radiological evaluation of dogs with heartworm disease as a preliminary screening to determine the need for further testing for the presence of PH. These cut-off values are different from the reference values established by previous authors for the CrPA/R4 ratio < 1.2 [28] and CdPA/R9 of 1 for healthy dogs [29,30], or those described to differentiate dogs with mitral regurgitation from healthy dogs (CrPA/R4 ($\geq$0.95) and CdPA/R9 $\geq$ 1.32) [17].

Other authors have already demonstrated the usefulness of these radiological indices to detect PH caused by other thoracic pathologies, such as Oui et al. (2015) [17], who used these indices, among others, to differentiate between dogs with mitral regurgitation and healthy dogs. In addition, other authors have demonstrated the usefulness of other similar radiological indicators to predict the presence of PH caused by different pathologies, such as the ratio of the area of the pulmonary artery crossing the ninth rib to the area of the ninth thoracic vertebra (areaPA/areaT9), the ratio of the width of the pulmonary artery crossing the ninth rib to the width of the ninth thoracic vertebra (widthPA/widthT9), or the caudal pulmonary artery to vein ratio [11,31].

As a limitation of the study, it must be remembered that, in veterinary medicine, the measurement of PH is based on indirect echocardiographic measurements and that they have not been confirmed with direct measurements through right heart catheterisation.

5. Conclusions

In conclusion, an increase in the cardiac silhouette does not appear to be useful in assessing the presence or absence of PH in dogs with heartworm disease. However, the results obtained for the CrPA/R4 and CdPA/R9 ratios seem to show cut-off values with quite acceptable sensitivity and specificity, which could suggest the evaluation of these ratios when carrying out a preliminary evaluation of the thoracic radiographs of a dog as a preliminary screening when assessing whether to perform complementary tests to evaluate the presence of PH. Moreover, additional studies with a larger number of animals, to enable a more robust statistical analysis, are necessary to further evaluate these radiological indicators.

Author Contributions: J.A.M.-A. and E.C. designed the study. S.F.-C., Y.F.-C., and E.C. wrote the manuscript. S.F.-C., Y.F.-C., A.C.-V., and N.C.-R. performed the fieldwork, collected the data, and performed the experiments. All authors participated in the discussion of the results, corrected, read, and approved the final manuscript. All authors have read and agreed to the published version of the manuscript.

Funding: The presented study was supported by own funds from the Internal Medicine Service FULP/ULPGC (SD-240/030/0026). N.C.-R. was supported by the "Grants for the predoctoral training program for researchers" programme of the Government of the Canary Islands (TESIS2021010010).

Institutional Review Board Statement: Ethical review and approval were not required for the animals in this study. All radiographs and echocardiographic measures were routinely collected for prescribed diagnostic purposes or official monitoring studies and subsequently made available to this study. All the dog owners were informed about the present study and consented to participate. The study was carried out in accordance with the current Spanish and European legislation on animal protection (Spanish Royal Decree 53/2013 and 2010/63/UE Directive).

Informed Consent Statement: Written informed consent has been obtained from the owner of the animals involved in this study.

Data Availability Statement: All data generated or analysed during this study are included in this article. The datasets used and/or analysed during the present study are available from the corresponding author upon reasonable request.

Acknowledgments: The authors would like to thank Uranovet for kindly providing the Urano test Dirofilaria®. Also, they would like to thank the Hospital Clínico Veterinario of the Universidad de Las Palmas de Gran Canaria (ULPGC) for their support.

Conflicts of Interest: The authors declare no conflicts of interest.

References

1. McCall, J.W.; Genchi, C.; Kramer, L.H.; Guerrero, J.; Venco, L. Heartworm disease in animals and humans. *Adv. Parasitol.* **2008**, *66*, 193–285. [CrossRef] [PubMed]
2. Kellihan, H.B.; Stepien, R.L. Pulmonary hypertension in dogs: Diagnosis and therapy. *Vet. Clin. N. Am. Small Anim. Pract.* **2010**, *40*, 623–641. [CrossRef] [PubMed]
3. Visser, L.C.; Im, M.K.; Johnson, L.R.; Stern, J.A. Diagnostic Value of Right Pulmonary Artery Distensibility Index in Dogs with Pulmonary Hypertension: Comparison with Doppler Echocardiographic Estimates of Pulmonary Arterial Pressure. *J. Vet. Intern. Med.* **2016**, *30*, 543–552. [CrossRef] [PubMed] [PubMed Central]
4. Venco, L.; Mihaylova, L.; Boon, J.A. Right pulmonary artery distensibility index (RPAD index). A field study of an echocardiographic method to detect early development of pulmonary hypertension and its severity even in the absence of regurgitant jets for doppler evaluation in heartworm-infected dogs. *Vet. Parasitol.* **2014**, *206*, 60–66. [CrossRef] [PubMed]
5. Bahr, R.J. Canine and feline cardiovascular system. In *Textbook of Veterinary Diagnostic Radiology*, 7th ed.; Thrall, D.E., Ed.; Elsevier: Amsterdam, The Netherlands, 2017; pp. 692–706.

6. Robert, J.B. Heart and pulmonary vessels. In *Textbook of Veterinary Diagnostic Radiology*, 5th ed.; Thrall, D.E., Ed.; Saunders Elsevier: Philadelphia, PA, USA, 2007; pp. 576–582.
7. Kellum, H.B.; Stepien, R.L. Sildenafil citrate therapy in 22 dogs with pulmonary hypertension. *J. Vet. Intern. Med.* **2007**, *21*, 1258–1264. [CrossRef] [PubMed]
8. Tudor, N.; Ionita, L.; Tapaloaga, D.; Tudor, P.; Ionita, C.; Vlagioiu, C. Radiographic cardiopulmonary changes in dogs with heartworm disease. *Rom. Biotechnol. Lett.* **2014**, *19*, 9918–9992.
9. Reinero, C.; Visser, L.C.; Kellihan, H.B.; Masseau, I.; Rozanski, E.; Clercx, C.; Williams, K.; Abbott, J.; Borgarelli, M.; Scansen, B.A. ACVIM consensus statement guidelines for the diagnosis, classification, treatment, and monitoring of pulmonary hypertension in dogs. *J. Vet. Intern. Med.* **2020**, *34*, 549–573. [CrossRef] [PubMed] [PubMed Central]
10. Adams, D.S.; Marolf, A.J.; Valdés-Martínez, A.; Randall, E.K.; Bachand, A.M. Associations between thoracic radiographic changes and severity of pulmonary arterial hypertension diagnosed in 60 dogs via Doppler echocardiography: A retrospective study. *Vet. Radiol. Ultrasound* **2017**, *58*, 454–462. [CrossRef] [PubMed]
11. Chanroon, K.; Tuntipas, T.; Sukpanich, P.; Punyarit, P.; Surachetpong, S. Novel methods to diagnose pulmonary hypertension in dogs assessed by thoracic radiography. *Thai J. Vet. Med.* **2018**, *48*, 385–391. [CrossRef]
12. Montoya-Alonso, J.A.; Carretón, E.; Morchón, R.; Silveira-Viera, L.; Falcón, Y.; Simón, F. The impact of the climate on the epidemiology of Dirofilaria immitis in the pet population of the Canary Islands. *Vet. Parasitol.* **2016**, *216*, 66–71. [CrossRef] [PubMed]
13. Montoya-Alonso, J.A.; Morchón, R.; Costa-Rodríguez, N.; Matos, J.I.; Falcón-Cordón, Y.; Carretón, E. Current Distribution of Selected Vector-Borne Diseases in Dogs in Spain. *Front. Vet. Sci.* **2020**, *7*, 564429. [CrossRef] [PubMed] [PubMed Central]
14. Montoya-Alonso, J.A.; Morchón, R.; García-Rodríguez, S.N.; Falcón-Cordón, Y.; Costa-Rodríguez, N.; Matos, J.I.; Rodríguez Escolar, I.; Carretón, E. Expansion of Canine Heartworm in Spain. *Animals* **2022**, *12*, 1268. [CrossRef] [PubMed] [PubMed Central]
15. Buchanan, J.W.; Bücheler, J. Vertebral scale system to measure canine heart size in radiographs. *J. Am. Vet. Med. Assoc.* **1995**, *206*, 194–199. [CrossRef] [PubMed]
16. Costanza, D.; Greco, A.; Piantedosi, D.; Bruzzese, D.; Pasolini, M.P.; Coluccia, P.; Castiello, E.; Baptista, C.S.; Meomartino, L. The heart to single vertebra ratio: A new objective method for radiographic assessment of cardiac silhouette size in dogs. *Vet. Radiol. Ultrasound* **2023**, *64*, 378–384. [CrossRef] [PubMed]
17. Oui, H.; Oh, J.; Keh, S.; Lee, G.; Jeon, S.; Kim, H.; Yoon, J.; Choi, J. Measurements of the pulmonary vasculature on thoracic radiographs in healthy dogs compared to dogs with mitral regurgitation. *Vet. Radiol. Ultrasound* **2015**, *56*, 251–256. [CrossRef] [PubMed]
18. Serrano-Parreño, B.; Carretón, E.; Caro-Vadillo, A.; Falcón-Cordón, Y.; Falcón-Cordón, S.; Montoya-Alonso, J.A. Evaluation of pulmonary hypertension and clinical status in dogs with heartworm by Right Pulmonary Artery Distensibility Index and other echocardiographic parameters. *Parasites Vectors* **2017**, *10*, 106. [CrossRef] [PubMed] [PubMed Central]
19. Venco, L.; Genchi, C.; Vigevani Colson, P.; Kramer, L. Relative utility of echocardiography, radiography, serologic testing and microfilariae counts to predict adult worm burden in dogs naturally infected with heartworms. In *Recent Advances in Heartworm Disease. Symposium'01*; Seward, R.L., Knight, D.H., Eds.; American Heartworm Society: Batavia, IL, USA, 2003; pp. 111–124.
20. Serrano-Parreño, B.; Carretón, E.; Caro-Vadillo, A.; Falcón-Cordón, S.; Falcón-Cordón, Y.; Montoya-Alonso, J.A. Pulmonary hypertension in dogs with heartworm before and after the adulticide protocol recommended by the American Heartworm Society. *Vet. Parasitol.* **2017**, *236*, 34–37. [CrossRef] [PubMed]
21. Falcón-Cordón, Y.; Montoya-Alonso, J.A.; Caro-Vadillo, A.; Matos-Rivero, J.I.; Carretón, E. Persistence of pulmonary endarteritis in canine heartworm infection 10 months after the eradication of adult parasites of Dirofilaria immitis. *Vet. Parasitol.* **2019**, *273*, 1–4. [CrossRef] [PubMed]
22. Duler, L.; Visser, L.C.; Jackson, K.N.; Phillips, K.L.; Pollard, R.E.; Wanamaker, M.W. Evaluation of radiographic predictors of left heart enlargement in dogs with known or suspected cardiovascular disease. *Vet. Radiol. Ultrasound* **2021**, *62*, 271–281. [CrossRef] [PubMed]
23. Johnson, L.; Boon, J.; Orton, E.C. Clinical characteristics of 53 dogs with Doppler-derived evidence of pulmonary hypertension: 1992-1996. *J. Vet. Int. Med.* **1999**, *13*, 440–447. [CrossRef] [PubMed]
24. Losonsky, J.M.; Thrall, D.E.; Lewis, R.E. Thoracic radiographic abnormalities in 200 dogs with spontaneous heartworm infestation. *Vet. Radiol.* **1983**, *24*, 120–123. [CrossRef]
25. Litster, A.; Atkins, C.; Atwell, R.; Buchanan, J. Radiographic cardiac size in cats and dogs with heartworm disease compared with reference values using the vertebral heart scale method: 53 cases. *J. Vet. Cardiol.* **2005**, *7*, 33–40. [CrossRef] [PubMed]
26. American Heartworm Society. *Current Canine Guidelines for the Prevention, Diagnosis and Management of Heartworm (Dirofilaria immitis) Infection in Dogs*; American Heartworm Society: Holly Springs, NC, USA, 2024.
27. Kim, S. Radiographic evaluation of canine heartworm disease. *Korean J. Vet. Clin. Med.* **1995**, *12*, 265.e75.
28. Thrall, D.E.; Losonsky, J.M. A method for evaluating canine pulmonary circulatory dynamics from survey radiographs. *J. Am. Anim. Hosp. Assoc.* **1976**, *12*, 457–462.
29. Bahr, R. The heart and pulmonary vessels. In *Textbook of Veterinary Diagnostic Radiology*, 6th ed.; Thrall, D., Ed.; Elsevier: St. Louis, MO, USA, 2013; pp. 585–607.

30. Losonsky, J.M. The pulmonary vasculature. In *Textbook of Veterinary Diagnostic Radiology*, 4th ed.; Thrall, D., Ed.; Saunders: St. Louis, MO, USA, 2002; pp. 420–430.
31. Lee, S.K.; Choi, J. Caudal pulmonary artery to vein ratio on radiography can predict pulmonary hypertension in dogs with mitral regurgitation. *Vet. Radiol. Ultrasound* **2023**, *64*, 18–27. [CrossRef] [PubMed]

animals

MDPI

Article

Echocardiographic Documentation of Dilated Cardiomyopathy Development in Dogs Naturally Infected with *Trypanosoma cruzi*

Eduardo E. Avalos-Borges [1], Carlos M. Acevedo-Arcique [2], Jose C. Segura-Correa [1], Matilde Jiménez-Coello [3], Nisha J. Garg [4] and Antonio Ortega-Pacheco [1,*]

[1] Departamento de Salud Animal y Medicina Preventiva, Facultad de Medicina Veterinaria y Zootecnia, Universidad Autónoma de Yucatán, Km 15.5 Carretera Mérida-Xmatkuil, Apdo. Postal 4-116 Itzimná, Mérida 97000, Yucatán, Mexico; animal.health@hotmail.com (E.E.A.-B.); jose.segura@correo.uady.mx (J.C.S.-C.)

[2] Hospital Veterinario para Perros y Gatos, Universidad Autónoma de Yucatán, Av. Itzaes No. 490 x 29, C. 18 No. 271, San José Vergel, Mérida 97000, Yucatán, Mexico; carlos.acevedo@correo.uady.mx

[3] Laboratorio de Microbiologia, Centro de Investigaciones Regionales "Dr. Hideyo Noguchi", Universidad Autónoma de Yucatán, Av. Itzaes No. 490 x 29, Mérida 97000, Yucatán, Mexico; mjcoello@correo.uady.mx

[4] Department of Microbiology & Immunology, University of Texas Medical Branch, Galveston, TX 77555-1070, USA; nigarg@utmb.edu

* Correspondence: opacheco@correo.uady.mx; Tel.: +52-9991508936

Simple Summary: *Trypanosoma cruzi* is a protozoan parasite with high tropism to the cardiac tissue, causing severe cardiopathies. This study determined the changes in the cardiac structure and myocardial parameters using real-time ultrasonography in dogs naturally infected with *T. cruzi* while living in an endemic region. A major indicator of cardiac involvement in infected dogs was changes in the left ventricular internal diameter (LVID) during systole and diastole. Changes in the intraventricular septum and LV posterior wall (LVPW) thickness at systole and diastole, as well as findings of fractional shortening and an E/A ratio above or below the normal range, may be used to predict dilated cardiomyopathy (DCM) progression.

Abstract: We aimed to characterize the echocardiographic alterations in dogs from an endemic region that were naturally infected with *T. cruzi*. Dogs (n = 130) seropositive for antibodies against *T. cruzi* and/or with acute parasitemia were enrolled in the study. Indicators of changes in the structure and systolic and diastolic functions of the left ventricle (LV) and blood flow patterns were evaluated by echocardiography. The frequency and extent of alterations in these indicators were associated with the severity of the disease. Briefly, 15 (11.54%) dogs were diagnosed with dilated cardiomyopathy (DCM), and 115 (88.46%) dogs were diagnosed as being without DCM. Infected dogs with DCM exhibited structural features of LV dysfunction, e.g., a significant ($p < 0.05$) increase in the LV internal diameter at systole and diastole (LVID-s, LVID-d) and a decline in the LV posterior wall (LVPW-d) thickness at diastole. Despite an increase in stroke volume and cardiac output indicating contraction force, DCM resulted in a decreased ejection fraction, affecting systolic function. Dogs that were diagnosed as DCM-negative but were positive for *T. cruzi* by PCR exhibited a high frequency of an increase in the thickness of the interventricular septum in systole (IVS-s) and the LV posterior wall in diastole (LVPW-d), a decline in the LV inner diameter (LVID-d, LVID-s), and fractional shortening (FS). The thinning of the LVPW at systole was the most defining feature observed in chronically infected dogs. In summary, this is the first study reporting the echocardiographic changes occurring in dogs naturally infected with *T. cruzi* and developing DCM. Our data suggest that changes in LVID are a major indicator of risk of cardiac involvement, and the observation of changes in the IVS, LVPW, and FS have predictive value in determining the risk of DCM development in infected dogs.

Keywords: echocardiographic alterations; left ventricle; dog; *Trypanosoma cruzi*; flow patterns; dilated cardiomyopathy; Chagas disease

1. Introduction

Chagas disease, also referred to as American Trypanosomiasis (AT), is an illness caused by a flagellated protozoan that is transmitted by hematophagous vectors. The etiological agent, *Trypanosoma cruzi*, is distributed across a large part of the American continent, and it infects several species of wild and domestic mammals and humans [1]. AT is endemic in Latin America, and an increase in the incidence of *T. cruzi* infection in dogs and humans has been noted in Mexico [2]. In Yucatan, Mexico, AT in dogs has been detected in rural and urban areas [3]. In the capital city of Yucatan, Merida, *T. cruzi* is frequently detected with a prevalence of 12.2% in apparently healthy domiciled dogs [4].

Chronic Chagas disease presents in humans as myocardial abnormalities, ranging from mild forms, such as apical aneurysm and left ventricular (LV) diastolic dysfunction only, to significant cardiac chamber dilatation, coupled with severe systolic dysfunction. Dilatation of the left and right ventricles is the most recognizable alteration of the heart in dogs and humans with Chagas disease [1,2,5–8]. Dogs are important reservoirs of *T. cruzi*, and due to the similar progression of the disease to that occurring in humans, canine AT can serve as a surrogate model for studying the course of human Chagas disease.

Electrocardiographic and echocardiographic studies in experimentally infected dogs have noted a decline in the ejection fraction is associated with thinning of the LV walls, mural thrombi, hypokinesis, and thickening of the septum [9]. More recently, echocardiography examination of seven dogs that were experimentally infected with *T. cruzi* eight years earlier identified an inversion of the E/A index, indicating a delayed relaxation pattern or mild dysfunction and hypomotility of the interventricular septum, but cardiomegaly signs were not found [10]. A detailed evaluation of the changes in cardiac structure and function in dogs naturally infected with *T. cruzi* isolates circulating in Mexico has not been carried out yet.

In this study, we aimed to characterize the echocardiographic alterations in dogs naturally infected with *T. cruzi* with ultrasound equipment normally used in routine veterinary practice. Our primary objective was to determine whether dogs naturally exposed to *T. cruzi* isolates circulating in the state of Yucatan could develop dilated cardiomyopathy and to describe the cardiac findings before the clinically severe form of the heart disease appeared in infected dogs. Our secondary objective was to obtain a comprehensive view of the changes in a range of myocardial parameters in infected dogs and determine whether these variabilities are reflective of human Chagas disease.

2. Materials and Methods

2.1. Bioethics and Study Area

All the animal studies were performed according to the protocol approved by the bioethics committee (CB-CCBA-I-2017-003) at the Facultad de Medicina Veterinaria y Zootecnia, Universidad Autónoma de Yucatán. Informed consent was obtained from the dog owners before the dogs were enrolled in the study. The study was conducted with pet dogs referred by veterinary clinics in Merida, Yucatan, Mexico (19°30″ and 21°35″ N; 87°30″ and 90°24″ W). The climate of the area is tropical sub-humid, with a well-defined rainy season during the months of May–June and October–November.

2.2. Selection of Animals

A cross-sectional study was carried out where 130 dogs with physical symptoms of cardiac involvement were randomly chosen. The inclusion criteria were that the dogs resided in Merida, Yucatan, with a pet owner and were seropositive for *T. cruzi* according to an ELISA. The enrolled dogs were later confirmed in terms of their *T. cruzi* exposure by Western blotting and/or PCR diagnostic approaches and had not received any treatment. The enrolled dogs (both males and females) were older than two years old and of variable sizes, weights, and breeds. Dogs were considered to have a cardiopathy when their medical history and physical exams were compatible with heart disease [11]. Electrocardiographic abnormalities, not attributable to electrolyte imbalance, were also recorded [12]. In some cases, radiographic studies or determination of blood pressure [13] were also conducted to confirm the cardiopathy. Healthy dogs (n = 16) that were seronegative for anti-*T. cruzi* antibodies and PCR-negative for *T. cruzi* DNA and exhibited no physical symptoms of cardiac involvement were used as the controls.

2.3. Blood Samples

From each dog, two blood samples were obtained via cephalic or jugular vein puncture. Half of each blood sample was collected in a PAXgene Blood DNA Tube (BD-QIAGEN) to preserve the DNA for later purification. A second aliquot of each blood sample was collected in a BD Vacutainer and centrifuged at 400 rpm for 15 min at room temperature to obtain the serum.

2.4. Serology

A serological diagnosis of AT was made according to the detection of immunoglobulins (IgGs) against *T. cruzi* by using the Chagatest ELISA recombinant v.3.0 kit (Wiener, Rosario, Argentina). The assay was carried out following the manufacturer's recommendations, except that the 2nd antibody was replaced with horseradish peroxidase (HRP)-conjugated goat anti-dog IgGs. The details of the protocol were previously described by us [14].

2.5. Western Blotting

Epimastigotes of the H4 strain of *T. cruzi* were lysed in Laemmli sample buffer containing a protease inhibitor cocktail (Sigma-Aldrich, St. Louis, MO, USA). The protein samples (20 μg) were resolved on 10% polyacrylamide gels and transferred onto nitrocellulose membranes. The membranes were probed with serum samples of the dogs, and color was developed using the standard methods. A serum sample was considered positive when at least five antigenic bands were recognized [4,14].

2.6. PCR Detection of *T. cruzi*

Total DNA was extracted from whole blood samples according to a previously published protocol [4]. Alternatively, the DNeasy Blood and Tissue Kit (69504, QIAGEN, Hilden, Germany) was used to isolate the genomic DNA from the blood samples by following the manufacturer's instructions. Total DNA was examined in terms of quality (OD_{260}/OD_{280} ratio of 1.7–2.0) and quantity ($[OD_{260} - OD_{320}] \times 50$-μg/mL) by using a DU 800 UV/visible spectrophotometer. To detect the presence of *T. cruzi* DNA in the blood, a PCR assay was carried out, as described previously [3].

2.7. Echocardiographic Evaluation of Cardiac Structures and Function

The dogs were prepared, positioned, and scanned according to the conventional technique [15–17]. Mindray M5 real-time ultrasound equipment (Mindray Electronics®, Shenzhen, China) equipped with a cardiac transducer probe with a frequency range of 2 to 4 MHz was employed for echocardiography. The parameters of the cardiac structure and function were obtained as follows: In the right parasternal window with a short-axis view of the LV at the level of the papillary muscles, the thickness of the interventricular septum and the free walls and the diameter of the LV were measured with the M mode in systole and diastole [15–17]. To compare these parameters between dogs of variable sizes, weights, and breeds, the measurements of the structures were normalized according to the allometric scale formula for cardiac M-mode measurements in adult dogs [18]. Using the Teichholtz method, the software calculated the fractional shortening (FS), ejection fraction (EF), stroke volume (SV), and cardiac output (CO), which together provide an indication of LV systolic function. In the left apical position with a view of four chambers, the trans-mitral flow was measured with a pulse Doppler instrument, placing the probe at the tip of the mitral valve [19]. The peaks of the E and A waves were marked, the E/A index (a marker of LV diastolic function) was calculated, and the flow patterns were identified as normal, delayed relaxation, or restrictive, as described previously [20]. When increased blood flow velocities were observed with a normal pattern, they were surmised to be pseudo-normal, as further confirmation with a tissue Doppler was not feasible.

Dogs were considered positive for DCM according to the ultra-sonographic criteria proposed by the European Society of Veterinary Cardiology (major criteria: dilatation of the LV in systole or diastole, ventricular spheroid structure, thinning of the septum, and a reduction in fractional shortening; minor criteria: increased space between point E and the septum, incongruent values for fractional shortening and left or bilateral atrial dilatation) [21]. The reference values were based on a prediction interval of 95%, as described in the published literature [18].

2.8. Statistical Analysis

The normal distribution of the variables of interest was confirmed by the Shapiro–Wilk test. The evaluated animals were first grouped according to the results of the serologic and molecular tests and the presence and absence of DCM as follows: group A, serology-negative, PCR-positive, DCM-negative; Group Aw, serology-negative, PCR-positive, DCM-positive; group B, serology-positive, PCR-negative, DCM-negative; group Bw, serology-positive, PCR-negative, DCM-positive; group C, serology-positive, PCR-positive, DCM-negative; and group Cw, serology-positive, PCR-positive, DCM-positive. The control group included seronegative, PCR-negative, DCM-negative, healthy dogs (n = 16).

Echocardiographic, age, and body size data showing a normal distribution were analyzed using a one-way ANOVA, followed by Tukey's post hoc test. The non-normally distributed data were analyzed using the Kruskal–Wallis test with Bonferroni correction, and the frequencies of the alterations in each indicator of cardiac structure and function were determined. Binary data were analyzed using the Chi^2 test or the exact Fisher's test to establish the association between sex, size, age, and transmitral flow patterns and a DCM-positive vs. DCM-negative disease status. All the statistical analyses were performed using Statgraphics v.19.0 software.

3. Results

3.1. Characteristics of T. cruzi Infection

All the dogs enrolled in the study were in the age range of 6.94 to 9.83 years old and of a small to medium size (weight range: 8.41–13.06 kg). PCR detection of *T. cruzi* DNA in the blood is indicative of circulating parasites and is mostly noted in response to acute infection or repeat infection. Seropositivity for anti-*T. cruzi* antibodies is generally noted in all infected cases irrespective of the stage of disease development. Thus, a lack of anti-*T. cruzi* IgGs in *T. cruzi* DNA-positive dogs indicates the very early stage of acute infection before adaptive humoral immunity has been elicited by the host. Our data showed that most of the dogs enrolled in the study were positive in both diagnostic tests, i.e., anti-*T. cruzi* antibodies by ELISA and Western blot analysis and parasite DNA detection by PCR (group C: 97 out of 130 dogs, 75% of total), indicating their exposure to repeat infections. Some dogs were PCR-positive only (group A: 26 out of 130, 20% of total), indicating the first, early phase of acute infection, and the fewest dogs were in the chronic phase (group B: 7 out of 130, 5% of total), indicated by their seropositive status only (Table 1).

Table 1. Categorization of 130 dogs based on serological and molecular tests for *Trypanosoma cruzi* infection and the presence of dilated cardiomyopathy.

Dilated Cardiomyopathy	Group A Seronegative, PCR-Positive	Group B Seropositive, PCR-Negative	Group C Seropositive, PCR-Positive	Total
Absent	23	6	86	115
Present	3	-	12	15
Total	26	6	98	130

Dogs were screened for anti-*T. cruzi* antibodies using two serological tests, including an enzyme-linked immunosorbent assay (ELISA) and Western blot analysis, and for circulating parasites using *T. cruzi*-specific PCR, as described in Materials and Methods. Echocardiography was performed to measure the changes in cardiac structure and function, and the animals in each group were further identified to have or not have dilated cardiomyopathy. No association was found between sex, age, or size and the presence of DCM according to the Chi^2 or Fisher's test.

3.2. Echocardiographic Findings

Dogs were analyzed by echocardiography to measure the changes in left ventricular (LV) structure and function and characterize the presence and absence of DCM. Figure 1 shows a representative image in B mode of the cardiac characteristics of DCM in a dog positive for anti-*T. cruzi* antibodies. Note the dilatation of the LV (ventricular sphericity) with the thinning of the interventricular septum and its bowing towards the right ventricle. A greater diameter of the right atrium compared to the left atrium is also noticeable.

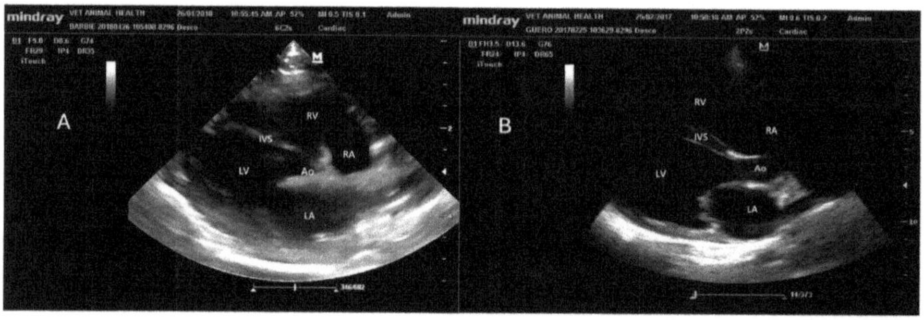

Figure 1. (**A**). Ultrasound (B-mode) image of the heart of a dog negative for *T. cruzi*. Chambers and septum are within the normal patterns for a healthy dog. (**B**). Ultrasound (B-mode) image of the heart of a dog with dilated cardiomyopathy (DCM) and seropositive for anti-*T. cruzi* antibodies. Shown is a right parasternal window with long-axis view of the four chambers. Dilatation of the left ventricle (spheroid shape) can be seen with thinning of the septum and its bowing towards the right ventricle. A greater diameter of the right atrium with respect to the left atrium is also noticeable. Abbreviations: RA = right atrium, LA = left atrium, RV = right ventricle, LV = left ventricle, Ao = aorta artery, IVS = interventricular septum.

3.3. M-Mode Ultrasonographic Findings

Representative M-mode ultrasonographic images and measurements of the LV structure from normal and infected dogs are shown in Figure 2. In comparison to non-infected, seronegative dogs (Figure 2A), the infected dogs with DCM (Figure 2C) and advanced DCM (Figure 2D) exhibited the highest values in terms of LV diameter, indicating the dilatation of the LV. Dogs that were infected but without DCM exhibited minor to no changes in LV diameter. Based on these structural features, we noted that 3 and 12 dogs within groups A and C, respectively, exhibited the features of DCM, while the remaining 115 infected dogs did not present with DCM (Table 1). Thus, overall, 15 out of the 130 infected dogs (i.e., 11.54%) exhibited DCM.

Next, we performed one-way ANOVA or Kruskall–Wallis analyses to determine whether the infection and seropositivity status were significantly correlated with any of the echocardiographic parameters of LV structure in dogs (Table 2A). In general, we did not observe any major changes in the IVS thickness at diastole (IVS-d) or systole (IVS-s) between any of the infected groups and the non-infected, healthy controls. The dogs exhibited an average IVS thickness of 0.4–0.5 cm at diastole and 0.64–0.70 cm at systole. The mean values of the internal diameter of the LV at diastole (LVID-d) and systole (LVID-s) were also not significantly changed in the DCM-negative dogs in groups A, B, or C (without DCM) compared to the healthy controls. However, the Aw (acute infection) and Cw (chronic and parasitemic dogs) groups with DCM exhibited a 23–40% increase in LVID-d and a 30–43% increase in LVID-s in comparison with the infected groups without DCM and the healthy controls. The mean LV posterior wall thickness at diastole (LVPW-d) was decreased by 14.9–17% in the Aw and Cw (vs. control) groups. The dogs in group B (chronic infection) exhibited a 22–28% decline in their LVPW-d and LVPW-s values ($p < 0.05$). Together, these results suggest that an increase in the LVID and thinning of the LVPW were major features of DCM in the infected dogs, and the risk of LVPW thinning was increased with chronic infection (group B).

Echocardiographic parameters of global LV function were also measured in all enrolled dogs (Table 2B). These results showed that the mean values for LV systolic function, including stroke volume (SV), heart rate (HR), and cardiac output (CO), were increased by 39–80%, 11.2–21.4%, and 62–110%, respectively, in the Aw and Cw groups compared to those noted in the infected groups (A, B, C) without DCM and the normal, healthy controls. The maximal increase in the SV, HR, and CO values was noted in the dogs in the Cw group (chronic-stage and parasitemic dogs). Despite their increased SV, the dogs in the Cw group exhibited a 15.9% decline in the ejection fraction (EF). Fractional shortening (FS) was decreased by 15% in the dogs in the Cw group. These findings indicate the early stage of DCM in the majority of the dogs in the Aw and Cw groups, as was also noted in the echocardiography imaging (Figure 2C).

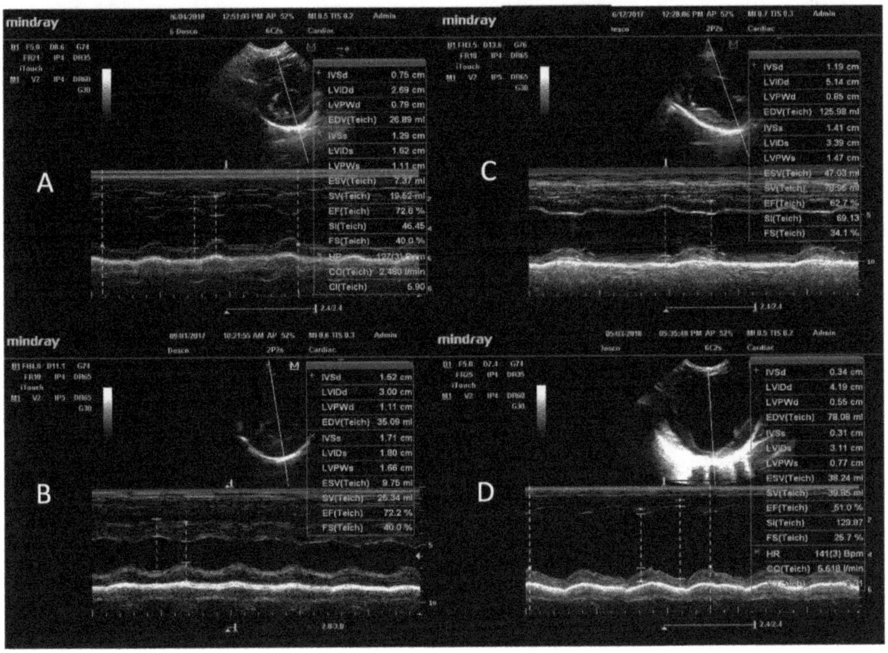

Figure 2. Representative images of the M-mode ultrasonographic measurement of the structures of the left ventricle in right parasternal window, short-axis view at the level of the papillary muscles, shown for infected and non-infected dogs. (**A**) Healthy, non-infected. (**B**) Seropositive for anti-*T. cruzi* antibodies and no DCM. (**C**) Seropositive with incipient DCM. (**D**) Seropositive with advanced DCM. Abbreviations: IVS = interventricular septum, LVID = left ventricle inner diameter, LVPW = left ventricle posterior wall, d = diastole, s = systole, FS = fractional shortening, EDV = end diastolic volume, ESV = end systolic volume, SV = stroke volume, EF = ejection fraction, SI= stroke index, CO = cardiac output, CI = cardiac index, HR = heart rate.

Table 2. Echocardiographic, age, and body size parameters in *Trypanosoma cruzi*-positive dogs.

Parameters	Mode	A (n = 23)	Aw (n = 3)	B (n = 6)	C (n = 86)	Cw (n = 12)	Controls (n = 16)
A: Structural features							
IVS diastole (IVS-d) *[1]	M mode	0.5 ± 0.07 ^{a}	0.51 ± 0.05	0.4 ± 0.08	0.49 ± 0.08 ^{b}	0.4 ± 0.08 ^{a,b}	0.44 ± 0.07
IVS systole (IVS-s) *[1]	M mode	0.69 ± 0.01	0.66 ± 0.05	0.67 ± 0.17	0.7 ± 0.15	0.64 ± 0.18	0.64 ± 0.09
LVID diastole (LVID-d) *[2]	M mode	1.44 ± 0.24 ^{a}	2.02 ± 0.16 ^{a,b}	1.57 ± 0.20 ^{b,c}	1.35 ± 0.25 ^{b,c,d}	2.05 ± 0.15 ^{a,c,d,e}	1.55 ± 0.1 ^{b,d,e}
LVID systole (LVID-s) *[2]	M mode	0.83 ± 0.18 ^{a}	1.15 ± 0.10 ^{a,b}	0.91 ± 0.13 ^{c}	0.76 ± 0.2 ^{b}	1.23 ± 0.19 ^{a,c}	0.86 ± 0.1 ^{b}
LVPW diastole (LVPW-d) *[2]	M mode	0.51 ± 0.09 ^{a}	0.39 ± 0.16 ^{a}	0.34 ± 0.03 ^{a,b}	0.49 ± 0.1 ^{b,c,d}	0.4 ± 0.09 ^{a,c,d,e}	0.47 ± 0.06 ^{b,d,e}
LVPW systole (LVPW-s) *[2]	M mode	0.71 ± 0.09 ^{a}	0.62 ± 0.18	0.53 ± 0.11 ^{a,b}	0.73 ± 0.12 ^{b}	0.69 ± 0.19 ^{b}	0.68 ± 0.08 ^{b}
EPSS [1] (cm)	M mode	0.39 ± 0.21 ^{a}	0.93 ± 0.13 ^{b}	0.5 ± 0.2	0.31 ± 0.21 ^{b,c}	0.96 ± 0.09 ^{a,c,d}	0.27 ± 0.11 ^{b,d}
LA/Ao [1] ratio	B mode	1.39 ± 0.51 ^{a}	1.69 ± 0.51	1.38 ± 0.17	1.25 ± 0.24 ^{b}	1.86 ± 0.43 ^{a,b,c}	1.19 ± 0.14 ^{c}
SI [1] ratio	B and M mode	1.99 ± 0.21 ^{a}	1.54 ± 0.1 ^{b}	1.87 ± 0.19	1.97 ± 0.24 ^{b,c}	1.53 ± 0.08 ^{a,c}	1.75 ± 0.07 ^{a,c}
B: Functional features							
Fractional shortening (FS), % **[1]	M mode	40.35 ± 6.66	40.33 ± 5.69	38.83 ± 5.03	41.39 ± 11.2	37 ± 10.72	41.75 ± 5.36
Ejection fraction (EF), % **[1]	M mode	72.61 ± 7.69	71.2 ± 7.30	68.33 ± 7.51	72.75 ± 12.58	64.17 ± 14.1	74.09 ± 6.68
Stroke volume (SV), mL **[1]	M mode	20.85 ± 14.37 ^{a}	33.52 ± 14.77	27.49 ± 26.35	22.17 ± 15.5 ^{b}	51.18 ± 28.4 ^{a,b}	24.01 ± 10.97
Heart rate [1]	M mode	102.43 ± 17.98	121 ± 8.54	95.33 ± 8.71	102.5 ± 31.57	121.25 ± 41.29	108.75 ± 23.95
Cardiac output (CO), L/min **[1]	M mode	2.1 ± 1.31 ^{a}	4 ± 1.53	2.63 ± 2.52	2.22 ± 1.55 ^{b}	6.6 ± 5.13 ^{a,b}	2.46 ± 0.81
E/A ratio [1]	PW Doppler	1.53 ± 0.34 ^{a}	1.24 ± 0.24	1.35 ± 0.26	1.37 ± 0.34	1.41 ± 0.61	1.1 ± 0.29 ^{a}
Age, years [1]	...	9.83 ± 3.46 ^{a}	7.33 ± 3.78	7.33 ± 3.33	8.52 ± 4.01	9.83 ± 2.25	5.94 ± 3.23 ^{a}
Weight, kg [1]	...	8.41 ± 6.31	9.53 ± 7.33	13.06 ± 11.86	12.77 ± 10.03	16.03 ± 18.61	9.56 ± 6.84

The Mindray© M5 ultrasound system was used to perform transthoracic echocardiography in the B and M modes and pulse wave Doppler echocardiography used to assess the changes in left ventricular diastolic function. Group A: seronegative by ELISA and Western blotting and PCR-positive, indicative of acute infection; group Aw: seronegative and PCR-positive with DCM; group B: seropositive and PCR-negative, indicative of chronic infection; group C: seropositive and PCR-positive, indicative of chronic-stage and repeat infections; group Cw: seropositive and PCR-positive for repeat infection with DCM. Controls included seronegative and PCR-negative dogs with no indication of cardiac involvement. Parameters of structural features in 1A are presented in cm. Data are presented as mean values ± standard deviation, calculated using Statgraphics software version 19. Significance ($p < 0.05$) is plotted with superscript letters, where different letters indicate statistical difference; * values were normalized as defined in Cornell et al., 2004 [18]. ** Values calculated with Teichholz formula. [1] Compared by Kruskal–Wallis test with Bonferroni correction, [2] compared by ANOVA and Tukey's test; Abbreviations: IVS, interventricular septum; LVID, left ventricular (LV) internal diameter; LVPW, LV posterior wall; EPSS, E point septal separation; LA/Ao, left atrium/aorta; SI, sphericity index; ELISA, enzyme-linked immunosorbent assay; PCR, polymerase chain reaction; DCM, dilated cardiomyopathy.

3.4. Pulse Doppler Ultrasonographic Findings

The transmitral flow was assessed and calculation of the E/A index was carried out using ultrasonic recording with a pulse Doppler instrument. Representative images of the four chambers in the left apical window view of a seronegative healthy dog and infected dogs are presented in Figure 3, and the mean E/A values for each group are presented in Table 2. Most of the infected dogs without DCM exhibited a normal transmitral flow pattern, as was noted in the healthy controls (Figure 3A,C). A pattern of delayed relaxation (E < A) was observed in some of the infected dogs without DCM and in all the dogs with early signs of DCM (Figure 3B,D).

An increase in the velocity and time of the E and A values (Figure 3D) in *T. cruzi*-positive dogs with DCM could correspond to a pseudo-normal pattern of blood flow, but tissue Doppler imaging (TDI) is required to confirm this. At the population level, 20% of the dogs with DCM (i.e., 3 out of 15) exhibited a restrictive pattern, while a normal pattern was observed in 60% and a delayed relaxation flow was observed in 20% of the infected DCM-positive dogs. Among the infected DCM-negative dogs, 75.44% had a normal pattern, 4.39% exhibited a delayed relaxation pattern, 5.26% exhibited some features of a pseudo-normal pattern, and 14.91% exhibited a restrictive flow pattern. Though all the infected dogs exhibited an increase in the E/A ratio and the maximal increase was noted in group A (acutely infected), the average E/A values in the infected dogs were within the normal range of 0.98–1.7, and no association of the flow patterns with infection or disease status was noted using the Chi^2 test (probability of association between type of flows and groups: 0.82). These results suggest that changes in the blood flow patterns are not directly correlated with infection and/or disease status.

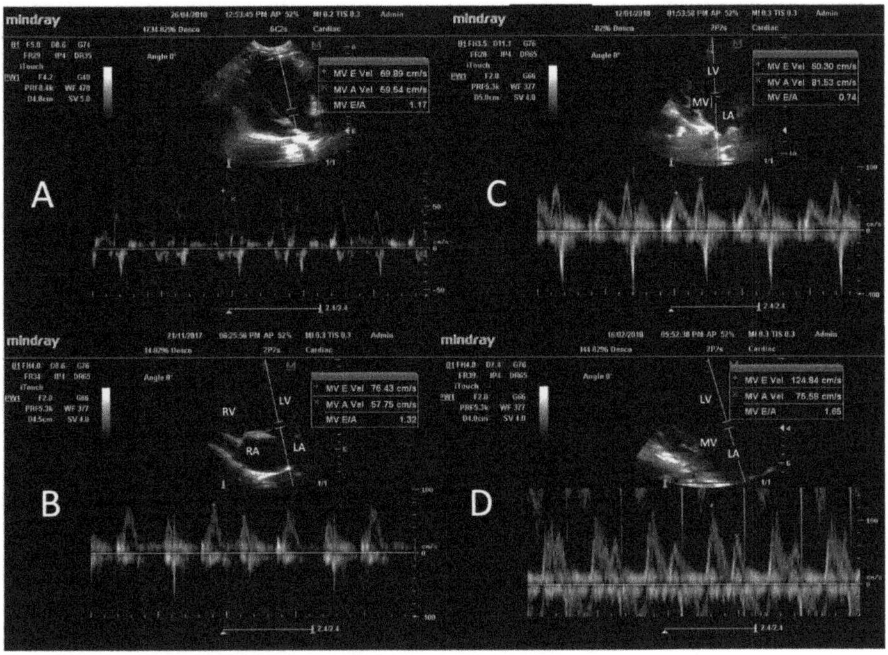

Figure 3. Ultrasonic recording with pulse Doppler of the transmitral blood flow and calculation of the E/A index in dogs naturally infected with *T. cruzi*. Shown is the left apical window view of four chambers. (**A**) Normal, healthy, seronegative for anti-*T. cruzi* antibodies; (**B**) pattern of delayed relaxation in a seropositive dog without DCM; (**C**) normal pattern in a seropositive dog without DCM; (**D**) flow of a restrictive pattern in a *T. cruzi*-positive dog with DCM. LA = left atrium, LV = left ventricle, RA = right atrium, RV = right ventricle, MV = mitral valve. MV E vel = velocity of peak E, MV A vel = velocity of peak A, MV E/A = index E/A.

We examined the frequency of alterations in the heart structure and function of the LV in infected dogs without DCM to determine whether status of infection is associated with particular features of LV dysfunction (Table 3). In group A (acute infection), 26.09% (6 out of 23) of the dogs exhibited an increase in the E/A ratio, and 21.74% of dogs exhibited a reduction in the LVID in both stages of the cardiac cycle. In group B (chronic infection), 33.3% of cases showed thinning of the LVPW at systole. In group C, the most frequent alterations were noted in LVID-s and LVID-d, which were decreased in 42.3% and 38.8% of dogs, respectively. These results suggest that changes in LVID-s and LVID-d and an increased E/A ratio are important indicators of the risk of DCM development in dogs that are acutely infected (group A) or have been repeatedly exposed to *T. cruzi* infection (group C) and are positive for *T. cruzi* by PCR. In comparison, thinning of the LV posterior wall was the most defining feature of the risk of DCM development in the chronically infected dogs (group B).

Table 3. Frequency of cardiac alterations in dogs that were positive for *Trypanosoma cruzi* exposure and negative for dilated cardiomyopathy.

Parameters	Mode	A (n = 23)		B (n = 6)		C (n = 85)		Reference Range
		Increase (%)	Decrease (%)	Increase (%)	Decrease (%)	Increase (%)	Decrease (%)	
IVS, at diastole (IVS-d)	M mode	4.35	...	...	...	8.24	1.18	0.29–0.59 [1]
IVS, at systole (IVS-s)	M mode	13.04	...	16.67	...	22.35	2.35	0.43–0.79 [1]
LVID, at diastole (LVID-d)	M mode	...	21.74	...	16.67	...	38.82	1.27–1.85 [1]
LVIDW, at systole (LVID-s)	M mode	...	21.74	...	...	...	42.35	0.71–1.26 [1]

Table 3. *Cont.*

Parameters	Mode	A (n = 23)		B (n = 6)		C (n = 85)		Reference Range
		Increase (%)	Decrease (%)	Increase (%)	Decrease (%)	Increase (%)	Decrease (%)	
LVPW, at diastole (LVPW-d)	M mode	13.04	…	…	…	10.59	1.18	0.29–0.60 [1]
LVPW, at systole (LVPW-s)	M mode	…	…	…	33.33	9.41	1.18	0.48–0.87 [1]
Fractional shortening (FS)	M mode	8.70	13.04	…	16.67	22.35	20.00	33.6–49.9 [2]
Ejection fraction (EF)	M mode	8.70	4.35	…	16.67	22.35	12.94	58.9–82.9 [3]
E/A ratio	PW Doppler	26.09	8.70	16.67	…	11.76	3.53	0.98–1.7 [4]

The Mindray© M5 ultrasound system was used to perform transthoracic echocardiography in the B and M modes and pulse wave Doppler echocardiography used to assess the left ventricular diastolic function. Please see Table 2 for units of measurement for each parameter: Group A: seronegative by ELISA and Western blot analysis and PCR-positive, indicative of acute infection; group B: seropositive and PCR-negative, indicative of chronic infection; group C: seropositive and PCR-positive, indicative of chronic and repeat infections. Reference cited: [1] Cornell et al., 2004 [18]; [2] Brown et al., 2003 [22]; [3] Serres et al., 2008 [23]; [4] Schober and Fuentes, 2001 [24].

4. Discussion

The presence of circulating *T. cruzi* DNA in seropositive dogs (group C) may indicate a low but persistent parasitemia in chronic Chagas disease. This trend is similar to the observation made in seropositive people from endemic areas that exhibit persistence of circulating *T. cruzi* using a PCR approach [25,26]. Yucatan State, the site of this study, is a *T. cruzi*-endemic zone with abundance of the vector *Triatoma dimidiata* [27] and vector-borne transmission. Curtis-Robles et al. [28] found 25% of dogs included in their study from southern Texas were PCR-positive/seropositive for *T. cruzi*. In the present study, a larger number of cases were positive according to both tests (97 out of 130 dogs), which cannot be justified only by the presence of low, occasional parasitemia. Thus, we surmise that repeat exposure to infection contributed to the detection of circulating parasites in the seropositive dogs.

The published literature has not addressed the prevalence of DCM and heart failure in dogs. Our in-depth analysis of changes in LV structure and function in the infected dogs in this study identified a prevalence of DCM of 11.54% in the infected dogs, and the DCM features in the dogs were consistent with the clinical characterization of chronic AT made by other investigators [10,29]. The majority of the dogs included in the study were without DCM or in the early stages of DCM development. These observations do not mean that DCM is uncommon in dogs with AT but that it is a terminal feature of Chagas disease [1,2,7,30]. It is also possible that most animals do not present with this clinical condition and perhaps many may die due to arrhythmias, blockages, endocarditis, valvular endocardiosis, congestive heart failure, or other conditions that may manifest before DCM develops [8,30]. In this study, it was also found that three dogs with DCM were negative in the serological tests but positive in the PCR test, which indicates that acute *T. cruzi* infection can also result in DCM and heart failure in dogs, as is noted in 5% of acutely infected human patients. Another possibility is that these dogs were exposed to *T. cruzi* infection after they had already developed DCM from other etiologies.

Ventricular dilatation is the main feature of DCM [21], and it is recognized as one of the most well-known alterations of AT in dogs and humans [1,2,5–7,30]. As expected, the groups of infected dogs with DCM showed higher mean values for LVID-s and LVID-d, and these values were statistically different when compared to the other groups (Table 2). In comparison, many of the PCR-positive dogs in group A and group C exhibited a reduction in their LVID-s and LVD-d values (Table 3). This pattern is similar to that reported [31] in puppies during the acute stage of Chagas disease.

Chetboul [21] has documented the thinning of the IVS and/or the LVPW in dogs with DCM. In this study, some of the infected DCM-negative dogs exhibited increased thickening of the IVS at systole (Table 3); however, no significant changes in IVS thickness were observed in the dogs with Chagas disease (vs. healthy controls) (Tables 2 and 3). Barr et al. [9] also documented septal thickening during systole in dogs with Chagas disease, which exceeded the maximum reference value for septal thickness. The reduction in LVPW-s seen in group B is consistent with the decrease in the size of the cardiac wall in dogs with AT reported in other studies [9].

The frequency of reduction in the LVID at systole and diastole, along with the observation of the thickening of the IVS and LVPW at systole, particularly in the PCR-positive dogs (group A and group C), suggests a tendency towards a progressive decline in the ventricular diameter. This condition has been reported in patients with acute AT [9,10,29,31,32] but not in the chronic phase of Chagas disease in experimental models and human patients. This acute left ventricular diameter reduction and septum/wall thickening of seropositive/PCR-positive dogs is possibly due to an increase in ventricular mass, as reported in mice with myocarditis [33], which in turn is produced by persistent parasitemia [26,34]. In the chronic stages, fibrosis caused by the loss of myocardial cells and their replacement by collagen fibers [35–37] may contribute to this reduction before DCM develops over time due to compensatory cardiovascular mechanisms [5,26]. It is also possible that there was a difference in the DTU of *T. cruzi* circulating in Merida at the time of this study with respect to previous studies [9,10,29] The *T. cruzi* DTU (TcI) circulating in this area [38] is documented to be cardiotropic and highly virulent [39,40].

Although the Teichholtz method is not optimal for determining EF, SV, and CO, it allows us to obtain values that can be obtained even using low-end ultrasound machines without specific software and by operators with little experience in echocardiography, which offers a great advantage when working in small veterinary clinics.

As noted in other studies [9,29], we recorded systolic dysfunction (decreased FS and EFs) in some of the infected animals without DCM (Table 3), though the average values for these parameters in the infected dogs (groups A, B, C) were not significantly different from those observed in the healthy controls (Table 2). This observation indicates that though they were identified as DCM-negative with normal myocardial contractility, some of the infected dogs were progressing towards the clinical development of DCM. The EF in the animals with DCM was less efficient than that in the animals without DCM despite the observation that SV and CO were significantly greater due to the dilated ventricular space. Yet there is a possibility that the increase in the systolic function parameters in the PCR-positive groups, especially in group C (Table 2), occurred due to the stress effect of the ultrasonographic procedure, which was conducted without anesthesia. Stress is documented to increase the heart rate and the ventricular contractibility transiently in small animals [41]. Further studies comparing sedated and non-sedated dogs will be needed to address the effect of stress on LV function in infected dogs.

The mean diastolic function (E/A index) was found within the reference ranges [24] in all groups (Table 2), though the E/A ratio tended to increase in all the infected groups. Blood flow patterns were not directly correlated with AT or its severity (with or without DCM) because there were similar numbers of cases with delayed relaxation and restrictive and normal patterns, with no predominance of any one parameter, in both the DCM-positive and DCM-negative infected dogs. A pattern of delayed relaxation in dogs was reported by Pascon et al. [10] in *T. cruzi*-infected dogs without DCM, but it has not been described in *T. cruzi*-positive dogs with DCM, although this and the other patterns found in the infected dogs are consistent with the pathophysiology of DCM [21].

It is particularly important to consider those indicators that were statistically different between the DCM-positive and DCM-negative groups (i.e., IVS-s, LVID-d, LVID-s, LVPW-d, and EF) because they allow for an approximation in the evaluation of these variables in dogs that are at risk of developing DCM. It should be noted that the alterations found here, when applied to an uncontrolled, randomized population, may not be associated with *T. cruzi* infection only since it is possible that there are other diseases or chronic degenerative processes that may affect the heart in a similar manner as was observed for the *T. cruzi*-infected dogs.

5. Conclusions

Changes in the heart structure were evident in infected dogs with DCM. Some of the infected dogs exhibited a decreased left ventricular diameter with the presence of IVS and LVPW thickening, indicating the risk of DCM development. Transmitral flow patterns were not associated with the severity of cardiac damage in the DCM-positive infected dogs. The results of the present study also suggest that changes in the LVID at systole and diastole are a major indicator of cardiac involvement in *T. cruzi*-infected dogs, and observation of changes in the thickness of the IVS and LVPW at systole or diastole, fractional shortening, and an E/A ratio above or below the normal range will provide high predictive efficacy in determining the risk of DCM progression.

Author Contributions: E.E.A.-B. and A.O.-P. conceived the study and analyzed the data; E.E.A.-B. and C.M.A.-A. carried out the echocardiography; C.M.A.-A., funding acquisition; M.J.-C., supervision and provision of reagents for serology and PCR experiments; J.C.S.-C. performed the data curation and analysis; A.O.-P. and M.J.-C. wrote the original draft; N.J.G. provided intellectual input and conceptualization and edited the manuscript. All authors have read and agreed to the published version of the manuscript.

Funding: This research received no external funding. N.J.G. is supported by grants from the National Institute of Allergy and Infectious Diseases (R01AI054578; R01AI136031) of the National Institutes of Health.

Institutional Review Board Statement: The study was conducted in accordance with the bioethics committee (CB-CCBA-I-2017-003) at the Facultad de Medicina Veterinaria y Zootecnia, Universidad Autónoma de Yucatán.

Informed Consent Statement: Informed consent was obtained from the dog owners before the dogs were enrolled in the study.

Data Availability Statement: All the data are presented in the study. Raw data are available from the corresponding author upon reasonable request.

Acknowledgments: The authors would like to thank Manuel Marquez-Solis for his kind collaboration during the ultrasonographic evaluation.

Conflicts of Interest: The authors declare no conflicts of interest.

References

1. Taylor, M.A.; Coop, R.L.; Wall, R.L. *Veterinary Parasitology*, 3rd ed.; Wiley-Blackwell: New Delhi, India, 2016; pp. 356–358.
2. Weese, J.S.; Peregrine, A.S.; Anderson, M.E.C.; Fulford, M.B. Parasitic Diseases. In *Companion Animal Zoonoses*; Weese, J.S., Fulfords, M.B., Eds.; Wiley-Blackwell: Ames, IA, USA, 2011; pp. 3–108.
3. Jiménez-Coello, M.; Poot-Cob, M.; Ortega-Pacheco, A.; Guzma-Marin, E.; Ramos-Ligonio, A.; Sauri-Arceo, C.H.; Acosta-Viana, K.Y. American trypanosomiasis in dogs from an urban and rural area of Yucatan, Mexico. *Vector Borne Zoonotic Dis.* **2008**, *8*, 755–761. [CrossRef]
4. Jiménez-Coello, M.; Acosta-Viana, K.Y.; Guzmán-Marín, E.; Bárcenas-Irabién, A.; Ortega-Pacheco, A. American trypanosomiasis and associated risk factors in owned dogs from the major city of Yucatan, Mexico. *J. Venom. Anim. Toxins Incl. Trop. Dis.* **2015**, *21*, 37. [CrossRef]
5. Camacho, A.A. Miocardiopatía chagásica en caninos. In *Afecciones Cardiovasculares en Pequeños Animales*; Belerenian, G.C., Mucha, C.J., Camacho, A.A., Eds.; Inter-Médica: Buenos Aires, Argentina, 2007; pp. 180–183.
6. Guedes, P.M.M.; Veloso, V.M.; Alfonso, L.C.C.; Caliari, M.V.; Carneiro, C.M.; Diniz, L.F.; Marques-da-Silva, E.A.; Caldas, I.S.; do Valle Mata, M.A.; Souza, S.M.; et al. Development of chronic cardiomyopathy in canine Chagas disease correlates whith high IFN-gamma, TN-alpha, and low IL-10 production during the acute infection phase. *Vet. Immunol. Immunopathol.* **2009**, *130*, 43–52. [CrossRef] [PubMed]
7. Rassi, A., Jr.; Rassi, A.; Marin-Neto, J.A. Chagas disease. *Lancet* **2010**, *375*, 1388–1402. [CrossRef] [PubMed]
8. Pereira-Nunes, M.C.; Dones, W.; Morillo, C.A.; Justiniano-Encina, J.; Ribeir, A.L. Chagas disease an overview of clinical and epidemiological aspects. *J. Am. Coll. Cardiol.* **2013**, *62*, 767–776.
9. Barr, S.C.; Holmes, R.A.; Klei, T.R. Electrocardiographic and echocardiographic features of trypanosomiasis in dogs inoculated with North American *Trypanosoma cruzi* isolates. *Am. J. Vet. Res.* **1992**, *53*, 521–527. [CrossRef] [PubMed]
10. Pascon, J.P.D.; Neto, G.B.P.; Sousa, M.G.; Sousa, M.G.; Paulino, D.; Camacho, A.A. Clinical characterization of chronic chagasic cardiomyopathy in dogs. *Pesq. Vet. Bras.* **2010**, *30*, 115–120. [CrossRef]
11. Gompf, R.E. History and physical examination. In *Manual of Canine and Feline Cardiology*, 5th ed.; Smith, F.W.K., Tilley, L.P., Oyama, M., Sleeper, M.M., Eds.; Elsevier: St. Louis, MO, USA, 2016; pp. 3–25.
12. Tilley, L.; Burtnick, N. *ECG for the Small Animal Practitioner*; Teton New Media: Jackson, WY, USA, 2009; pp. 64–84.
13. Kittleson, M.D.; Kienle, R.D. Pulmonary arterial and systemic arterial hypertension. In *Small Animal Cardiovascular Medicine*; Kittleson, M.D., Kienle, R.D., Eds.; Mosby: St. Louis, MO, USA, 1998; pp. 1472–1845.

14. Torres-Vargas, J.; Jimenez-Coello, M.; Guzman-Marin, E.; Acosta-Viana, K.Y.; Yadon, Z.E.; Gutierrez-Blanco, E.; Guillermo-Cordero, J.L.; Garg, N.J.; Ortega-Pacheco, A. Quantitative and histological assessment of maternal-fetal transmission of *Trypanosoma cruzi* in guinea pigs: An experimental model of congenital Chagas disease. *PLOS Negl. Trop. Dis.* **2018**, *12*, e0006222. [CrossRef] [PubMed]
15. Boon, J.A. *Veterinary Echocardiography*, 2nd ed.; Wiley-Blackwell: Ames, IA, USA, 2011; pp. 50–132.
16. de Madron, E. Normal views: 2D, TM, spectral and color doppler. In *Clinical Echocardiography of the Dog and Cat*; de Madron, E., Chetboul, V., Bussadori, C., Eds.; Elsevier: St Louis, MO, USA, 2015; pp. 21–37.
17. Boon, J.A. *Two-Dimensional and M-Mode Echocardiography for the Small Animal Practitioner*, 2nd ed.; Wiley-Blackwell: Ames, IA, USA, 2016; pp. 75–83.
18. Cornell, C.C.; Kittleson, M.D.; Della Torre, P.; Häggström, J.; Lombard, C.W.; Pedersen, H.D.; Vollmar, A.; Wey, A. Allometric scaling of M-mode cardiac measurements in normal adult dogs. *J. Vet. Int. Med.* **2004**, *18*, 311–321.
19. Gallay-Lepoutre, J.; Bélanger, M.C.; Nadeau, M.E. Prospective evaluation of Doppler echocardiography, tissue Doppler imaging and biomarkers measurement for the detection of doxorubicin-induced cardiotoxicity in dogs: A pilot study. *Res. Vet. Sci.* **2016**, *105*, 153–159. [CrossRef]
20. de Madron, E. Assessment of distolic function. In *Clinical Echocardography of the Dog and Cat*; de Madron, E., Chetboul, V., Bussadori, C., Eds.; Elsevier: St Louis, MO, USA, 2015; pp. 3–19.
21. Chetboul, V. Dilated cardiomyopathy and other cardiomyopathies in dogs. In *Clinical Echocardiography of the Dog and Cat*; de Madron, E., Chetboul, V., Bussadori, C., Eds.; Elsevier Masson: St Louis, MO, USA, 2016; pp. 181–205.
22. Brown, D.J.; Rush, J.E.; MacGregor, J.; Ross, J.N., Jr.; Brewer, B.; Rand, W.M. M-mode echocardiographic ratio indices in normal dogs, cats, and horses: A novel quantitative method. *J. Vet. Int. Med.* **2003**, *17*, 653–662.
23. Serres, F.; Chetboul, V.; Tissier, R.; Poujol, L.; Gouni, V.; Sampedrano, C.; Pouchelon, J.L. Comparison of 3 ultrasound methods for quantifying left ventricular systolic function: Correlation with disease severity and prognostic value in dogs with mitral valve disease. *J. Vet. Int. Med.* **2008**, *22*, 566–577. [CrossRef] [PubMed]
24. Schober, K.E.; Fuentes, V.L. Effects of age, body weight, and heart rate on transmitral and pulmonary venous flow in clinically normal dogs. *Am. J. Vet. Res.* **2001**, *62*, 1447–1454. [CrossRef]
25. Salomone, O.A.; Juri, D.; Omelianiuk, M.O.; Sembaj, A.; Afiguerri, A.M.; Carriazo, C.; Barral, J.M.; Madoery, R. Prevalence of circulating Trypanosoma cruzi detected by polymerase chain reaction in patients with Chagas cardiomyopathy. *Am. J. Cardiol.* **2000**, *85*, 1274–1276. [CrossRef]
26. Marin-Neto, J.A.; Cunha-Neto, E.; Maciel, B.C.; Simões, M.V. Pathogenesis of chronic Chagas heart disease. *Circulation* **2007**, *115*, 1109–1123. [CrossRef]
27. Dumonteil, E.; Ramirez-Sierra, M.J.; Ferral, J.; Euan-Garcia, M.E.; Chavez-Nuñez, L. Usefulness of community participation for the fine temporal monitoring of house infestation by non-domiciliated triatomines. *J. Parasitol.* **2009**, *95*, 469–471. [CrossRef] [PubMed]
28. Curtis-Robles, R.; Snowden, K.F.; Dominguez, B.; Dinges, L.; Rodgers, S.; Mays, G.; Hamer, S.A. Epidemiology and molecular typing of *Trypanosoma cruzi* in naturally infected hound dogs and associated triatomine vectors in Texas, USA. *PLOS Negl. Trop. Dis.* **2017**, *11*, e0005298. [CrossRef] [PubMed]
29. Meurs, K.M.; Anthony, M.A.; Slater, M.; Miller, M.W. Chronic *Trypanosoma cruzi* infection in dogs: 11 cases (1987–1996). *J. Am. Vet. Med. Assoc.* **1998**, *213*, 497–500. [CrossRef]
30. Graiff, D.S. Relación Entre Perros Seropositivos a *Trypanosoma cruzi* y Alteraciones Electrocardiográficas Compatibles Con Miocardiopatía Chagásica Canina en la Localidad de La Para (Córdoba-Argentina). Master's Thesis, Universidad Nacional del Litoral, Rojas, Argentina, 2010.
31. Vitt, J.P.; Saunders, A.B.; O'Brien, M.T.; Mansell, J.; Ajithdoss, D.K.; Hamer, S.A. Diagnostic features of acute Chagas myocarditis with sudden death in a family of boxer dogs. *J. Vet. Intern. Med.* **2016**, *30*, 1210–1215. [CrossRef]
32. Rassi, A.; de Rezende, A.J.M. American trypanosomiasis (Chagas disease). *Infect. Dis. Clin. North. Am.* **2012**, *26*, 275–291. [CrossRef]
33. Steinl, D.C.; Xu, L.; Khanicheh, E.; Ellertsdottir, E.; Ochoa-Espinosa, A.; Mitterhuber, M.; Glatz, K.; Kuster, G.M.; Kaufmann, B.A. Noninvasive contrast-enhanced ultrasound molecular imaging detects myocardial inflammatory response in autoimmune myocarditis. Circ. Cardiovasc. *Imaging* **2016**, *9*, e004720.
34. Gironès, N.; Fresno, M. Etiology of Chagas disease myocarditis: Autoimmunity, parasite persistence, or both? *Trends. Parasitol.* **2003**, *19*, 19–22. [CrossRef] [PubMed]
35. Guedes, P.M.M.; Veloso, V.M.; Caliari, M.V.; Carneiro, C.M.; Souza, S.M.; de Lana, M.; Chiari, E.; Bahia, M.T.; Galvao, L.M.C. *Trypanosoma cruzi* high infectivity in vitro is related to cardiac lesions during long-term infection in Beagle dogs. *Mem. Inst. Oswaldo Cruz* **2007**, *102*, 141–147. [PubMed]
36. Garzoni, L.R.; Adesse, D.; Soares, M.J.; Rossi, M.I.D.; Borojevic, R.; de Meirelles, M.D.L. Fibrosis and hypertrophy induced by *Trypanosoma cruzi* in a three-dimensional cardiomyocyte-culture system. *J. Infect. Dis.* **2008**, *197*, 906–915. [CrossRef] [PubMed]
37. Cunha-Neto, E.; Teixeira, P.C.; Nogueira, L.G.; Kalil, J. Autoimmunity. In *Advances in Parasitology, Chagas Disease*; Weiss, L.M., Tanowitz, H.B., Eds.; Elsevier Academic Press Inc.: San Diego, CA, USA, 2011; Volume 76, pp. 129–152.

38. Jiménez-Coello, M.; Shelite, T.; Castellanos-Gonzalez, A.; Saldarriaga, O.; Rivero, R.; Ortega-Pacheco, A.; Acevedo-Arcique, C.; Amaya-Guardia, K.; Garg, N.; Melby, P.; et al. Efficacy of recombinase polymerase amplification to diagnose *Trypanosoma cruzi* infection in dogs with cardiac alterations from an endemic area of Mexico. *Vector Borne Zoonotic Dis.* **2018**, *18*, 417–423. [CrossRef] [PubMed]
39. Miles, M.A.; Llewellyn, M.S.; Lewis, M.D.; Yeo, M.; Baleela, R.; Fitzpatrick, S.; Gaunt, M.W.; Mauricio, I.L. The molecular epidemiology and phylogeography of *Trypanosoma cruzi* and parallel research on Leishmania: Looking back and to the future. *Parasitology* **2009**, *136*, 1509–1528. [CrossRef]
40. Espinoza, B.; Rico, T.; Sosa, S.; Oaxaca, E.; Vizcaino-Castillo, A.; Caballero, M.L.; Martínez, I. Mexican *Trypanosoma cruzi* Tc I strains with different degrees of virulence induce diverse humoral and cellular immune responses in a murine experimental infection model. *J. Biomed. Biotechnol.* **2010**, *2010*, 890672. [CrossRef]
41. Sisson, D.D. Pathophysiology of heart failure. In *Textbook of Veterinary Internal Medicine*, 7th ed.; Ettinger, S.J., Feldman, E.C., Eds.; Saunders; Elsevier: St Louis, MO, USA, 2010; pp. 1143–1158.

animals

MDPI

Article

Improved Antibody Detection for Canine Leptospirosis: ELISAs Modified Using Local Leptospiral Serovar Isolates from Asymptomatic Dogs

Pannawich Boonciew [1], Watcharee Saisongkorh [2], Suppalak Brameld [2], Matsaya Thongpin [2], Alongkorn Kurilung [3], Pratomporn Krangvichian [4], Waree Niyomtham [1], Kanitha Patarakul [5,6], Thanmaporn Phichitraslip [7], David J. Hampson [8] and Nuvee Prapasarakul [1,9,*]

1 Department of Veterinary Microbiology, Faculty of Veterinary Science, Chulalongkorn University, Henri-Dunent Street, Pathumwan, Bangkok 10330, Thailand; pannawichb12@gmail.com (P.B.); pewwaree@yahoo.com (W.N.)
2 Laboratory of Immunology, National Institute of Health, Department of Medical Sciences, Ministry of Public Health, Nonthaburi 11000, Thailand; watcharee.s@dmsc.mail.go.th (W.S.); suppalak37@gmail.com (S.B.); matsaya.t@dmsc.mail.go.th (M.T.)
3 Siriraj Metabolomics and Phenomics Center, Faculty of Medicine Siriraj Hospital, Mahidol University, Bangkok 10700, Thailand; alongkornkurilung@gmail.com
4 Medical Microbiology, Interdisciplinary Program, Graduate School, Chulalongkorn University, Bangkok 10330, Thailand; p_krangvichian@hotmail.com
5 Department of Microbiology, Faculty of Medicine, Chulalongkorn University, Bangkok 10330, Thailand; kanitha.pa@chula.ac.th
6 Chula Vaccine Research Center (Chula VRC), Center of Excellence in Vaccine Research and Development, Chulalongkorn University, Bangkok 10330, Thailand
7 Faculty of Veterinary Technology, Kasetsart University, Bangkok 10900, Thailand; cvttpp@ku.ac.th
8 School of Veterinary Medicine, Murdoch University, Perth, WA 6150, Australia; d.hampson@murdoch.edu.au
9 Center of Excellence in Diagnostic and Monitoring of Animal Pathogens (DMAP), Chulalongkorn University, Bangkok 10330, Thailand
* Correspondence: nuvee.p@chula.ac.th

Simple Summary: Leptospirosis is an infectious zoonotic disease caused by pathogenic Leptospira and affects both humans and animals throughout the world. Domestic animals, such as dogs, can act as potential reservoirs of leptospires for human or animal infections and environmental contamination. Due to the different varieties of the predominant local circulating pathogenic serovars in several regions, serodiagnosis tests, such as conventional microscopic agglutination tests (MATs) or others that do not include local serovars and their protein antigens, have limitations, and may fail to detect the disease and carry out antibody surveillance in certain regions. This study aimed to develop a more accurate antibody detection tool for canine leptospirosis in our region by using an indirect enzyme-linked immunosorbent assay (ELISA) using relevant local isolates of leptospiral serovars from asymptomatic dogs. All the modified IgG-ELISAs with the local isolates gave positive results for all infected dogs, especially outer membrane protein (OMP)-based IgG-ELISAs that showed negative results for all dogs from non-endemic areas, demonstrating improved accuracy and reduced limitations over those of the standard MAT and providing an enhanced method for leptospirosis detection in the study area. This improvement is crucial when investigating the epidemiology of the disease and preventing its spread. The article highlights the requirement and significance of utilising local circulating isolates in serological approaches to accurately diagnose and monitor leptospirosis.

Abstract: Leptospirosis is a zoonotic disease of significant concern for human and animal health, with domestic animals, including dogs, acting as reservoirs for human infection. Serology is widely used for leptospirosis diagnosis, even though the standard microscopic agglutination test (MAT) using a panel of serovars lacks specificity and can lead to detection limitations in certain regions. In this study, we aimed to develop an antibody detection tool for dogs using an indirect enzyme-linked immunosorbent assay (ELISA) with a set of local serovar isolates, including Paidjan, Dadas, and Mini, to enhance the accuracy of leptospirosis surveillance in our region. The specificity and sensitivity of various antigen preparations, namely leptospiral whole-cell protein (WCP), total membrane protein (TMP), and outer membrane protein (OMP), were assessed using sera from infected and non-infected dogs, as well as negative puppy sera. Leptospirosis diagnosis was supported using a genus-specific nested polymerase chain reaction test on all collected sera. Protein preparations were validated using SDS-PAGE and Western blotting analysis. In the results, the standard MAT failed to detect antibodies in any of the dogs confirmed as being infected using PCR and isolation, highlighting its limitations. In contrast, the OMP-based ELISAs using local isolates of *Leptospira* serovars gave positive results with sera from all infected dogs, and negative results with sera from all dogs from non-endemic areas. IgG titres of infected and unvaccinated dogs from endemically affected areas were significantly higher than those in non-endemic regions. Using the OMP-based IgG/ELISAs with the local serovar Dadas resulted in higher specificity and lower sensitivity than when using the WCP- and TMP-based IgG/ELISAs. Agreement analysis revealed fair and moderate concordance between OMP-based IgG/ELISAs and PCR results, whereas slight and fair agreement was observed between OMP-based ELISAs and the MAT. Overall, the modified OMP-based IgG/ELISAs, utilising relevant local serovar isolates from dogs, demonstrated improved accuracy in detecting leptospirosis in the study area, overcoming the limitations of the MAT. This study highlights the importance of identifying and incorporating these local circulating serovar isolates into serological techniques for leptospirosis diagnosis and surveillance.

Keywords: canine leptospirosis; diagnosis; serological test; microscopic agglutination test; enzyme-linked immunosorbent assay; local serovars or isolates

1. Introduction

Leptospirosis is considered a worldwide significant infectious zoonosis for humans and animals, particularly in developing countries, including Thailand [1–4]. The disease is caused by pathogenic spirochetes in the genus *Leptospira*, which currently contains 68 species, in which pathogenic groups can be divided into over 24 pathogenic serogroups and 250 pathogenic serovars based on the surface epitopes of the lipopolysaccharide (LPS) antigens on their outer membrane [2,5–7]. As various pathogenic serovars can infect humans and animals, it is crucial to identify the serovar involved and develop improved serological diagnostic techniques to understand their epidemiology and prevent the spread of disease [5,8–13].

Mammals, especially domestic animals (dogs and livestock), can play a significant role as reservoirs for disease maintenance and transmission, contaminating the environment via their urine [1,14]. The four pathogenic serovars Canicola, Icterohaemorrhagiae, Grippotyphosa, and Pomona can be found in dogs, as assessed by antibody prevalence or isolation, and have been included in several commercial canine vaccines and diagnostic methods [15,16]. Despite this, studies on antibody prevalence have suggested that the serovars circulating in dogs vary among locations.

According to studies previously conducted in Thailand, various leptospiral serovars, including Autumnalis, Australis, Bataviae, Bratislava, Canicola, Copenhageni, Grippotyphosa, Icterohaemorrhagiae, Javanica, Mini, Pomona, Sejroe, Shermani, and Tarassovi have been shown to have high seroprevalence in dogs; moreover, these serovars also have been described as the predominant circulating serovars found in humans in Thailand [17–21]. Previous reports demonstrate that dogs may spread and increase the probability of human infection with the circulating serovars, where dogs and humans share the same areas and environment [22–25].

In Thailand, *L. interrogans* serovars Paidjan, Dadas, and Batavaie, and *L. weilii* serogroup Mini were the leptospires most frequently detected in asymptomatic dogs via a cross-agglutination test, whereas evidence for their presence could not be detected via the MAT. Phylogenetic analysis indicated that they were genetically related to *L. interrogans* isolates from the urine of asymptomatic humans with a previous history of symptoms and close contact with unvaccinated animals in their backyards (*L. interrogans* serovars Paidjan strain CUDO5 and Dadas strain CUDO8) [26,27]. In addition, two local canine isolates of *L. weilii* belonged to sequence type (ST) 94 (containing an undesignated serovar and strains CUDO6 and CUDO13), and this ST was closely related to ST183 (containing an undesignated serovar and strain LNT1234) and ST193 (containing serovar Hekou, strain H27 of *L. weilii*), isolated from humans in Laos and China [26–30]. Moreover, these serovars, especially Bataviae, have been reported to cause disease in humans and dogs and are carried by rodents [13,31–34]. Overall, dogs represent a significant potential source of human leptospirosis infection [26,32,35].

Current methods for diagnosing animal leptospirosis follow recommendations by the World Organisation for Animal Health (OIE), focusing either on detecting the agent or immune response indicating past exposure [36–38]. PCR is the gold-standard method for detecting early and chronic infection in agent detection, while serum antibody detection using the MAT with live leptospires or an ELISAs with protein antigens is suggested for surveillance [38,39]. The serovars used in the test should reflect the serovars circulating in the investigated populations [36,37].

Recent studies in Thailand have identified issues with the use of serological and PCR detection methods for guiding the country's strategic approach to leptospirosis. The local Leptospiral isolates, specifically serovars Paidjan and Dadas for *L. interrogans* and serogroup Mini for *L. weilii*, are not included in the 24 serovars used in the MAT panel by Thailand's National Institute of Health. The MAT panel includes 23 pathogenic serovars and 1 non-pathogenic serovar [40,41]. Challenges arise due to differences in strains and serovars, impacting diagnostic accuracy. Discrepancies between PCR and MAT results have been observed, impacting the application of serological detection for surveillance in Thailand [26]. To address these limitations, several studies have developed ELISAs to detect IgM and IgG antibodies against leptospiral antigen proteins, which are cost-effective, offer good specificity and sensitivity, and are suitable for the large-scale surveillance of animal populations [42–44].

This study aimed to assess the effectiveness of using locally sourced leptospiral isolates from dogs in Thailand for modified indirect ELISAs by comparing these with the standard MAT test, which utilises conventional serovars, to enhance the accuracy of leptospirosis detection in canine sera. Different protein antigen preparations comprising whole-cell, total membrane, and outer membrane protein were evaluated to provide recommendations for optimal antigen selection to improve test sensitivity and specificity.

2. Materials and Methods

2.1. Bacterial Cultures and Growth Conditions

Nine *Leptospira* isolates were used for antibody detection in the MAT and for protein extraction in developing modified ELISAs. These comprised five local leptospiral isolates that were found in the urine of asymptomatic dogs in Thailand, including three from *Leptospira interrogans*: serogroup Bataviae, serovar Paidjan, strain CUDO5; serogroup Grippotyphosa, serovar Dadas, strain CUDO8; and serogroup Bataviaem serovar Bataviae, strain D64. They also comprised two *Leptospira weilii* strains: serogroup Mini strains CUDO6 and CUD13 (non-identified serovars) [26,32], and four isolates of *L. interrogans* that are commonly used in commercial leptospirosis vaccines in dogs, comprising serovars Icterohaemorrhagiae, Pomona, Grippotyphosa, and Canicola. The methods used to control biohazards associated with handling Leptospira were approved by the Institutional Biosafety Committee of the Faculty of Veterinary Science, Chulalongkorn University (IBC 2031004). All leptospiral isolates were cultivated and incubated in 25 mL of liquid Ellinghausen–McCullough–Johnson–Harris (EMJH) medium (Difco, Sparks, MD, USA) supplemented with enrichment EMJH (Difco, USA) and 3% (*v*/*v*) rabbit serum for two weeks in aerobic conditions at 28–30 °C. The other leptospiral standard serovars used in the MAT panel were provided by the National Institute of Health (NIH), Thailand [40,41]. The list of all 29 isolates of 27 serovars used is shown in Table 1.

Table 1. Names of 29 leptospiral isolates, including 24 representative reference isolates and 5 local Thai isolates, that were used in the microscopic agglutination test (MAT) and enzyme-linked immunosorbent assay (ELISA) in the study. These isolates were associated with 27 serovars (serovar Bataviae and Mini each had two different isolates).

Serovars/Isolates	MAT	ELISA
24 representative reference isolates (belonging to 24 serovars)		
Australis	Used	-
Aumtumnalis	Used	-
Ballum	Used	-
Bataviae	Used	-
Canicola	Used	Used
Cellidoni	Used	-
Cynopteri	Used	-
Djasiman	Used	-
Grippotyphosa	Used	Used
Hebdonadis	Used	-
Icterohaemorrhagiae	Used	Used
Javanica	Used	-
Louisaina	Used	-
Manhao	Used	-
Mini	Used	-
Panama	Used	-
Pomona	Used	Used
Pyrogenes	Used	-
Ranarum	Used	-
Sarmin	Used	-
Sejroe	Used	-
Shermani	Used	-
Tarasovi	Used	-
Semaranga	Used	-
5 local Thai isolates (belonging to 4 serovars)		
Paidjan strain CUDO5	Used	Used
Dadas strain CUDO8	Used	Used
Bataviae strain D64	Used	Used
Mini strain CUDO6	Used	Used
Mini strain CUD13	Used	Used

2.2. Serum Samples and Groups of Dogs

Whole canine sera were collected from 260 dogs and divided into five groups based on their status and origin. The collection of serum samples from dogs was approved by the Chulalongkorn University Animal Care and Use Committee (CU-ACUC; Protocol NO. 1531075, groups 1 and 2). Moreover, the remaining canine sera from another study were authorised by the owners with written informed consent for the participation of their dogs in this study (groups 3, 4, and 5). Group 1 (n = 6) included leptospiral-infected dogs in an area endemic to local isolates of leptospiral serovars (Nan Province) from a previous study. Leptospirosis infection was confirmed by testing urine using a nested PCR targeting a genus-specific region on the leptospiral 16S ribosomal RNA (*rrs*) gene. The four leptospiral serovar isolates were also obtained from the urine of these dogs [26]. The serogroups and serovars of these isolates were identified by using the microscopic agglutination test (MAT) with polyclonal and monoclonal antibodies raised against *Leptospira* isolates at the OIE National Collaborating Centre for Reference and Research on Leptospirosis, The Netherlands [27,30]. The isolates were identified as *L. interrogans* serovars Paidjan and Dadas and *L. weilii* serogroup Mini, as mentioned earlier. Group 2 (n = 21) included unvaccinated dogs from the same endemic area as that of the positive leptospiral-infected dog group from a previous study [26]. The urines of these dogs were all negative for *Leptospira*, as determined via nested PCR and isolation. Group 3 (n = 112) was a set of one-year-old dogs from Bangkok with a history of a complete vaccination program from the Blood Bank Unit, Small Animal Hospital, Faculty of Veterinary Science, Chulalongkorn University. Each dog had received various combinations of vaccine isolates from the multivalent leptospiral vaccine (bivalent, trivalent, or quadrivalent leptospiral vaccine) when they were vaccinated at different hospitals or small animal clinics before blood donation. The quadrivalent vaccine for leptospirosis contained four leptospiral serovars from the species *Leptospira interrogans*, including serovars Icterohaemorrhagiae, Canicola, Grippotyphosa, and Pomona (trivalent vaccine: serovars Icterohaemorrhagiae, Canicola, and Grippotyphosa; bivalent vaccine: serovars Icterohaemorrhagiae and Canicola). Group 4 (n = 108) comprised unvaccinated dogs from non-endemic areas for the isolated serovars (Bangkok, Samut Prakan, and Chonburi provinces). Group 5 (n = 13) included two-month-old unvaccinated puppies from the same non-endemic areas as those represented by Group 4, and these served as a negative control to determine the cut-off values. The sera from Groups 4 and 5 were provided by staff of the Faculty of Veterinary Technology, Kasetsart University. All the canine sera were tested for the *Leptospira* genus gene via nested PCR to help investigate any history of leptospirosis infection.

2.3. Detection of Leptospira in Sera Using a Genus-Specific Nested PCR Assay

DNA was extracted from 100 µL samples of each canine serum using Genomic DNA Extraction and Purification Kit (Thermo Scientific™, Thermo Fisher Scientific, Waltham, MA, USA), following the manufacturer's instructions. DNA extracts were stored at −20 °C before use. A single-tube nested PCR was used to detect a genus-specific region on the leptospiral 16S ribosomal RNA (*rrs*) gene found in pathogenic and intermediate pathogenic *Leptospira* spp., as previously described [45]. A 25 µL nested PCR was performed using two sets of primers, including rrs-outer-F (5′-CTCAGAACTAACGCTGGCGGCGCG-3′), rrs-outer-R (5′-GGTTCGTTACTGAGGGTTAAAACCCCC-3′), rrs-inner-F (5′-CTGGCGGCGCGTCTTA-3′), and rrs-inner-R (5′-GTTTTCACACCTGACTTACA-3′). The sizes of the final amplicon were 547 and 443 bp. The genomic DNA extracted from the local isolates of the leptospiral serovars from *L. interrogans* and *L. weilii* was used as the positive control. In contrast, the negative control was the reaction mixture without a DNA template. Both positive and negative controls were included in each run of the nested PCR assay. The resulting amplicons were separated via 1.5% agarose gel electrophoresis and visualised under ultra-violet light following staining with ethidium bromide. The DNA extracts of a serum sample showing bands of resulting amplicons were recorded as positive PCR results.

2.4. A Preliminary Study Using the Microscopic Agglutination Test (MAT)

Fifty serum samples including all of those in Group 1 (6 sera) and Group 2 (21 sera), and a subset of 23 serum samples from Group 3 were selected for use in a preliminary study of the MAT. This was conducted to test the hypothesis that the use of the approved standard MAT might not detect antibody titres in the sera samples from dog groups in this study, especially leptospiral-infected dogs. In the MAT panel, 29 isolates of 27 serovars were used, including 24 representative reference isolates of *Leptospira* serovars [40,41] and 5 local isolates of *Leptospira* serovars isolated from Thai asymptomatic dogs [26,32]. The MAT procedure was carried out as previously described [46]. Individual serum samples were tested via two-fold dilutions from 1:20 to 1:10,240. The threshold of MAT cut-off for reactivity was defined as ≥1:20 [32]. The titre, the maximum dilution at which 50% of leptospires agglutinated with the antibody from the serum dilution, was used to interpret the MAT results.

2.5. Leptospiral Protein Preparations for Modified ELISAs

2.5.1. Whole-Cell Protein Using Sonicated Leptospiral Preparations

All leptospiral protein preparations were extracted in duplicate using 50 mL of *Leptospira* EMJH culture (one replicate; 25 mL). After protein preparations were completed, the extracted proteins from each replicate were pooled in a single tube. Whole-cell protein (WCP) extraction was modified from a previously described method [47]. In total, of 2.1–2.9 × 10^8 cells/mL of each *Leptospira* isolate cultured in EMJH medium was washed three times with phosphate-buffered saline (PBS, pH 7.4) and centrifuged at 13,000× *g* at 4 °C for 10 min. The cells were subjected to 10 cycles of −80 °C freezing and 10 min of thawing at room temperature. Each *Leptospira* preparation was then sonicated for 30 min at 4 °C in an ice bath. Cells containing PBS were mixed using SiLibeads Type S with an equivalent volume of PBS, and ten cycles of shaking for 1 min and cooling for 2 min were conducted. The WCP extracts were then centrifuged at 10,000× *g* for 5 min to elute them, after which each was aliquoted and kept at −20 °C until needed. The protein concentrations of the extracts ranged from 502 to 641 µg/mL for the five local isolates and 707 to 872 µg/mL for the four vaccine isolates of *Leptospira* serovars.

2.5.2. Total Membrane Protein Fraction Obtained Using Lysis Buffer

Total membrane protein (TMP) extraction utilising lysis buffer was modified from a previously described method [48]. For each isolate, 2.1–2.9 × 10^8 *Leptospira* cells cultured in EMJH medium were centrifuged three times at 13,000× *g* for 10 min at 4 °C using PBS for washing. The final *Leptospira* pellets were dissolved in 1 mL of lysis buffer (20 mM Tris (pH 8), 150 mM NaCl, 2 mM EDTA, and 2 mg of lysozyme per mL). Each live *Leptospira* serovar suspension was then sonicated after standing for 30 min in an ice bath at 4 °C. Finally, the *Leptospira* suspension was centrifuged at 13,000× *g* for 10 min at 4 °C to separate the complete membrane protein pellet from the soluble protein supernatant. For the ELISAs and protein concentration measurements, the TMP pellets were resuspended in PBS and stored at −20 °C. The concentrations of the TMPs for the five local isolates were 308 to 429 µg/mL, and these were 534 to 663 µg/mL for the four vaccine isolates of *Leptospira* serovars.

2.5.3. Outer Membrane Protein Isolation Using Triton X-114

Outer membrane protein (OMP) extraction using Triton X-114 was modified from previously described methods [48–51]. Leptospiral isolates containing 2.1–2.9 × 10^8 cells/mL in EMJH medium were washed in triplicate with PBS via centrifugation at 13,000× *g* and 4 °C for 10 min. Each *Leptospira* pellet was immersed in Triton X-114 buffer (20 mM Tris (pH 8), 150 mM NaCl, 2 mM EDTA, and 2% Triton X-114) for 1–2 h at 4 °C to extract the OMPs. The suspension was centrifuged at 17,000× *g* for 45 min to remove the insoluble pellet and preserve the supernatant. Phase separation was then conducted on the supernatant by adding PBS containing 20 mM $CaCl_2$, subjecting it to warming at 37 °C for 1 h, and centrifuging it at 6000× *g* for 10 min at 25 °C. The protein phase of each *Leptospira* isolate was divided into three phases: an aqueous phase, a detergent phase that contained outer membrane proteins, and pellets. The detergent phase was dissolved in PBS and stored at −20 °C for use in the ELISAs and measurements of protein concentration. The protein concentrations of all the isolates ranged from 113 to 227 g/mL.

2.6. Confirmation of Protein Components via SDS-PAGE and Western Blotting

SDS-PAGE and Western blotting techniques were modified from the methods described in previous studies [48,52]. The protein components of the three extractions from the five local isolates of *Leptospira* serovars were visualised via 12.5% polyacrylamide gel electrophoresis and staining with Coomassie Brilliant Blue R-250. For Western blotting, proteins in the polyacrylamide gels were transferred onto polyvinylidene difluoride (PVDF) membranes, which were equilibrated using absolute methanol. After protein transfer, the membranes were blocked with 5% skim milk in TBS buffer with 0.1% Tween (TBST) and washed three times with TBST after being incubated for 2 h at room temperature. Each membrane was incubated at 4 °C overnight with pooled leptospirosis sera from dogs confirmed as infected from Nan Province (Group 1) diluted 1:200 in TBST containing 5% Bovine serum albumin (TBST-BSA). The membranes were washed with TBST and incubated with 1:1000 horseradish peroxidase (HRP)-conjugated goat anti-dog IgG antibody in TBST at room temperature for 2 h. After incubation, the membranes were washed with TBST and visualised via chromogenic detection using a 3,3′-diaminobenzidine tetrahydrochloride (DAB) substrate.

2.7. Indirect Immunoglobulin G (IgG) ELISAs

2.7.1. Optimisation of Indirect ELISAs and Cut-Off OD Values

The optimum single-working protein concentrations and serum dilutions were established using checkerboard titration with the indirect ELISA technique, modified according to previously described studies [53,54]. The protein obtained from each extraction technique and isolate was suspended in 2 µg/mL of 0.05 M carbonate-bicarbonate buffer (pH 9.6), serially diluted two-fold to a final dilution of 0.015625 µg/mL, and applied to the wells of 96 well-microplates with a total volume of 100 µL of diluted protein in coating buffer. Three different groups of dog sera were used to optimise the indirect ELISAs, including (1) pooled serum samples from six asymptomatic dogs with *Leptospira* successfully isolated from their urine or who tested antigen-positive via PCR (Group 1), (2) pooled serum samples from six vaccinated dogs older than one year and with a history of complete vaccination (Group 3), and (3) pooled serum samples from 13 non-vaccinated 2-month-old dogs without a history of vaccination (Group 5). Each pooled sera preparation was added to PBS buffer with 0.05% Tween 20 containing 1% Bovine serum albumin (PBST-BSA) at a starting dilution of 1:80,serially diluted two-fold to a final dilution of 1:81,920 and then added to the wells across the columns of the microplates. After the optimisation of protein concentrations and serum dilutions, all 13 sera from the unvaccinated 2-month-old dogs (Group 5) were individually used to detect the optical density (OD) value and define the cut-off OD value for each modified indirect ELISA using the optimal protein concentration and serum dilution with the same conditions as those of the indirect ELISA technique used before. Each individual serum sample was examined for the OD value in triplicate. All OD values from 13 sera in triplicate were then computed as the single optimal OD cut-offs using the sum of the mean ODs plus four standard deviations (Mean + 4SD) to enhance the specificity of the modified ELISA tests for distinguishing between positive and negative results. The use of 4SD aimed to minimise false positives while maintaining sensitivity, as in other field studies [55–58].

2.7.2. IgG Antibody Detection

The optimal protein concentration for all three extraction methods (1 µg/mL) in 0.05 M carbonate-bicarbonate buffer (pH 9.6) was used to coat the wells (100 µL/well) overnight at 4 °C. After rinsing them three times with PBS buffer (pH 7.4), the plates were blocked with PBST-BSA blocking solution (100 µL/well) for 1 h before being washed three times with PBS buffer. The optimal serum sample dilutions for WCP, TMP, and OMP preparations were used at 1:1280, 1:640, and 1:640 with PBST-BSA, respectively. In total, 260 serum samples from the 5 dog groups were used. Each diluted serum sample was added to the well of the microplate (100 µL/well) and incubated at 37 °C for 1 h. After protein and serum incubation, plates were washed three times with PBS buffer. The bound IgG antibody was detected using a 1:10,000 dilution of goat anti-dog IgG conjugated with HRP enzyme in PBST-BSA buffer (100 µL/well) and incubated for 1 h at 37 °C. The microplates were washed three times with PBS buffer before 3,3′,5,5′-Tetramethylbenzidine (TMB) substrate (50 µL/well) was added to the wells, and the plates were then left at room temperature for 5–10 min to allow the development of a blue-green colour. Then, 2M sulphuric acid (50 µL/well) was used to terminate the colour development reaction, under which the colour changed to yellow. The IgG levels in individual serum samples were examined in duplicate. The reactivity was assessed by measuring the OD value at a 450 nm wavelength with a microplate ELISA reader. The OD of the blank control with PBST-BSA without serum dilution was subtracted from the OD of each well.

2.8. Statistical Analysis

The effectiveness of the modified indirect ELISAs was preliminarily assessed. The results of the modified ELISAs, PCR, and MAT assays were compared by utilising the first 50 serum samples selected from dog groups 1, 2, and 3. The evaluation included calculating the values for sensitivity, specificity, positive predictive value (PPV), negative predictive value (NPV), intra-assay (repeatability), inter-assay (reproducibility), and accuracy. The OD values obtained via IgG detection for all 260 serum samples from all serum groups, including the first 50 serum samples from dog groups 1, 2, and 3, that were used to evaluate the diagnostic performance of the modified ELISAs were calculated and analysed using nonparametric statistics via the Kruskal–Wallis test and Dunn's post hoc test in GraphPad prism software v.9 to compare the differences in the levels of IgG antibody between groups of dog sera within each modified indirect ELISA. A receiver–operator curve (ROC) analysis was determined and used to establish the appropriate cut-off OD values for the specificity and sensitivity of each modified ELISA test using MedCalc software v.20.006. The association between the antibody titre of the MAT and the antibody level of the modified ELISAs was determined using the 14 serum samples with MAT titres above a 1:20 dilution from dog groups 1, 2, and 3 via Pearson correlation analysis. Using agreement analysis and Cohen's kappa statistics, the degree of concordance among the three diagnostic methods was determined by comparing the results for the 50 serum samples (groups 1, 2, and 3) with the modified ELISAs against PCR and the MAT assays.

3. Results

3.1. Detection of *Leptospira* in Serum via Nested PCR

The results of the nested PCR for the 260 serum samples are shown in Table 2. Only the six serum samples of infected dogs from the endemic area (Group 1) tested positive.

3.2. Antibody Titres in the Microscopic Agglutination Test

The 50 serum samples, which were selected from the three dog groups, Group 1, 2 and 3, were tested via the MAT against 29 isolates (27 serovars), including 24 representative reference isolates and 5 local isolates of *Leptospira* serovars. Tables 2 and 3 record the number of evaluated sera and antibody titres against the serovars. Of the 50 serum samples, only 11 (11/21; 52%) sera from unvaccinated dogs in the endemic area (Group 2) and 3 (3/23; 13%) sera from vaccinated dogs in the non-endemic areas (Group 3) had antibody titres equal to or greater than a 1:20 dilution. In contrast, no sera (0%) from positive leptospiral-infected dogs in the endemic areas (Group 1) had antibody titres equal to or above the 1:20 dilution. The positive MAT serum samples from groups 2 and 3 were in the 1:20 to 1:80 dilution range, with Group 2 having higher antibody titres than group 3. In addition, antibody titres were only detected for four *Leptospira* serovars, including serovars Hebdonadis, Sejroe, Shermani, and Paidjan, with the numbers of positive sera to these serovars in MAT being four, one, nine, and one, respectively. One serum sample from an unvaccinated dog in the endemic area (Group 2) had antibody titres against the local isolate of serovar Paidjan, previously recovered from asymptomatic dogs. One serum sample from the same group displayed antibody titres with serovars Hebdonadis and Shermani, which are reference serovars used in standard MAT panels in Thailand.

Table 2. Information about the five dog groups, and the number and percentage of 260 sera from the groups that tested positive via urine isolation (agent detection), urine and serum PCR (agent detection), the MAT (antibody detection), and modified ELISAs (antibody detection) using a local leptospiral serovar isolate.

Groups	Group 1 Infected Dogs from Nan Province That Were Confirmed as Infected via Positive PCR and Isolation	Group 2 Unvaccinated Dogs from Nan Province	Group 3 Vaccinated Dogs from Non-Endemic Areas		Group 4 Unvaccinated Dogs from Non-Endemic Areas	Group 5 Unvaccinated Puppies from Non-Endemic Areas	Total
Number of samples	6	21	23 *	89 **	108	13	260
			112				
Area	Endemic area: Nan Province	Endemic area: Nan Province	Non-endemic area: Bangkok, Samut Prakan, and Chonburi provinces		Non-endemic area: Bangkok, Samut Prakan, and Chonburi provinces	Non-endemic area: Bangkok, Samut Prakan, and Chonburi provinces	-
Age	>1 year	>1 year	>1 year		>1 year	Two months	-
Vaccination status	No vaccination	No vaccination	Complete vaccination		No vaccination	No vaccination	-
Methods (Nested PCR, Isolation, and MAT assays)							
Nested PCR from urine	6 (6/6; 100%)	0 (0/21; 0%)	N/A		N/A	N/A	6 (6/27; 22%)
Isolation from urine	4 (4/6; 67%)	0 (0/21; 0%)	N/A		N/A	N/A	4 (4/27; 15%)
Nested PCR from sera	6 (6/6; 100%)	0 (0/21; 0%)	0 (0/112; 0%)		0 (0/108; 0%)	0 (0/13; 0%)	6 (6/260; 2%)
MAT from sera	0 (0/6; 0%)	0 (0/21; 0%)	0 (0/23; 0%)	N/A	N/A	N/A	14 (14/50; 28%)
Methods (modified ELISAs), protein preparation, and local isolates of serovars used in the ELISAs							
WCP-Dadas/IgG-ELISA	6 (6/6; 100%)	14 (14/21; 67%)	12 (12/23; 52%)	25 (25/89; 28%)	0 (0/108; 0%)	0 (0/13; 0%)	57 (57/260; 22%)
			37 (37/112; 33%)				
TMP-Dadas/IgG-ELISA	6 (6/6; 100%)	13 (13/21; 62%)	6 (6/23; 26%)	12 (12/89; 13%)	0 (0/108; 0%)	0 (0/13; 0%)	37 (37/260; 14%)
			18 (18/112; 16%)				

Table 2. *Cont.*

<table>
<tr><th rowspan="2">Groups</th><th>Group 1</th><th>Group 2</th><th colspan="2">Group 3</th><th>Group 4</th><th>Group 5</th><th rowspan="2">Total</th></tr>
<tr><th>Infected Dogs from Nan Province That Were Confirmed as Infected via Positive PCR and Isolation</th><th>Unvaccinated Dogs from Nan Province</th><th colspan="2">Vaccinated Dogs from Non-Endemic Areas</th><th>Unvaccinated Dogs from Non-Endemic Areas</th><th>Unvaccinated Puppies from Non-Endemic Areas</th></tr>
<tr><td rowspan="2">OMP-Dadas/IgG-ELISA</td><td rowspan="2">6 (6/6; 100%)</td><td rowspan="2">15 (10/21; 48%)</td><td>0
(0/23; 0%)</td><td>0
(0/89; 0%)</td><td rowspan="2">0 (0/108; 0%)</td><td rowspan="2">0 (0/13; 0%)</td><td rowspan="2">21 (21/260; 8%)</td></tr>
<tr><td colspan="2">0 (0/112; 0%)</td></tr>
</table>

N/A: Not applicable. WCP: whole-cell protein; TMP: total membrane protein; OMP: outer membrane protein; *: Twenty-three serum samples from group 3 were examined via both the MAT and ELISAs, and used for the preliminary evaluation of the effectiveness of diagnostic performance of the modified ELISAs. **: Another 89 serum samples were examined only via ELISAs and were used to screen for leptospiral IgG antibody and confirm the usefulness of the modified ELISAs.

Table 3. Frequency distribution of serum antibody titres in the microscopic agglutination test (MAT) with 29 different isolates of *Leptospira* using 50 sera from three groups of dogs (groups 1, 2, and 3).

<table>
<tr><th rowspan="2">Groups</th><th rowspan="2">Number</th><th rowspan="2">MAT</th><th colspan="5">Serovars and Serological Titres in MAT</th></tr>
<tr><th>Hebdonadis</th><th>Sejroe</th><th colspan="2">Shermani</th><th>Paidjan (CUDO5)</th></tr>
<tr><td>Isolation and PCR-confirmed infected dogs from Nan Province (Group 1)</td><td>6</td><td>0
(0/6; 0%)</td><td>-</td><td>-</td><td colspan="2">-</td><td>-</td></tr>
<tr><td rowspan="3">Unvaccinated dogs from Nan Province (Group 2)</td><td rowspan="3">21</td><td rowspan="3">11
(11/21; 52%)</td><td>1 *</td><td>1</td><td colspan="2">9 *</td><td rowspan="3">1</td></tr>
<tr><td rowspan="2">1:20</td><td rowspan="2">1:80</td><td>4</td><td>5</td></tr>
<tr><td>1:20</td><td>1:40</td></tr>
<tr><td rowspan="2">Vaccinated dogs from Bangkok (Group 3)</td><td rowspan="2">23</td><td rowspan="2">3
(3/23; 13%)</td><td rowspan="2">-</td><td rowspan="2">-</td><td colspan="2">3</td><td rowspan="2">-</td></tr>
<tr><td colspan="2">1:20</td></tr>
<tr><td>Total</td><td>50</td><td>14 (28%)</td><td>1</td><td>1</td><td colspan="2">12</td><td>1</td></tr>
</table>

* One sample from an unvaccinated dog from Nan Province (Group 2) had an MAT titre more than or equal to 1:20 for the two leptospiral serovars Hebdonadis and Shermani. The MAT was performed using 29 different isolates of *Leptospira* consisting of 24 representative reference isolates and 5 local Thai isolates. These isolates belonged to 27 serovars, of which 2 serovars, Bataviae and Mini, each included 2 different isolates.

3.3. The Protein Components Confirmed via SDS-PAGE and Western Blot Analysis Using Serum from a Dog with Leptospirosis

The protein components of the three extractions from the five local leptospiral serovar isolates were assessed via SDS-PAGE and Western blotting before use in modified ELISAs, and the results are shown in Supplementary Figure S1. The proteins from the preparations of WCP, TMP, and OMP were shown to react to pooled sera from a dog confirmed as positive for leptospirosis via isolation and PCR, and displayed different staining intensities and patterns.

3.4. Determination of the Optimal OD Cut-Off Values for Indirect ELISAs

Based on the optimal protein concentration and serum dilution for each antigen protein preparation, all 39 OD values from 13 unvaccinated 2-month-old dogs (Group 5) from non-endemic areas that acted as a negative control due to their history of no vaccination and the fact that they were PCR-negative for the leptospiral 16S ribosomal RNA (rrs) gene were examined in triplicate and computed to define the optimal OD cut-off values using the sum of the mean ODs plus four standard deviations (Mean + 4SD). The cut-off OD values for the ELISA using WCP, TMP, and OMP at a protein concentration of 0.1 μg/well (1 μg/mL) and serum dilutions of 1:1280, 1:640, and 1:640 were 0.593, 0.816, and 0.898, respectively. Serum samples that displayed an OD value equal to or greater than the values of the optimal OD cut-off were considered positive in the modified ELISAs that used the leptospiral serovar isolates in this study.

3.5. Detection of Antibody Levels via the Modified ELISAs

For the 260 serum samples, the number and percentage that tested positive in the modified ELISAs using five local isolates of *Leptospira* serovars and four *Leptospira* serovars commonly used in the leptospirosis vaccine in dogs are shown in Table 3 and Supplementary Table S1. Sera from all infected dogs with positive infection status confirmed via urine PCR and isolation (group 1) had positive antibody results under all ELISAs modified using local isolates of *Leptospira* serovars. At the same time, they did not exhibit positive titres under the MAT or in the ELISAs using the common *Leptospira* serovars used in canine vaccines. However, some sera from all unvaccinated dogs from the concurrent endemic area (Group 2) and vaccinated dogs from Bangkok (Group 3) that were negative under serum PCR tested positive under the MAT and modified ELISAs. In the WCP and TMP-based ELISAs using local isolates of *Leptospira* serovars, 52% to 71% and 0% to 33% of sera from groups 2 and 3, respectively, were positive. Using the OMP-based ELISAs, 43 to 62% of the sera from Group 2 were positive, but none of the sera from Group 3 were positive. Moreover, all sera from unvaccinated dogs from non-endemic areas (Group 4) and unvaccinated puppies from non-endemic areas (Group 5) were negative in all ELISAs utilising local isolates of *Leptospira* serovars. On the other hand, the OMP-based ELISAs coated with the set of serovars used in the vaccine gave strongly positive results for the vaccinated dogs (Group 3), for 66% to 76% of the sera. In contrast, infected dogs (group 1) and unvaccinated dogs (Group 2) were negative. Additionally, some of the sera, 21% to 31%, from the unvaccinated dogs in Group 4 were positive, whereas none of the sera from Group 5 were positive in any modified ELISAs utilising the *Leptospira* serovars commonly used in the canine leptospirosis vaccine.

The distribution of IgG antibody levels against the OMP of five local isolates of *Leptospira* and four common *Leptospira* serovars used in vaccines in the modified ELISAs are shown in Figure 1, and Supplementary Figures S2 and S3. All OMP-based ELISAs using local isolates of *Leptospira* serovars identified IgG antibodies in dogs from the endemic area (groups 1 and 2), with a highly significant difference (p-value < 0.05) from dogs in non-endemic areas (groups 3, 4, and 5). On the other hand, some of the WCP and TMP-based ELISAs using local isolates of *Leptospira* serovars did not significantly differentiate (p-value > 0.05) between the groups of dogs from the endemic and non-endemic areas (Figure 1, and Supplementary Figure S2). In contrast, all the modified ELISAs using the common *Leptospira* serovars used in vaccines showed that only the IgG antibody levels of the vaccinated dogs from non-endemic areas (Group 3) were significantly higher (p-value < 0.05) than those in the four other dog groups from the endemic and non-endemic areas (Supplementary Figure S3).

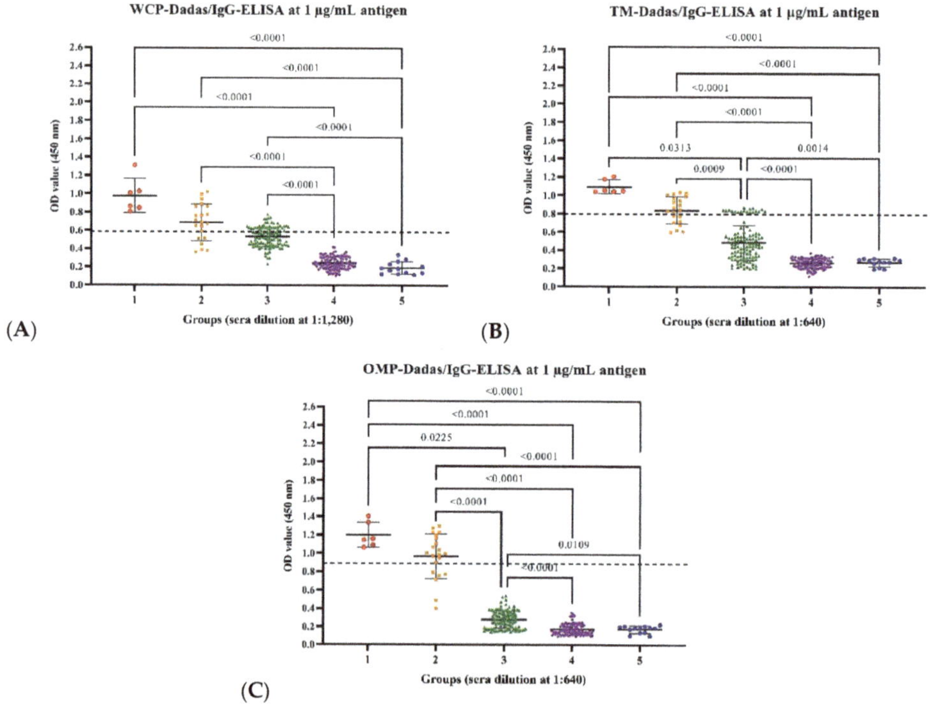

Figure 1. The levels of IgG antibody detected in modified ELISAs against whole-cell protein (WCP), total membrane protein (TMP), and outer membrane protein (OMP) from the local Thai isolate of *Leptospira* serovar, including serovar Dadas, at 1:1280 (**A**), 1:640 (**B**), and 1:640 (**C**) sera dilutions. Comparisons among 260 sera from five groups consisting of dogs from Nan Province confirmed as infected via PCR and isolation (Group 1), unvaccinated dogs from Nan Province (Group 2), vaccinated dogs from Bangkok (Group 3), unvaccinated dogs from non-endemic areas (Group 4), and unvaccinated puppies from non-endemic areas (Group 5). The significant differences of the IgG antibody levels between dog sera group were analyzed using Kruskal–Wallis test and Dunn's post hoc test (p-value < 0.05). (**A–C**) demonstrated the modified ELISAs against WCP, TMP and OMP from serovar Dadas identified that dogs from endemic area (groups 1 and 2) have the higher level of IgG antibody than dogs in non-endemic areas (groups 3, 4, and 5), with a highly significant difference.

3.6. Diagnostic Performance of All Modified ELISAs

The sensitivity, specificity, and intra- and inter-assay comparisons for all the modified ELISAs are shown in Supplementary Table S2. The diagnostic performance of the modified ELISAs was calculated using the results of the serum PCR, MAT, and ELISAs with 50 sera from three dog groups (groups 1, 2, and 3). All the modified ELISAs had the highest sensitivity compared with PCRs from both urine and sera. At the same time, OMP-based ELISAs showed the highest specificity, followed by TMP-based ELISAs and WCP-based ELISAs. For the comparison of the ELISA with the MAT, the highest sensitivity was for WCP-based ELISAs, whereas the highest specificity was for OMP-based ELISAs. In comparing the sensitivity among all modified ELISAs to that of the MAT, all ELISAs from the serovar Dadas produced the highest sensitivity, at 64.3 to 85.7%. However, the highest specificity was mostly found with OMP-based ELISAs, mainly where serovar Mini06 and Bataviae were used. The intra-assay (repeatability) and inter-assay (reproducibility) values of all three ELISA protein platforms were assessed for diagnostic precision. The percentage of the coefficient of variation (% CV) determined via the intra-assay analysis of the three modified ELISA platforms were 6.3, 4.1, and 2.6 for WCP, TMP, and OMP-based ELISAs, respectively. Furthermore, the inter-assay assessment showed that the percentage of the coefficient of variation for the three ELISA platforms based on WCP, TMP, and OMP were 11.0, 9.3, and 8.5, respectively.

3.7. Receiver–Operator Curve (ROC) Analysis, Correlation Analysis, and Agreement Analysis

The receiver–operator curve (ROC) analysis and area under the ROC curve (AUC) for the determination of the cut-off OD value for the specificity and sensitivity of each modified ELISA are shown in Figure 2 and Supplementary Figure S4. The AUC of the modified ELISAs against WCP/IgG-ELISAs, TMP/IgG-ELISAs, and OMP/IgG-ELISAs ranged from 0.610 to 0.676, 0.607 to 0.696, and 0.585 to 0.691, respectively. The cut-off OD values of each modified ELISA were 0.564–0.744, 0.413–0.865, and 0.2695–0.899 for the WCP/IgG-ELISAs, TMP/IgG-ELISAs, and OMP/IgG-ELISAs, respectively. Among the cut-off OD values of all the modified ELISAs, the modified ELISA tests for the WCP/IgG-ELISA, TMP/IgG-ELISA, and OMP/IgG-ELISA using the serovar Dadas isolate yielded the highest specificity, while maintaining good sensitivity in the ROC analysis.

The results of the Pearson correlation of MAT antibody titres and the IgG antibody levels under modified ELISAs in dog sera with MAT titres above a 1:20 dilution are shown in Supplementary Figure S5. Overall, the antibody levels in dogs with MAT titres above a 1:20 dilution from the MAT and ELISA positively correlated with those under all modified ELISA platforms. IgG antibodies from the WCP-Mini13/IgG-ELISA, TMP-Mini06/IgG-ELISA, and OMP-Paidjan/IgG-ELISA had the most positive correlation to the MAT antibody titre among the IgG-ELISAs.

The agreement analysis of serum PCR and the MAT against all modified ELISAs is shown in Supplementary Table S3. The degree of concordance from a comparison of a PCR assay from urine and all modified ELISAs demonstrated that only the WCP-/IgG-ELISA using serovars Dadas, Bataviae, and Mini13, and the OMP-Dadas/IgG-ELISA showed slight agreement (κ = 0.10–0.20), while the other modified ELISAs represented fair agreement (κ = 0.21–0.40). On the other hand, a comparison of the serum PCR results with those of all modified ELISAs demonstrated that the TMP-Mini06/IgG-ELISA and OMP/IgG-ELISA from serovar Bataviae and Mini06 had moderate agreement (κ = 0.41–0.60), whereas the others showed fair agreement—except for the WCP-Dadas/IgG-ELISA, which only had slight agreement. In contrast, a comparison of the MAT assay with all modified ELISAs showed that the seven modified ELISAs, including the WCP/IgG-ELISA using the serovars Dadas, Mini06, and Mini13, the TMP/IgG-ELISA using the serovars Paidjan and Bataviae, and the OMP/IgG-ELISA from serovars Dadas and Paidjan showed fair agreement with the MAT, while the remaining modified ELISAs displayed slight agreement.

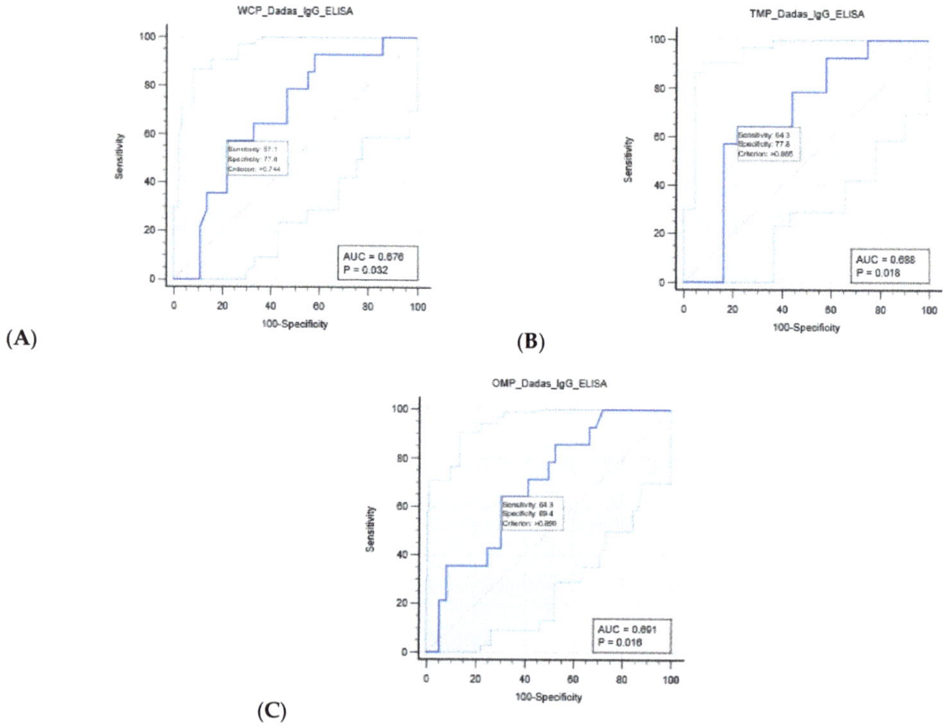

Figure 2. Receiver–operator curve (ROC) and area under the curve of ROC (AUC) of modified ELISAs against whole-cell protein (WCP), total membrane protein (TMP), and outer membrane protein (OMP) from the local Thai isolates of *Leptospira* serovars, including serovar Dadas, at 1:1280 (**A**), 1:640 (**B**), and 1:640 (**C**) sera dilutions, with 50 sera from three groups consisting of dogs from Nan Province confirmed as infected via PCR and isolation (Group 1), unvaccinated dogs from Nan Province (Group 2), and vaccinated dogs from Bangkok (Group 3). All the ROC curve and AUC of the ROC of modified ELISAs were analyzed by MedCalc software. (**A**–**C**) displayed the sensitivity and specificity of the modified ELISAs against WCP, TMP and OMP from serovar Dadas using the cut-off OD values that were set by ROC analysis showed the moderate sensitivity, specificity and AUC, with p-value < 0.05.

4. Discussion

Leptospirosis is a zoonotic disease that poses a threat to both human and animal health. Dogs can act as asymptomatic carriers of the pathogen, making them potential sources of infection for humans [1,32]. To effectively combat leptospirosis, it is essential to have reliable diagnostic tools and vaccines for surveillance and prevention, respectively. Serological techniques, including ELISAs, have been developed to diagnose and screen for leptospirosis in dogs [42,53,59–61]. This study aimed to improve the sensitivity and specificity of antibody detection in leptospirosis in our region by proposing modified indirect ELISAs incorporating local isolates of *Leptospira* serovars isolated from asymptomatic dogs in Thailand.

The leptospiral serum antibody content of animals reflects exposure to immunodominant *Leptospira* antigens during infection and vaccination. Leptospiral surface proteins are often cross-reactive, while the surface lipopolysaccharide (LPS) components are mostly serovar-specific and generate antibodies specific for those serovars [62–66]. Previous studies on antibody detection in dogs have used sera from animals that have either been infected or have received serovar-specific vaccines [42,53,57,65–67]. In the current study, the canine sera used were chosen from various types of dogs from areas where there were different circulating serovars. This allowed an analysis of serovar-specific and non-serovar-specific antibodies generated against local isolates of leptospiral serovars. The five groups of dogs represented those from an area endemic for local isolates (groups 1 and 2) and those where these isolates have not been recorded (groups 3–5). This allowed us to evaluate and confirm the modified ELISAs' diagnostic performance using local leptospiral serovars for detecting corresponding antibodies in dogs.

Group 5, including unvaccinated puppies from non-endemic regions, was chosen as the negative control group for a serological investigation since the dogs had not been vaccinated and tested negative under PCR, confirming the absence of the pathogen in them. This group provides a clear baseline for seronegative samples, which is crucial for determining the cut-off in serological tests. The MAT was not used for these sera because the MAT is typically reserved for cases with uncertain pathogen presence or to measure antibody response, and these puppies were already confirmed as negative under PCR, a method that directly detects a pathogen's genetic material with high sensitivity and specificity, rendering further MAT testing unnecessary [68,69]. The reliability of the PCR in confirming the absence of the pathogen justifies its use over the MAT, which, despite its high specificity, has lower sensitivity and can yield false negatives in early infection stages or in patients with low antibody levels. Therefore, the PCR-negative status of Group 5 puppies ensures a valid negative control for establishing serological test cut-offs [68–71].

Determining an optimal cut-off value for ELISA tests is essential for accurately distinguishing between positive and negative results in detecting antibodies against *Leptospira* spp. The method of setting the cut-off value at the average optical density (OD) plus four standard deviations (SD) is highlighted for its ability to improve the specificity of the test, ensuring that positive results truly reflect the presence of the disease [55–58]. This approach is supported by a study in which use of the ELISA gave high specificity (95.6%) and sensitivity (100%) compared with those of the MAT for detecting leptospirosis in dogs, by adopting the mean OD + 4SD as the cut-off value [57]. This method effectively minimises the risk of false positives, which is crucial for reliable diagnostics in a clinical setting [72].

Using local isolates of *Leptospira* serovars as protein antigens in the modified ELISAs significantly improved the accuracy of serological detection for leptospirosis in dogs in Thailand. Previous studies also have demonstrated the efficacy of ELISAs using local serovars, such as the *L. interrogans* serovar Canicola, in detecting leptospiral antibodies and increasing sensitivity and specificity compared with those of the standard MAT [42,53]. Similar results have been observed in leptospirosis studies in dogs and cattle using local serovars and isolates from endemic areas, such as *L. fainei*, serovar Hurstbridge, strain BUT 6T, and *L. interrogans*, serovar Hardjo [60,73]. These findings and the present study indicate that using known or local serovars in ELISA testing enhances the diagnostic accuracy of serological antibody detection.

The ELISAs modified using local isolates of *Leptospira* serovars demonstrated the ability to detect IgG antibodies and differentiate their levels in sera between different groups of dogs, particularly in infected dogs (group 1) that did not produce the antibody titres detected in the MAT but showed positive results via PCR on urine and serum. This is consistent with previous studies that have shown the effectiveness of modified ELISAs using local isolates of *Leptospira* serovars in detecting antibodies in negative sera and in sera shown to be positive under the MAT [42,53,54,60,73]. Unvaccinated dogs from endemic areas (Group 2) and vaccinated dogs from non-endemic areas (Group 3) exhibited different ELISA results and antibody levels compared with infected dogs (Group 1). While serum PCR results were negative for groups 2 and 3, MAT and ELISAs showed positive titres and antibody levels in some animals. The positive animals in Group 2 might have developed antibodies from previous exposure to local leptospiral isolates in the endemic area. In contrast, antibodies in Group 3 could have been generated via exposure to serogroup cross-reactive antigen proteins from vaccine serovars. In diagnosing leptospirosis, it is well recognised that there is no direct correlation between the results of tests that detect the agent, such as PCR, and serological testing that provides evidence of past exposure [38,44]. However, combining PCR results with serological test results can improve the effectiveness of leptospirosis diagnosis [74–77]. Using ELISAs with local serovars and isolates may enhance the diagnostic efficacy of serological tests, especially in samples that are PCR-negative and negative for conventional antibody detection under the MAT.

The modified ELISAs employed different protein components as antigens, including WCP, TMP, and OMP preparations. WCPs of leptospiral cells contain many proteins derived from the whole and outer membranes. On the other hand, TMPs consist of cytoplasmic membrane and outer membrane proteins. The heat shock proteins GroEL and DnaK, which are primarily present in cytoplasmic membranes, are potential immunoreactive protein antigens in these preparations. The OMPs, which include numerous possible highly immunogenic proteins, such as LipL32, LipL45/31, and LipL41, are the most conserved components across all leptospiral serovars and species. In contrast, lipopolysaccharides have a wide range of carbohydrate side chains, influencing antigenic diversity across several leptospiral serovars [47–52]. Among these, the ELISAs based on WCPs exhibited higher sensitivity, followed by those based on TMPs and OMPs. The WCP-based ELISAs are better suited for antibody surveillance due to their high sensitivity; nevertheless, this enhanced sensitivity can frequently result in false-positive results. Although ELISAs based on total membrane proteins (TMPs) and outer membrane proteins (OMPs), especially OMP-based ELISAs, demonstrated a high level of specificity and are therefore suitable for antibody screening and diagnosis to differentiate between infection and no infection or vaccination in dogs, they tend to yield a high percentage of false-negative results. False-positive and false-negative serological reactivities reflect the sensitivity and specificity of each method via the synthesis of protein antigens and serovars used in the ELISAs. These findings can be attributed to the differences in protein components and the impact of various serovars on the sensitivity and specificity of the ELISAs [42,48,53,61].

Agreement analysis between the ELISAs modified using local isolates of *Leptospira* serovars and the PCR test in this study indicated a slight to moderate degree of concordance, suggesting a reliable correlation between the two methods. The agreement between the ELISAs and MAT ranged from slight to fair, highlighting more satisfactory agreement between PCR-based antigen detection and ELISA-based antibody detection using local serovars and isolates compared with the MAT using reference serovars combined with local serovars and isolates [73,78,79]. To enhance the accuracy of serological diagnostic testing for antibody detection and monitoring in canine leptospirosis, the utilisation of local serovars or isolates in combination with appropriate antigen preparations is crucial. Thus, the identification of the local serovars and isolates in each area, as well as the serovars identified via serotyping, is essential and should be explored.

The discrepancies observed between the ELISA and the MAT results in diagnosing leptospirosis are primarily due to the different antigenic profiles that each test targets and the stages of disease at which they are most effective. The ELISA is designed to detect antibodies against specific antigens or epitopes, which may be highly specific to certain strains of a pathogen, while the MAT detects antibodies against a broader range of antigens presented by live bacteria, including surface proteins not targeted by the ELISA [80]. This can lead to discrepancies, particularly if the ELISA and MAT do not include antigens and leptospires from the strains circulating in each region, potentially resulting in false negatives. The World Health Organization recommends using a locally optimised MAT panel that represents the currently circulating strains to improve sensitivity [81]. The discrepancies between these tests underscore the importance of the regional customisation of diagnostic tests to enhance accuracy, which can be achieved by including antigens from local strains in ELISAs or by optimizing MAT panels with these strains [80,81].

The research highlights the importance of finding and using local strains of *Leptospira* from dogs to improve diagnostic tests in regions with high prevalence like Thailand where leptospirosis presents significant public health and economic difficulties. The CDC stresses the significance of national surveillance for controlling leptospirosis, a condition that must be reported nationally in several countries. This study proposes that creating ELISA plates specific to the most common local serovars could enhance diagnostic precision, given the current diagnostic methods' shortcomings in accurately detecting the disease due to their low specificity and the risk of cross-reactivity with other tropical diseases. A detailed plan for consistent monitoring and readiness for diagnosis, which involves creating a collection of serovars, regularly assessing the composition of ELISA plates, improving surveillance programs, developing ELISAs based on recombinant antigens, and fostering collaboration and data exchange, is suggested to direct future diagnostic methods and improve public health outcomes in the area [32,44,82].

The introduction of new serovars in tests for leptospirosis has the potential to greatly improve the accuracy of diagnostic tests, which are now limited by the specificity and sensitivity of current assays. Integrating the new serovars with the authorised set may enhance diagnostic accuracy and epidemiological surveillance, perhaps establishing them as the standard for leptospirosis diagnosis and surveillance, pending additional research and validation studies. This method would require revisions to current regulations and guidelines to include the new genotypes and serovars. If the new set is not officially adopted, it can still be used in conjunction with the approved set to improve surveillance and research, helping to understand the genetic diversity and evolution of *Leptospira* spp. Adopting new serovars requires regulatory compliance, integration with current processes, improved epidemiological data gathering, heightened public health awareness, and perhaps revised immunisation approaches. Both sets of serovars can be used together, with the new serovars being utilised for more accurate and quicker diagnosis. This development offers a chance to enhance the comprehension and control of leptospirosis by improving diagnostic precision and epidemiological monitoring [83–88].

Overall, the findings of this study contribute to the advancement of leptospirosis surveillance and emphasise the importance of using local isolates of *Leptospira* serovars in the design of effective serological techniques. Further research and validation studies are warranted to confirm the efficacy and applicability of these indirect ELISAs modified using local *Leptospira* serovar isolates in diverse geographical locations and dog populations.

5. Conclusions

In conclusion, this study demonstrates the effectiveness of ELISAs modified using local isolates of leptospiral serovars in improving the sensitivity and specificity of antibody detection for leptospirosis in dogs. Using protein antigens extracted from local isolates of leptospiral serovars enhanced the diagnostic accuracy of the serological tests, particularly in samples that were PCR-negative and negative for conventional antibody detection under the MAT. The choice of protein antigen preparation, such as WCPs, TMPs, or OMPs, affects the sensitivity and specificity of the ELISAs. The WCP-based ELISAs offer higher sensitivity, while the OMP-based ELISAs provide higher specificity. The ELISA modified with the combination of local isolates of leptospiral serovars with appropriate antigen preparations can improve the accuracy of serological diagnostic testing for antibody detection and monitoring in canine leptospirosis. The importance of the identification of local circulating leptospiral serovars in certain regions needs to be determined, and has practical implications for achieving more accurate diagnosis and management of the disease in endemic areas to prevent leptospirosis in animals and humans.

Supplementary Materials: The following supporting information can be downloaded at: https://www.mdpi.com/article/10.3390/ani14060893/s1, Figure S1: The confirmation of the protein components from the five local Thai leptospiral serovar isolates using SDS-PAGE and Western blotting; Table S1: The number and percentage of dog sera that tested positive by modified ELISAs using other four local isolates of leptospiral serovars and four common leptospiral serovars used in the leptospirosis vaccine for dogs; Figure S2: The levels of IgG antibody by modified ELISAs with WCP, TMP, and OMP from the other four local Thai leptospiral serovar isolates; Table S2: The evaluation of diagnostic test performance from the modified ELISAs using the five local Thai leptospiral serovar isolates compared to PCR and MAT; Figure S3: The levels of IgG antibody by modified ELISAs with WCP, TMP, and OMP from four common leptospiral serovars used in the leptospirosis vaccine for dogs; Table S3: Cohen's kappa statistics of modified ELISAs against PCR and MAT; Figure S4: The analysis of ROC and AUC of ROC from the modified ELISAs with WCP, TMP, and OMP from four local Thai leptospiral serovar isolates; Figure S5: Pearson correlation of antibody titer in MAT and antibody level in modified ELISAs using sera from dogs in groups 1, 2, and 3 that had MAT titers above 1:20 dilution.

Author Contributions: All authors have made valuable contributions to the research process in this study. Conceptualisation and design of the experimental study, and data analysis: P.B., A.K., W.N., K.P., D.J.H. and N.P.; methodology: P.B., A.K., P.K., W.N., K.P., T.P., D.J.H. and N.P.; investigation of the experiments: P.B., W.S., S.B., M.T., A.K., P.K., W.N., K.P., T.P., D.J.H. and N.P.; validation: P.B., A.K. and W.N.; software: P.B., A.K. and P.K.; formal data analysis: P.B. and A.K.; resources and data curation: P.B., A.K. and N.P.; writing—original draft preparation: P.B., A.K. and N.P.; writing—review and editing: K.P., D.J.H. and N.P.; visualisation: P.B. and N.P.; supervision; N.P.; project administration: N.P.; funding acquisition: P.B., A.K. and N.P. All authors have read and agreed to the published version of the manuscript.

Funding: This original research project was supported by grants from the Secondary Century Fund (C2F) for Doctoral Scholarship, the Ratchadapisek Sompoch Endowment Fund (grant No. CUGR63403101), and the 90th Anniversary of Chulalongkorn University Fund. No potential competing interests are reported by the authors.

Institutional Review Board Statement: All canine serum samples used in this study were obtained from the remaining sera from other studies. The experimental protocols of the blood collection for canine serum samples from groups 1 and 2 were approved by Chulalongkorn University Animal Care and Use Committee (CU-ACUC; Protocol NO. 1531075). All procedures were carried out under the relevant guidelines and regulations. Furthermore, canine sera from groups 3, 4, and 5 were acquired from another study with authorisation from the owners provided via written informed consent to the participation of their dogs in this study. However, the utilisation of all canine sera in this study was considered for a decision on the Animal Care and Use protocol by the Faculty of Veterinary Science Animal Care and Use Committee, Chulalongkorn University, which concluded that the protocol was exempt from needing a certificate of project approval, following the memorandum of the committee's decision on the protocol.

Informed Consent Statement: The study did not involve any subjects from humans.

Data Availability Statement: The data presented in this study are available on request from the corresponding author. The data are not publicly available due to [some data containing information that could compromise the privacy of research participants, including another ongoing study].

Acknowledgments: The authors would like to thank Suwarat Wadeerat, head of the Blood Bank Unit, Small Animal Hospital, Faculty of Veterinary Science, Chulalongkorn University, Bangkok, Thailand, for providing the vaccinated dog sera and Nattakarn Naimon from the Department of Veterinary Technology, Faculty of Veterinary Technology, Kasetsart University, Bangkok, Thailand, for assisting in the preparation of the unvaccinated dog and unvaccinated puppy sera from non-endemic areas.

Conflicts of Interest: The authors declare that the original research was conducted in the absence of any commercial or financial relationships that could be construed as a potential conflict of interest and that the scientific writing was not performed using generative AI.

References

1. Adler, B.; de la Peña Moctezuma, A. *Leptospira* and Leptospirosis. *Vet. Microbiol.* **2010**, *140*, 287–296. [CrossRef]
2. Costa, F.; Hagan, J.E.; Calcagno, J.; Kane, M.; Torgerson, P.; Martinez-Silveira, M.S.; Stein, C.; Abela-Ridder, B.; Ko, A.I. Global Morbidity and Mortality of Leptospirosis: A Systematic Review. *PLoS Neglected Trop. Dis.* **2015**, *9*, e0003898. [CrossRef] [PubMed]
3. Tangkanakul, W.; Smits, H.L.; Jatanasen, S.; Ashford, D.A. Leptospirosis: An Emerging Health Problem in Thailand. *Southeast Asian J. Trop. Med. Public Health* **2005**, *36*, 281–288.
4. Wuthiekanun, V.; Sirisukkarn, N.; Daengsupa, P.; Sakaraserane, P.; Sangkakam, A.; Chierakul, W.; Smythe, L.D.; Symonds, M.L.; Dohnt, M.F.; Slack, A.T.; et al. Clinical Diagnosis and Geographic Distribution of Leptospirosis, Thailand. *Emerg. Infect. Dis.* **2007**, *13*, 124–126. [CrossRef]
5. Cerqueira, G.M.; Picardeau, M. A Century of *Leptospira* Strain Typing. *Infect. Genet. Evol.* **2009**, *9*, 760–768. [CrossRef]

6. Lehmann, J.S.; Matthias, M.A.; Vinetz, J.M.; Fouts, D.E. Leptospiral Pathogenomics. *Pathogens* **2014**, *3*, 280–308. [CrossRef]
7. Arent, Z.; Pardyak, L.; Dubniewicz, K.; Płachno, B.; Kotula-Balak, M. *Leptospira* Taxonomy: Then and Now. *Med. Weter.* **2022**, *78*, 2022–6694. [CrossRef]
8. Kmety, E.; Dikken, H. *Classification of the Species Leptospira interrogans and History of Its Serovars*; University Press Groningen: Groningen, The Netherlands, 1993.
9. Feresu, S.B.; Ann Bolin, C.; van de Kemp, H.; Korver, H. Identification of a Serogroup Bataviae *Leptospira* Strain Isolated from an Ox in Zimbabwe. *Zentralbl Bakteriol* **1999**, *289*, 19–29. [CrossRef]
10. Bourhy, P.; Collet, L.; Clément, S.; Huerre, M.; Ave, P.; Giry, C.; Pettinelli, F.; Picardeau, M. Isolation and Characterization of New *Leptospira* Genotypes from Patients in Mayotte (Indian Ocean). *PLoS Neglected Trop. Dis.* **2010**, *4*, e724. [CrossRef]
11. Benacer, D.; Zain, S.N.M.; Sim, S.Z.; Khalid, M.K.N.M.; Galloway, R.L.; Souris, M.; Thong, K.L. Determination of *Leptospira borgpetersenii* Serovar Javanica and *Leptospira interrogans* Serovar Bataviae as the Persistent *Leptospira* Serovars Circulating in the Urban Rat Populations in Peninsular Malaysia. *Parasit Vectors* **2016**, *9*, 117. [CrossRef]
12. Song, N.; Zhang, W.; Ding, Y.; Wu, D.; Dai, Z.; Xu, L.; Cao, Y. Preliminary Characterization of Dog Derived Pathogenic Strains of *Leptospira interrogans* Serovar Australis in Nanchang of Jiangxi Province, China. *Front. Vet. Sci.* **2020**, *7*, 607115. [CrossRef]
13. Rahman, S.A.; Khor, K.H.; Khairani-Bejo, S.; Lau, S.F.; Mazlan, M.; Roslan, A.; Goh, S.H. Detection and Characterization of *Leptospira* spp. In Dogs Diagnosed with Kidney and/or Liver Disease in Selangor, Malaysia. *J. Vet. Diagn. Investig.* **2021**, *33*, 834–843. [CrossRef]
14. Narkkul, U.; Thaipadungpanit, J.; Srisawat, N.; Rudge, J.W.; Thongdee, M.; Pawarana, R.; Pan-Ngum, W. Human, Animal, Water Source Interactions and Leptospirosis in Thailand. *Sci. Rep.* **2021**, *11*, 3215. [CrossRef]
15. Eric Klaasen, H.L.; Adler, B. Recent Advances in Canine Leptospirosis: Focus on Vaccine Development. *Vet. Med.* **2015**, *6*, 245–260. [CrossRef]
16. Costa, A.; Colocho, R.A.B.; Pereira, C.R.; Lage, A.P.; Heinemann, M.B.; Dorneles, E.M.S. Canine Leptospirosis in Stray and Sheltered Dogs: A Systematic Review. *Anim. Health Res. Rev.* **2022**, *23*, 39–58. [CrossRef]
17. Meeyam, T.; Tablerk, P.; Petchanok, B.; Pichpol, D.; Padungtod, P. Seroprevalence and Risk Factors Associated with Leptospirosis in Dogs. *Southeast Asian J. Trop. Med. Public Health* **2006**, *37*, 148–153.
18. Niwetpathomwat, A.; Assarasakorn, S. Preliminary Investigation of Canine Leptospirosis in a Rural Area of Thailand. *Med. Weter.* **2007**, *63*, 59–61.
19. Jittapalapong, S.; Sittisan, P.; Sakpuaram, T.; Kabeya, H.; Maruyama, S.; Inpankaew, T. Coinfection of *Leptospira* spp. and Toxoplasma Gondii among Stray Dogs in Bangkok, Thailand. *Southeast Asian J. Trop. Med. Public Health* **2009**, *40*, 247–252. [PubMed]
20. Pumipuntu, N.; Suwannarong, K. Seroprevalence of *Leptospira* spp. in Cattle and Dogs in Mahasarakham Province, Thailand. *J. Health Res.* **2016**, *30*, 223. [CrossRef]
21. Ngasaman, R.; Saechan, V.; Prachantasena, S.; Yingkajorn, M.; Sretrirutchai, S. Investigation of *Leptospira* Infection in Stray Animals in Songkhla, Thailand: Leptospirosis Risk Reduction in Human. *Vector Borne Zoonotic Dis.* **2020**, *20*, 432–435. [CrossRef] [PubMed]
22. Rubel, D.; Seijo, A.; Cernigoi, B.; Viale, A.; Wisnivesky-Colli, C. *Leptospira interrogans* in a Canine Population of Greater Buenos Aires: Variables Associated with Seropositivity. *Rev. Panam. Salud Publica* **1997**, *2*, 102–105. [CrossRef]
23. Suárez Hernández, M.; Martínez Sánchez, R.; Posada Fernández, P.; Bustelo Aguila, J.; Carrera Nodal, O.; Fleite, F.B.; García, I.V.; Sánchez Sibello, A. Leptospirosis in Children in Ciego De Avila Province, Cuba. *Rev. Soc. Bras. Med. Trop.* **1999**, *32*, 145–150. [CrossRef]
24. Patil, D.; Dahake, R.; Roy, S.; Mukherjee, S.; Chowdhary, A.; Deshmukh, R. Prevalence of Leptospirosis among Dogs and Rodents and Their Possible Role in Human Leptospirosis from Mumbai, India. *Indian J. Med. Microbiol.* **2014**, *32*, 64–67. [CrossRef] [PubMed]
25. Lelu, M.; Muñoz-Zanzi, C.; Higgins, B.; Galloway, R. Seroepidemiology of Leptospirosis in Dogs from Rural and Slum Communities of Los Rios Region, Chile. *BMC Vet. Res.* **2015**, *11*, 31. [CrossRef] [PubMed]
26. Kurilung, A.; Chanchaithong, P.; Lugsomya, K.; Niyomtham, W.; Wuthiekanun, V.; Prapasarakul, N. Molecular Detection and Isolation of Pathogenic *Leptospira* from Asymptomatic Humans, Domestic Animals and Water Sources in Nan Province, a Rural Area of Thailand. *Res. Vet. Sci.* **2017**, *115*, 146–154. [CrossRef]
27. Kurilung, A.; Keeratipusana, C.; Suriyaphol, P.; Hampson, D.J.; Prapasarakul, N. Genomic Analysis of *Leptospira interrogans* Serovar Paidjan and Dadas Isolates from Carrier Dogs and Comparative Genomic Analysis to Detect Genes under Positive Selection. *BMC Genom.* **2019**, *20*, 168. [CrossRef]
28. Boonsilp, S.; Thaipadungpanit, J.; Amornchai, P.; Wuthiekanun, V.; Bailey, M.S.; Holden, M.T.G.; Zhang, C.; Jiang, X.; Koizumi, N.; Taylor, K.; et al. A Single Multilocus Sequence Typing (Mlst) Scheme for Seven Pathogenic *Leptospira* Species. *PLoS Neglected Trop. Dis.* **2013**, *7*, e1954. [CrossRef] [PubMed]
29. Voronina, O.L.; Kunda, M.S.; Aksenova, E.I.; Ryzhova, N.N.; Semenov, A.N.; Petrov, E.M.; Didenko, L.V.; Lunin, V.G.; Ananyina, Y.V.; Gintsburg, A.L. The Characteristics of Ubiquitous and Unique *Leptospira* Strains from the Collection of Russian Centre for Leptospirosis. *Biomed Res. Int.* **2014**, *2014*, 649034. [CrossRef]
30. Kurilung, A.; Perreten, V.; Prapasarakul, N. Comparative Genomic Analysis and a Novel Set of Missense Mutation of the *Leptospira* Weilii Serogroup Mini from the Urine of Asymptomatic Dogs in Thailand. *Front. Microbiol.* **2021**, *12*, 731937. [CrossRef]

31. Amran, F.; Khalid, M.K.N.M.; Mohamad, S.; Ripen, A.M.; Ahmad, N.; Goris, M.G.A.; Muhammad, A.H.; Halim, N.A.N. Draft Genome Sequence of *Leptospira interrogans* Serovar Bataviae Strain Lepimr 22 Isolated from a Rodent in Johor, Malaysia. *Genome Announc.* **2016**, *4*, e00956-16. [CrossRef]
32. Altheimer, K.; Jongwattanapisan, P.; Luengyosluechakul, S.; Pusoonthornthum, R.; Prapasarakul, N.; Kurilung, A.; Broens, E.M.; Wagenaar, J.A.; Goris, M.G.A.; Ahmed, A.A.; et al. *Leptospira* Infection and Shedding in Dogs in Thailand. *BMC Vet. Res.* **2020**, *16*, 89. [CrossRef]
33. Booncieww, P.; Kurilung, A.; Altheimer, K.; Hartmann, K.; Prapasarakul, N. Draft Genome Sequence of *Leptospira interrogans* Serovar Bataviae Strain D64, Isolated from the Urine of an Asymptomatic Dog in Pathum Thani, Thailand. *Microbiol. Resour. Announc.* **2020**, *9*, e00361-20. [CrossRef] [PubMed]
34. Ramli, S.R.; Bunk, B.; Spröer, C.; Geffers, R.; Jarek, M.; Bhuju, S.; Goris, M.; Mustakim, S.; Pessler, F. Complete Genome Sequencing of *Leptospira interrogans* Isolates from Malaysia Reveals Massive Genome Rearrangement but High Conservation of Virulence-Associated Genes. *Pathogens* **2021**, *10*, 1198. [CrossRef]
35. Trevejo, R.T.; Rigau-Pérez, J.G.; Ashford, D.A.; McClure, E.M.; Jarquín-González, C.; Amador, J.J.; Reyes, J.O.d.L.; Gonzalez, A.; Zaki, S.R.; Shieh, W.; et al. Epidemic Leptospirosis Associated with Pulmonary Hemorrhage-Nicaragua, 1995. *J. Infect. Dis.* **1998**, *178*, 1457–1463. [CrossRef] [PubMed]
36. Musso, D.; La Scola, B. Laboratory Diagnosis of Leptospirosis: A Challenge. *J. Microbiol. Immunol. Infect.* **2013**, *46*, 245–252. [CrossRef] [PubMed]
37. Leptospirosis–Manual of Diagnostic Tests and Vaccines for Terrestrial Animals, 8th ed. 2021. Available online: https://www.woah.org/en/what-we-do/standards/codes-and-manuals/terrestrial-manual-online-access/ (accessed on 8 October 2021).
38. Pinto, G.V.; Senthilkumar, K.; Rai, P.; Kabekkodu, S.P.; Karunasagar, I.; Kumar, B.K. Current Methods for the Diagnosis of Leptospirosis: Issues and Challenges. *J. Microbiol. Methods* **2022**, *195*, 106438. [CrossRef] [PubMed]
39. Gökmen, T.G.; Soyal, A.; Kalayci, Y.; Önlen, C.; Köksal, F. Comparison of 16s Rrna-Pcr-Rflp, Lipl32-Pcr and Ompl1-Pcr Methods in the Diagnosis of Leptospirosis. *Rev. Inst. Med. Trop. Sao Paulo* **2016**, *58*, 64. [CrossRef] [PubMed]
40. Suwancharoen, D.; Chaisakdanugull, Y.; Thanapongtharm, W.; Yoshida, S. Serological Survey of Leptospirosis in Livestock in Thailand. *Epidemiol. Infect.* **2013**, *141*, 2269–2277. [CrossRef]
41. Dara, C.; Anunnatsiri, S.; Kessomboon, P.; Laha, T.; Brameld, S.; Saisongkorh, W.; Sripa, B.; Suttiprapa, S. *Leptospira* spp. Seroprevalence of Co-Infection with Opisthorchis Viverrini Infection in Northeast, Thailand. *Suranaree J. Sci. Technol.* **2022**, *29*, 070029.
42. Kumar, R.S.; Pillai, R.M.; Mukhopadhyay, H.K.; Antony, P.X.; Thanislass, J.; Srinivas, V.M.V.; Vishnupriya, S. Seroepidemiology of Canine Leptospirosis by Ielisa and Mat. *Vet. World* **2013**, *6*, 926–930. [CrossRef]
43. Al-orry, W.; Arahou, M.; Hassikou, R.; Mennane, Z. A Review of Laboratory Diagnosis and Treatment of Leptospirosis. *Int. J. Pharm. Pharm. Sci.* **2016**, *8*, 7–13. [CrossRef]
44. Sykes, J.E.; Reagan, K.L.; Nally, J.E.; Galloway, R.L.; Haake, D.A. Role of Diagnostics in Epidemiology, Management, Surveillance, and Control of Leptospirosis. *Pathogens* **2022**, *11*, 395. [CrossRef]
45. Boonsilp, S.; Thaipadungpanit, J.; Amornchai, P.; Wuthiekanun, V.; Chierakul, W.; Limmathurotsakul, D.; Day, N.P.; Peacock, S.J. Molecular Detection and Speciation of Pathogenic *Leptospira* spp. In Blood from Patients with Culture-Negative Leptospirosis. *BMC Infect. Dis.* **2011**, *11*, 338. [CrossRef]
46. Goris, M.G.; Hartskeerl, R.A. Leptospirosis Serodiagnosis by the Microscopic Agglutination Test. *Curr. Protoc. Microbiol.* **2014**, *32*, 12E-5. [CrossRef] [PubMed]
47. Adler, B.; Murphy, A.M.; Locarnini, S.A.; Faine, S. Detection of Specific Anti-Leptospiral Immunoglobulins M and G in Human Serum by Solid-Phase Enzyme-Linked Immunosorbent Assay. *J. Clin. Microbiol.* **1980**, *11*, 452–457. [CrossRef] [PubMed]
48. Guerreiro, H.; Croda, J.; Flannery, B.; Mazel, M.; Matsunaga, J.; Reis, M.G.; Levett, P.N.; Ko, A.I.; Haake, D.A. Leptospiral Proteins Recognized During the Humoral Immune Response to Leptospirosis in Humans. *Infect. Immun.* **2001**, *69*, 4958–4968. [CrossRef] [PubMed]
49. Haake, D.A.; Chao, G.; Zuerner, R.L.; Barnett, J.K.; Barnett, D.; Mazel, M.; Matsunaga, J.; Levett, P.N.; Bolin, C.A. The Leptospiral Major Outer Membrane Protein Lipl32 Is a Lipoprotein Expressed During Mammalian Infection. *Infect. Immun.* **2000**, *68*, 2276–2285. [CrossRef] [PubMed]
50. Noda, R.F.A.A.; Fleitas, O.; Rodriguez, I.; Beltran, J.F.; Falcon, R.; Almaguer, T.; TH, S. Triton X-100 Vs. Triton X-114: Isolation of Outer Membrane Proteins from *Leptospira* spp. *J. Vet. Sci.* **2017**, *1*, 7–12.
51. Thoduvayil, S.; Dhandapani, G.; Brahma, R.; Balaya, R.D.A.; Mangalaparthi, K.K.; Patel, K.; Kumar, M.; Tennyson, J.; Satheeshkumar, P.K.; Kulkarni, M.J.; et al. Triton X-114 Fractionated Subcellular Proteome of *Leptospira interrogans* Shows Selective Enrichment of Pathogenic and Outer Membrane Proteins in the Detergent Fraction. *Proteomics* **2020**, *20*, e2000170. [CrossRef]
52. Maneewatchararangsri, S.; Reamtong, O.; Kalambaheti, T.; Pumirat, P.; Vanaporn, M.; Limmathurotsakul, D.; Thavornkuno, C. Development of Enzyme-Linked Immunosorbent Assay for Human Leptospirosis Serodiagnosis Using *Leptospira* Secretome Antigen. *Southeast Asian J. Trop. Med. Public Health* **2017**, *48*, 576–584.
53. Sathiyamoorthy, A.; Selvaraju, G.; Palanivel, K.M.; Srinivasan, A.P. Development of Indirect Enzyme-Linked Immunosorbent Assay for Diagnosis of Canine Leptospirosis. *Vet. World* **2017**, *10*, 530–535. [CrossRef] [PubMed]
54. Araghian, A.; Elmi, A.; Farahbakhsh, M.; Hosseini, S.; Faezi, S. Seroepidemiology of Leptospirosis in Guilan Province, Northern Iran: Comparison between Mat and Igm-Elisa Techniques. *J. Infect. Dev. Ctries.* **2018**, *12*, 109–114. [CrossRef] [PubMed]

55. Gómez-Morales, M.A.; Selmi, M.; Ludovisi, A.; Amati, M.; Fiorentino, E.; Breviglieri, L.; Poglayen, G.; Pozio, E. Hunting Dogs as Sentinel Animals for Monitoring Infections with *Trichinella* spp. in Wildlife. *Parasites Vectors* **2016**, *9*, 154. [CrossRef] [PubMed]
56. Gottschalk, M.; Altman, E.; Charland, N.; De Lasalle, F.; Dubreuil, J.D. Evaluation of a Saline Boiled Extract, Capsular Polysaccharides and Long-Chain Lipopolysaccharides of Actinobacillus Pleuropneumoniae Serotype 1 as Antigens for the Serodiagnosis of Swine Pleuropneumonia. *Vet. Microbiol.* **1994**, *42*, 91–104. [CrossRef]
57. Ribotta, M.J.; Higgins, R.; Gottschalk, M.; Lallier, R. Development of an Indirect Enzyme-Linked Immunosorbent Assay for the Detection of Leptospiral Antibodies in Dogs. *Can. J. Vet. Res.* **2000**, *64*, 32–37.
58. Woo, P.C.; Leung, A.S.; Lau, S.K.; Chong, K.T.; Yuen, K.Y. Use of Recombinant Mitogillin for Serodiagnosis of Aspergillus Fumigatus-Associated Diseases. *J. Clin. Microbiol.* **2001**, *39*, 4598–4600. [CrossRef] [PubMed]
59. Ye, C.; Yan, W.; Xiang, H.; He, H.; Yang, M.; Ijaz, M.; Useh, N.; Hsieh, C.-L.; McDonough, P.L.; McDonough, S.P.; et al. Recombinant Antigens Rlipl21, Rloa22, Rlipl32 and Rligacon4-8 for Serological Diagnosis of Leptospirosis by Enzyme-Linked Immunosorbent Assays in Dogs. *PLoS ONE* **2014**, *9*, e111367. [CrossRef]
60. Penna, B.; Marassi, C.D.; Libonati, H.; Narduche, L.; Lilenbaum, W.; Bourhy, P. Diagnostic Accuracy of an in-House Elisa Using the Intermediate Species *Leptospira* Fainei as Antigen for Diagnosis of Acute Leptospirosis in Dogs. *Comp. Immunol. Microbiol. Infect. Dis.* **2017**, *50*, 13–15. [CrossRef]
61. Lizer, J.; Velineni, S.; Weber, A.; Krecic, M.; Meeus, P. Evaluation of 3 Serological Tests for Early Detection of Leptospira-Specific Antibodies in Experimentally Infected Dogs. *J. Vet. Intern. Med.* **2018**, *32*, 201–207. [CrossRef]
62. Tsuji, M.; Kawaoka, Y.; Naiki, M.; Yanagawa, R. Isolation of Antigenically Active Components from Leptospiral Serovar-Specific Lipopolysaccharide Antigen by Alkaline Treatment. *Microbiol. Immunol.* **1981**, *25*, 949–957. [CrossRef]
63. Ono, E.; Takase, H.; Naiki, M.; Yanagawa, R. Purification, Characterization and Serological Properties of a Glycolipid Antigen Reactive with a Serovar-Specific Monoclonal Antibody against *Leptospira interrogans* Serovar Canicola. *J. Gen. Microbiol.* **1987**, *133*, 1329–1336. [CrossRef]
64. Cinco, M.; Banfi, E.; Panfili, E. Leptospiral Lipopolysaccharide Presence in the Outer Envelope: Electrophoretic Evidence and Immunological Specificity. *Zentralblatt Für Bakteriol. Mikrobiol. Und Hyg. A* **1988**, *269*, 277–283. [CrossRef]
65. Surujballi, O.; Mallory, M. An Indirect Enzyme Linked Immunosorbent Assay for the Detection of Bovine Antibodies to Multiple *Leptospira* Serovars. *Can. J. Vet. Res.* **2004**, *68*, 1–6.
66. Cariou, C.; Herbet, G.; Ripart, P.; Martin-Cagnon, N.; Bouvet, J.; Schneider, M.; Guiot, A.-L.; Cupillard, L. Development of Antibody Elisa Specific of *Leptospira interrogans* Serovar Grippotyphosa, Canicola, and Icterohaemorrhagiae to Monitor Vaccine Immunogenicity. *Vet. Immunol. Immunopathol.* **2020**, *219*, 109960. [CrossRef]
67. Sripattanakul, S.; Prapong, T.; Kamlangdee, A.; Katzenmeier, G.; Haltrich, D.; Hongprayoon, R.; Prapong, S. *Leptospira borgpetersenii* Leucine-Rich Repeat Proteins and Derived Peptides in an Indirect Elisa Development for the Diagnosis of Canine Leptospiral Infections. *Trop. Med. Infect. Dis.* **2022**, *7*, 311. [CrossRef]
68. Garibyan, L.; Avashia, N. Polymerase Chain Reaction. *J. Investig. Dermatol.* **2013**, *133*, 1–4. [CrossRef]
69. Yang, S.; Rothman, R.E. Pcr-Based Diagnostics for Infectious Diseases: Uses, Limitations, and Future Applications in Acute-Care Settings. *Lancet Infect. Dis.* **2004**, *4*, 337–348. [CrossRef] [PubMed]
70. Chirathaworn, C.; Inwattana, R.; Poovorawan, Y.; Suwancharoen, D. Interpretation of Microscopic Agglutination Test for Leptospirosis Diagnosis and Seroprevalence. *Asian Pac. J. Trop. Biomed.* **2014**, *4* (Suppl. 1), S162–S164. [CrossRef] [PubMed]
71. Martin, E.A.; Heseltine, J.C.; Creevy, K.E. The Evaluation of the Diagnostic Value of a Pcr Assay When Compared to a Serologic Micro-Agglutination Test for Canine Leptospirosis. *Front. Vet. Sci.* **2022**, *9*, 815103. [CrossRef] [PubMed]
72. Chen, H.-W.; Tilley, D.H.; Canal, E.; Guevara, C.; Maves, R.; Hall, E.; Halsey, E.S.; Zhang, Z.; Ching, W.-M.; Kochel, T.J. Detection of Leptospira-Specific Antibodies Using a Recombinant Antigen-Based Enzyme-Linked Immunosorbent Assay. *Am. J. Trop. Med. Hyg.* **2013**, *89*, 1088–1094. [CrossRef]
73. Souza, M.A.; Castro, J.; Tavares, T.; Soares, P.; Santos, M.; Ganda, M.; Gomes, D.O.; Lima, A.M.C. Comparison of Microscopic Agglutination Test and Indirect Elisa in the Diagnosis of Bovine Leptospirosis. *Biosci. J.* **2014**, *30*, 833–838.
74. Ooteman, M.C.; Vago, A.R.; Koury, M.C. Evaluation of Mat, Igm Elisa and Pcr Methods for the Diagnosis of Human Leptospirosis. *J. Microbiol. Methods* **2006**, *65*, 247–257. [CrossRef]
75. Jose, L.R.; Mahadeviah, S.N.; Balamurugan, V.; Kini, K.R. Evaluation of an in-House Lipl32 Polymerase Chain Reaction for Diagnosis of Leptospirosis and Its Correlation with Various Serological Diagnostic Techniques. *Indian J. Med. Microbiol.* **2018**, *36*, 385–390. [CrossRef]
76. Philip, N.; Affendy, N.B.; Masri, S.N.; Yuhana, M.Y.; Than, L.T.L.; Sekawi, Z.; Neela, V.K. Combined Pcr and Mat Improves the Early Diagnosis of the Biphasic Illness Leptospirosis. *PLoS ONE* **2020**, *15*, e0239069. [CrossRef] [PubMed]
77. Shukla, S.; Mittal, V.; Singh, P.; Singh, A. Evaluation of Taqman Based Real-Time Pcr Assay Targeting Lipl32 Gene for Leptospirosis in Serologically Positive Human Urine Samples from North India. *Indian J. Med. Microbiol.* **2021**, *39*, 11–14. [CrossRef] [PubMed]
78. Bajani, M.D.; Ashford, D.A.; Bragg, S.L.; Woods, C.W.; Aye, T.; Spiegel, R.A.; Plikaytis, B.D.; Perkins, B.A.; Phelan, M.; Levett, P.N.; et al. Evaluation of Four Commercially Available Rapid Serologic Tests for Diagnosis of Leptospirosis. *J. Clin. Microbiol.* **2003**, *41*, 803–809. [CrossRef]
79. McHugh, M.L. Interrater Reliability: The Kappa Statistic. *Biochem. Med.* **2012**, *22*, 276–282. [CrossRef]
80. Mulla, S.; Chakraborty, T.; Patel, M.; Pandya, H.P.; Dadhaniya, V.; Vaghela, G. Diagnosis of Leptospirosis and Comparison of Elisa and Mat Techniques. *Indian J. Pathol. Microbiol.* **2006**, *49*, 468–470.

81. Jayasundara, D.; Gamage, C.; Senavirathna, I.; Warnasekara, J.; Matthias, M.A.; Vinetz, J.M.; Agampodi, S. Optimizing the Microscopic Agglutination Test (Mat) Panel for the Diagnosis of Leptospirosis in a Low Resource, Hyper-Endemic Setting with Varied Microgeographic Variation in Reactivity. *PLoS Neglected Trop. Dis.* **2021**, *15*, e0009565. [CrossRef]
82. Lee, H.S.; Levine, M.; Guptill-Yoran, C.; Johnson, A.J.; von Kamecke, P.; Moore, G.E. Regional and Temporal Variations of *Leptospira* Seropositivity in Dogs in the United States, 2000–2010. *J. Vet. Intern. Med.* **2014**, *28*, 779–788. [CrossRef]
83. Bertasio, C.; Boniotti, M.B.; Lucchese, L.; Ceglie, L.; Bellinati, L.; Mazzucato, M.; Furlanello, T.; D'Incau, M.; Natale, A.K. Detection of New *Leptospira* Genotypes Infecting Symptomatic Dogs: Is a New Vaccine Formulation Needed? *Pathogens* **2020**, *9*, 484. [CrossRef]
84. Di Azevedo, M.I.N.; Aymée, L.; Borges, A.; Lilenbaum, W. Molecular Epidemiology of Pathogenic *Leptospira* spp. Infecting Dogs in Latin America. *Animals* **2023**, *13*, 2422. [CrossRef] [PubMed]
85. Ko, A.I.; Goarant, C.; Picardeau, M. Leptospira: The Dawn of the Molecular Genetics Era for an Emerging Zoonotic Pathogen. *Nat. Rev. Microbiol.* **2009**, *7*, 736–747. [CrossRef] [PubMed]
86. Lauretti-Ferreira, F.; Silva, P.L.D.; Alcântara, N.M.; Silva, B.F.; Grabher, I.; Souza, G.O.; Nakajima, E.; Akamatsu, M.A.; Vasconcellos, S.A.; Abreu, P.A.E.; et al. New Strategies for *Leptospira* Vaccine Development Based on Lps Removal. *PLoS ONE* **2020**, *15*, e0230460. [CrossRef] [PubMed]
87. Matthias, M.A.; Lubar, A.A.; Acharige, S.S.L.; Chaiboonma, K.L.; Pilau, N.N.; Marroquin, A.S.; Jayasundara, D.; Agampodi, S.; Vinetz, J.M. Culture-Independent Detection and Identification of *Leptospira* Serovars. *Microbiol. Spectr.* **2022**, *10*, e0247522. [CrossRef]
88. Smith, A.M.; Stull, J.W.; Moore, G.E. Potential Drivers for the Re-Emergence of Canine Leptospirosis in the United States and Canada. *Trop. Med. Infect. Dis.* **2022**, *7*, 377. [CrossRef]

animals

Article

Enhancing Control of *Leishmania infantum* Infection: A Multi-Epitope Nanovaccine for Durable T-Cell Immunity

Clara Hurtado-Morillas [1], Abel Martínez-Rodrigo [2,3], José A. Orden [1], Laura de Urbina-Fuentes [1], Alicia Mas [1,*] and Gustavo Domínguez-Bernal [1]

1 INMIVET, Animal Health Department, School of Veterinary Medicine, Complutense University of Madrid, 28040 Madrid, Spain; clarahur@ucm.es (C.H.-M.)
2 INMIVET, Animal Science Department, School of Veterinary Medicine, Complutense University of Madrid, 28040 Madrid, Spain
3 Centro de Investigación en Sanidad Animal, Instituto Nacional de Investigación y Tecnología Agraria y Alimentaria, Consejo Superior de Investigaciones Científicas (CISA-INIA-CSIC), 28130 Madrid, Spain
* Correspondence: alimas@ucm.es

Simple Summary: Canine leishmaniosis is a potentially fatal disease in dogs caused by the *Leishmania* parasite. Vaccination seems to be the safest, most cost-effective, and long-lasting control strategy. Currently, there is still no vaccine that totally guarantees complete protection against *Leishmania* infection. Here, we designed and evaluated the effectiveness of a nanovaccine against *Leishmania infantum* infection in the murine model. This vaccine strategy, consisting of the HisDTC peptide encapsulated in polymeric nanoparticles, induced in vaccinated groups a lower parasite load in comparison to the control groups, which was correlated with the induction of a cellular immune response profile against *Leishmania infantum* measured throughout different cytokines, antibodies titers, and memory T cells. These results provide evidence that the HisDTC peptide encapsulated in polymeric nanoparticles is a potential vaccine strategy against canine leishmaniosis.

Abstract: Canine leishmaniosis (CanL) is a growing health problem for which vaccination is a crucial tool for the control of disease. The successful development of an effective vaccine against this disease relies on eliciting a robust and enduring T-cell immune response involving the activation of $CD4^+$ Th1 and $CD8^+$ T-cells. This study aimed to evaluate the immunogenicity and prophylactic efficacy of a novel nanovaccine comprising a multi-epitope peptide, known as HisDTC, encapsulated in PLGA nanoparticles against *Leishmania infantum* infection in the murine model. The encapsulation strategy was designed to enhance antigen loading and sustain release, ensuring prolonged exposure to the immune system. Our results showed that mice immunized with PLGA-encapsulated HisDTC exhibited a significant reduction in the parasite load in the liver and spleen over both short and long-term duration. This reduction was associated with a cellular immune profile marked by elevated levels of pro-inflammatory cytokines, such as IFN-γ, and the generation of memory T cells. In conclusion, the current study establishes that PLGA-encapsulated HisDTC can promote effective and long-lasting T-cell responses against *L. infantum* in the murine model. These findings underscore the potential utility of multi-epitope vaccines, in conjunction with appropriate delivery systems, as an alternative strategy for CanL control.

Keywords: canine leishmaniosis; PLGA nanoparticles; multi-epitope peptide; nanovaccine; LetiFend®; *Leishmania infantum*

1. Introduction

Canine leishmaniosis (CanL), a life-threatening vector-borne zoonotic disease caused by the protozoa of the genus *Leishmania*, is a significant health risk to both dogs and humans in endemic regions worldwide. This disease is present in all continents except Oceania, and it is endemic in more than fifty countries in the Mediterranean Basin, South America, and Central and Southwestern Asia [1]. Far from disease control, the incidence of CanL in non-autochthonous countries shas increased in recent years due to globalization, among other causes [2].

Among the twelve species described as capable of infecting dogs, *Leishmania (L.) infantum* is the principal factor responsible for CanL and the most widely distributed [3]. Dogs, beyond enduring the potential severity and fatality of this disease, constitute the principal reservoir for *L. infantum* in relation to human infections [1]. *L. infantum*, the etiological agent of visceral leishmaniosis (VL), represents the most severe form of this disease in humans, ranking second in mortality and fourth in morbidity among tropical diseases, with 50,000 to 90,000 cases annually [4]. Consequently, leishmaniosis emerges not only as an important disease in veterinary medicine but also as a global public health challenge for which effective control tools must be developed.

CanL's control is, however, challenging primarily due to the high cost, limited efficacy, and potential toxicity associated with available treatments [5,6]. Compounding this issue, there are currently no effective measures to prevent human infection, which makes it even more important to design and implement adequate control measures targeting canids. Vaccination has emerged as a crucial, safe, cost-effective, and long-lasting strategy for controlling leishmaniosis. While five vaccines have been licensed against CanL, only the following two are currently commercially available: LetiFend® (LETI Pharma, S.L.U., Spain) in Europe, composed of a chimera protein (protein Q) formed by five antigenic fragments from four different *L. infantum* ribosomal proteins LiP2a, LiP2b, and LiP0 and the histone H2A, and Leish-Tec® (Hertape, Brazil) in Brazil, which consists of a recombinant protein A2 from *L. donovani* with saponin as a vaccine adjuvant. A third recent vaccine was approved by the European Medicines Agency (EMA) for commercialization [7]. Despite these advancements, uncertainties persist regarding their efficacy. Although these vaccines may mitigate the risk of clinical signs and disease progression, they do not guarantee complete protection against the infection establishment, and infected dogs may still serve as potential reservoirs for the parasite [8,9]. Consequently, there is a pressing need for further studies, which are necessary to develop more effective strategies capable of preventing infection and controlling the spread of this disease among both canids and humans.

Peptide vaccines have become a focal point in vaccine development because of their easy and rapid manufacture, along with their inherent stability and safety [10]. While peptide-based vaccines can elicit a specific immune response against pathogens, their immunogenic potential may be limited, necessitating the use of adjuvants, specific delivery systems, or multiple doses to enhance their immunogenicity [10]. In this context, the utilization of Toll-like receptor ligands (TLRLs) as adjuvants has shown promising results as they can promote the initiation of a robust innate immune response during the early stages of *Leishmania* infection, paving the way for the development of an effective cell-mediated immune response [11,12]. Several licensed vaccines currently incorporate TLRLs as adjuvants, such as TLRL-4 [13,14] and TLRL-9 [15]. Ongoing studies are exploring other Toll-like receptor agonists, such as TLRL-2 (Pam3CSK4) and TLRL-3 (Polyinosinic-polycytidylic acid (poly(I:C)) as potential vaccine adjuvants against leishmaniosis [16–18].

In this context, in parallel to the use of adjuvants, nanotechnology has introduced innovative possibilities for designing vaccine delivery systems against various infectious diseases, including leishmaniosis [19–21]. Among different nanoparticles (NPs), poly(lactic-*co*-glycolic) acid (PLGA) nanoparticles, approved by the EMA, have gained considerable attention as carriers for antigen delivery in vaccine development due to their properties. The biodegradable and biocompatible nature of PLGA allows for the safe administration and controlled release of encapsulated antigens. Additionally, PLGA nanoparticles protect antigens from degradation, enhance antigen uptake through antigen-presenting cells, and facilitate co-encapsulation with adjuvants, thereby enhancing immune responses [22].

This study specifically explores the use of the HisDTC multi-epitope peptide vaccine delivered by PLGA NP in combination with TLRLs as adjuvants. This peptide has already been described to induce a specific cellular immune response in vaccinated mice that effectively controls parasite multiplication against *L. infantum* [23]. To enhance the long-term efficacy of the HisDTC-induced immune response against *L. infantum* infection in BALB/c mice, this study evaluates the co-encapsulation of the multi-epitope peptide with TLRL-2 and TLRL-3 as immunomodulators with PLGA nanoparticles. For this purpose, we analyzed the parasite load in target organs and characterized the specific immune response generated against *L. infantum* through this vaccine strategy in the murine model of VL.

The use of the murine model serves as a crucial initial step towards evaluating the vaccine's efficacy. Furthermore, this study highlights the promising role of nanotechnology-based vaccine delivery systems in overcoming challenges associated with vaccine development for complex infectious diseases such as leishmaniosis.

2. Materials and Methods

2.1. Mice, Parasites, and Soluble Leishmania Antigen Production

Studies were performed on 6–8-week-old female BALB/c mice (Janvier-Labs, Laval, France), which were housed in the Animal Facility of the Complutense University of Madrid under specific pathogen-free conditions. The study was approved (reference: PROEX 211/18) by the Animal Welfare Committee of the Community of Madrid, Spain. Animal care and procedures followed Spanish and EU legislation (Law 32/2007, R.D. 53/2013, and Council Directive 2010/63/EU).

Leishmania infantum BCN150 zymodeme MON-1 (M/CAN/ES/96/BCN150) promastigotes were cultured, as reported previously, at 26 °C in Schneider's medium (Sigma-Aldrich, Saint Louis, MO, USA) supplemented with 20% of inactivated fetal bovine serum (FBS, Sigma-Aldrich, Saint Louis, MO, USA), 200 U/mL of penicillin and 200 mg/mL of streptomycin (Lonza, Basel, Switzerland) [24]. All the parasites were previously passaged in BALB/c mice to maintain their virulence [25]. The specific soluble *Leishmania* antigen (SLA) was prepared from stationary-phase *L. infantum* promastigotes as described elsewhere, and it was stored at −20 °C until use [26].

2.2. Vaccine Preparation

2.2.1. Chimeric Multi-Epitope Peptide Formulation and TLRL Adjuvants Synthesis

The multi-epitope peptide vaccine HisDTC was previously designed, as detailed in [23]. The peptides were synthesized by GenScript Biotech (Leiden, The Netherlands) with purity $\geq$ 95% and stored at −20 °C until encapsulation.

TLRL-2 (Pam3CSK4) and TLRL-3 (Poly(I:C) HMW (VacciGrade™, InvivoGen, Toulouse, France) were used as adjuvants when needed.

2.2.2. PLGA Nanoparticles Preparation and Characterization

PLGA-NPs containing the chimeric peptide HisDTC and/or immunomodulators TLRL-2/3, when required, were performed using the double emulsion method (w/o/w) by Nanovex Biotechnologies (Llanera, Asturias, Spain). Briefly, a solution of PLGA (lactide/glycolide 50:50 molar ratio) was prepared in dichloromethane (20 mg/mL) and mixed with 1 mL of aqueous medium containing, when needed, the corresponding amount of each active ingredient to be encapsulated (peptide HisDTC; TLRL-2; TLRL-3) forming a primary solution of water in oil. The resulting solution was then added to a second aqueous solution containing polyvinyl acetate (PVA) 2%, and the mixture was emulsified via homogenization. The final double (w/o/w) emulsion was left under constant homogenization for 18 h to allow the organic solvent to evaporate. Nanoparticles were then purified by subjecting them to three consecutive cycles of centrifugation and redispersion. The final collected PLGA formulations were dispersed in a phosphate final medium pH 7 with 3% sucrose to stabilize PLGA nanoparticles for subsequent lyophilization, which were conserved at 4 °C until use.

The particle size and polydispersity index (PDI) of the synthesized nanoparticles were characterized using the Dynamic Light Scattering (DLS) technique and the Zetasizer ZS90

(Marlvern Instruments Ltd., Marlvern, UK). Zeta potential measurements were performed at 25 °C using the M3-PALS (Mixed Measurement Mode Phase Analysis) technique and the Zetasizer ZS90 (Marlvern Instruments Ltd.).

TLRL-3 and TLRL-2 concentrations were determined using the Quant-iT™ OliGreen ssDNA Reagent Assay Kit (Invitrogen, Molecular Probes, Eugene, OR, USA) and OPA (Fluoraldehyde) kit (Thermo Scientific, Waltham, MA, USA), respectively. The concentration of HisDTC was determined using the BCA kit (Thermo Scientific, Waltham, MA, USA). A Synergy™ HTX Multi-Mode Microplate Reader (Agilent Technologies, Thermo Scientific, Waltham, MA, USA) was used for fluorescence and absorbance measurements. All measurements were conducted using aqueous dispersions of PLGA-NPs prior to their lyophilization.

2.3. *Immunization, Infection and Sacrifice of Mice*

BALB/c mice were divided into five groups ($n = 9$ per group) and immunized subcutaneously on weeks 0 and 2, with a total volume of 100 µL of sterile phosphate-buffered saline (PBS) containing the following formulations: group 1 mice were immunized with 10 µg of HisDTC, 5 µg TLRL-2 and 38 µg of TLRL-3 encapsulated in PLGA-NPs (HisDTC-TLRLs-PLGA group); group 2 mice were immunized with 10 µg of HisDTC encapsulated (HisDTC-PLGA group); group 3 mice were and with 10 µg of protein Q (LetiFend group); groups 4 and 5 served as control mice which received 5 µg of TLRL-2 and 38 µg of TLRL-3 encapsulated in PLGA nanoparticles (TLRLs-PLGA group) and PLGA empty nanoparticles (PLGA group), respectively.

Animals were stratified into two analyses. Firstly, to assess the short-term immunoprophylactic capacity of the formulations on week 6, five mice per group were infected intravenously with 5×10^5 stationary-phase *L. infantum* promastigotes (short-term evaluation). Lately, with the aim of studying whether the effectiveness of the vaccination was prolonged in the long term, four mice per group were infected intravenously with 5×10^5 stationary-phase *L. infantum* promastigotes on week fourteen (long-term evaluation). In both cases, seven weeks post-infection, the animals were sacrificed, and serum samples, the liver, and spleen were collected to perform parasitological and immunological analyses.

2.4. *Evaluation of Delayed-Type Hypersensitivity (DTH)*

The cell-mediated immune response was examined by assessing the DTH reaction. Mice received an intradermal injection of 50 µg of SLA in a 35 µL volume of sterile PBS into the right footpad two days prior to infection for short and long-term evaluation. DTH response was assessed by measuring the difference in footpad swelling 48 h after SLA injection compared to the left footpad that received a control PBS injection, using a digital caliper (Sigma-Aldrich, Saint Louis, MO, USA).

2.5. *Quantification of Parasite Burden Using Limited Dilution Assay*

Parasite burdens in the liver and spleen were quantified using the limited dilution assay [27]. Briefly, the spleen and a portion of the liver that had been previously weighed were homogenized in 4 mL of Schneider's Insect Medium supplemented with 20% FBS. Subsequently, tissue homogenates were serially diluted (4-fold) in microtiter culture plates (Corning Costar, Thermo Scientific, Waltham, MA, USA) and incubated in quadruplicate at 26 °C. After eight days, the presence of motile promastigotes was examined using an inverted microscope. The parasite load was determined as the last positive dilution containing at least one viable parasite; it was expressed as the parasite number per gram of tissue, in the case of the liver, or per tissue, in the case of the spleen.

2.6. *Isolation and Co-Culture System of Bone Marrow-Derived Murine Dendritic Cells (BMDCs)*

Ten days before sacrifice, bone marrow stem cell progenitors were harvested from the femurs and tibias of naïve BALB/c mice ($n = 5$) and cultured in RPMI 1640 with L-glutamine and 25 mM of HEPES (Sigma-Aldrich, Saint Louis, MO, USA), supplemented with 10%

FBS and a mixture of antibiotics (100 U/mL penicillin, 100 mg/mL streptomycin, Lonza) in the presence of 20 ng/mL of the murine granulocyte-macrophage colony-stimulating factor (GM-CSF; BioLegend®, San Diego, CA, USA), modified from the protocol previously described [28]. Fresh medium containing GM-CSF was added to the cultures every 3 days. On day 6, cells were harvested, resuspended in RPMI medium with a freshly added growth factor, and cultured at 37 °C with 5% CO_2. Finally, on day 9, non-adherent naïve BMDCs were seeded in 24-well plates (5×10^5 cells/mL) and used for subsequent experiments.

2.7. Extracellular Cytokine Profile and Arginine Metabolism Elicited by Ex Vivo Co-Cultured Splenocytes Stimulations

To quantify cytokine production in immunized and infected mice, BMDCs obtained from naïve mice were plated (5×10^5 cells/mL) into 24-well plates and either stimulated or not with 25 µg/mL of SLA overnight. Following this, splenocytes obtained from immunized and infected mice were co-cultured at a ratio of 1:5 (BMDCs/splenocytes) at 37 °C and 5% CO_2. After 96 h, supernatants were collected and stored at –80 °C until processing. The levels of IFN-γ (ELISA MAX™ Deluxe Set Mouse, BioLegend, San Diego, CA, USA), TNF-α (Duoset ELISA kit, Development System R&D, Abingdon, UK), and IL-10 (BD OptEIA kit, Bioscience, San Diego, CA, USA) were quantified using commercially available ELISA kits following the instructions provided by the manufacturers.

In parallel, L-arginine metabolism was evaluated by quantifying nitric oxide (NO) production and arginase activity using the same experimental set. After 96 h, the co-culture of naïve BMDCs with splenocytes from the animals of different groups in the presence of 25 µg/mL of SLA at 37 °C and 5% CO_2. NO production was measured indirectly from the concentration of nitrites in supernatants using the Griess reaction as described [29]. Intracellular arginase activity was estimated using an enzymatic-colorimetric assay method, which detects urea production [30].

2.8. Flow Cytometric Characterization and Immunophenotyping of Cellular Immune Response

Seven weeks post-infection, spleen samples were collected to assess cytokine-producing T lymphocytes and characterize memory T cell subpopulations triggered by vaccination. Spleen samples from immunized and infected animals were macerated, the red blood cells lysed, and the isolated spleen cells were stimulated ex vivo with SLA (25 µg/mL) at 37 °C with 5% CO_2 for 12 h to characterize T lymphocyte subpopulations or for 48 h to assess intracellular cytokine staining. These simulations were performed in polypropylene tubes (Falcon, BD Pharmingen, San Diego, CA, USA) at a concentration of 1×10^6 cells/mL. The non-stimulated culture received only the culture medium.

Dead cells were excluded using the LIVE/DEAD Zombie RED Fixable Viability Kit (BioLegend, San Diego, CA, USA). The events (200,000 live cells) were acquired on a CytoFLEX S flow cytometer (Beckman Coulter, Life Science, Brea, CA, USA), and the CytExpert™ 2.4 software package (Beckman Coulter, Life Science, Brea, CA, USA) was utilized for data analysis.

2.8.1. Intracellular Cytokine Production

Upon stimulation and prior to harvest, isolated spleen cells were pre-incubated with brefeldin A (BioLegend, San Diego, CA, USA) for 4 h at 37 °C. To block non-specific antibody binding, the TruStain FcX™ antibody (Clone 96, BioLegend, San Diego, CA, USA) was then added. Subsequently, splenocytes were incubated with Zombie RED (Fixable Viability Kit, BioLegend, San Diego, CA, USA) for 10 min at room temperature, washed and immunostained with anti-CD4 FITC (clone GK1.5, BioLegend) (diluted 1:800) and anti-CD8 PE (clone QA17A07, BioLegend) (diluted 1:800) fluorophore-labeled antibodies. After two washes with PBS containing 2% FCS, cells were fixed and permeabilized using Cyto-FastTM Fix/Perm Solution (BioLegend) according to the manufacturer's protocol. Then, cells were stained with anti-IFN-γ PerCP (clone XMG1.2, BioLegend) and anti-IL-10 PCy7 (clone JES5-16E3, BioLegend) at room temperature for 30 min, washed twice, and analyzed. The production of intracellular cytokines was expressed as percentages, representing the ratio of the percentage of IFN-γ or IL-10-producing-T cells in SLA-stimulated cultures.

2.8.2. Specific T Lymphocyte Subsets

For immunophenotyping T lymphocyte memory subsets, the TruStain FcX™ antibody (Clone 96, BioLegend) was added to block the non-specific binding of antibodies. Subsequently, splenocytes were incubated with Zombie RED for 10 min at room temperature. Finally, cells were stained with the following fluorophore-labeled anti-mouse monoclonal antibodies: anti-CD4 FITC (clone GK1.5, BioLegend), anti-CD8 PCy7 (clone QA17A07, BioLegend), anti-CD44 PE (clone IM7, BioLegend) and anti-CD62L PerCP (clone MEL-14, BioLegend) for 20 min at 4 °C in the dark. The T lymphocyte subsets were characterized by the percentage values obtained from the SLA-stimulated culture.

2.9. Leishmania-Specific IgG1 and IgG2a Antibody Titration

Sera samples from immunized and infected mice were used for the titration of the IgG isotype antibody using the ELISA assay as previously described [31]. Briefly, mice serum was serially two-fold-diluted starting from 1:400 for 90 min in an ELISA 96-well microtiter plate (Nunc Immunoplate MaxiSorp, Thermo Scientific, Waltham, MA, USA) previously coated with 6 μg/mL of specific SLA antigens. Next, the IgG isotype goat anti-mouse antibody conjugated with peroxidase (HRP) (Southern Biotech, Birmingham, AL, USA) was added for 1 h. Finally, the reaction was detected using the TMB substrate (Thermo Scientific, Waltham, MA, USA). The optical density was measured at 450 nm, and the reciprocal end-point titer was determined as the highest serum dilution, providing an absorbance three times higher than the negative control (serum from uninfected and non-immunized mice).

2.10. Statistical Analysis

Statistical analyses were performed using GraphPad Prism software (version 8.3 for Windows, San Diego, CA, USA). First, the Shapiro–Wilk normality test was performed to assess the normality distribution of quantitative variables. Then, one or two-way ANOVA with multiple comparisons of Tukey's post hoc test was conducted to determine which means from the independent groups were significantly different. The antibody response of animals was analyzed using Kruskal–Wallis's test, as these data did not follow a normal distribution. Statistical significance in all models was considered for p-values ≤ 0.05.

3. Results

3.1. PLGA-Encapsulated HisDTC Induces a Robust and Durable DTH Response

The PLGA polymeric nanoparticles prepared using the w/o/w solvent evaporation method had a regular morphology and a smooth surface. The particle sizes used for this study ranged from 243 to 287 nm, presenting a PDI of 0.09 ± 0.02. The Zeta potential of these formulations used in the present work had an average of −29 mV.

PLGA-encapsulated HisDTC vaccine formulations were evaluated for their ability to induce a DTH response after the immunization of BALB/c mice. These formulations included the multi-epitope peptide HisDTC with or without TLRLs (HisDTC-TLRLs-PLGA and HisDTC-PLGA groups, respectively). Additionally, the LetiFend® vaccine (LetiFend group) was used as a reference, enabling a comparative assessment of the vaccine's effectiveness against CanL. Control groups included the administration of encapsulated TLRL-2/3 (TLRLs-PLGA group) and empty PLGA nanoparticles (PLGA group). All vaccine strategies were administered twice at 14-day intervals via the subcutaneous route. The DTH reaction was studied as an indicator of a specific cell-mediated immune response against *Leishmania* induced by vaccination prior to infection. A short-term (four weeks after the last immunization) evaluation of the DTH response revealed that mice immunized with HisDTC (both HisDTC-TLRLs-PLGA and HisDTC-PLGA groups) exhibited significantly ($p < 0.01$) higher swelling rates compared to the rest of the groups (Figure 1). In contrast, mice vaccinated with LetiFend® and control groups (TLRLs-PLGA and PLGA) showed similar magnitudes of DTH response, with no statistical differences between them.

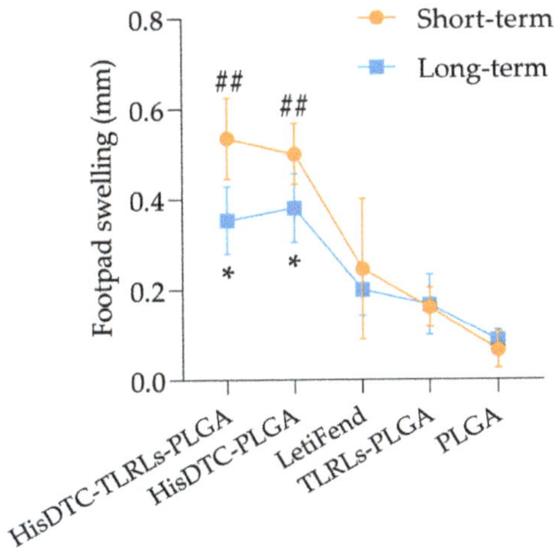

Figure 1. Delayed type of hypersensitivity response in immunized mice. BALB/c mice were immunized twice. Prior to infection, four weeks after the last immunization(short-term), and fourteen weeks post-immunization (long-term), DTH was assessed. After 48 h of SLA injection in the right footpad, inflammation was measured, and the difference with the control left footpad injected with PBS was calculated. Data are represented as the mean ± SEM. Statistical analysis was performed with one-way ANOVA followed by Turkey's post hoc test. Hashes indicate significant differences (##; $p < 0.01$) between immunized and control groups (TLRLs-PLGA and PLGA). Asterisks indicate differences (*; $p < 0.05$) between the immunized and PLGA control group.

Interestingly, for the long-term evaluation (fourteen weeks after the last immunization), HisDTC-vaccinated mice (with and without adjuvants) displayed milder DTH responses compared to the short-term evaluation but still demonstrated the most pronounced induration. They were the only groups that presented statistically significant ($p < 0.05$) differences compared to the PLGA control group (Figure 1).

3.2. *Immunization with HisDTC-Based PLGA Nanoformulations Confer Short and Long-Lasting Protection against L. infantum Infection*

To assess the efficacy and durability of protection conferred by the HisDTC peptide encapsulated in PLGA nanoparticles against *L. infantum* infection, immunized mice were infected with stationary phase *L. infantum* promastigotes either four weeks (short-term evaluation) or fourteen weeks (long-term evaluation) after the last immunization. Subsequently, seven weeks post-infection, the animals were euthanized, and parasite load quantification was performed in target organs using a limiting dilution assay.

For the short-term evaluation (Figure 2a), the control groups (PLGA and TLRLs-PLGA) exhibited the highest parasite loads in both the spleen and liver organs. Conversely, mice immunized with HisDTC, w/o TLRLs (HisDTC-TLRLs-PLGA and HisDTC-PLGA groups), as well as LetiFend-vaccinated animals, displayed a lower parasite load compared to both control groups. This decrease was statistically significant in mice immunized with the HisDTC peptide alone ($p < 0.05$). Specifically, of these two HisDTC immunized groups, HisDTC without immunomodulators was the most effective at controlling *L. infantum* multiplication, presenting the statistically significant lowest parasite burdens among all groups ($p < 0.01$).

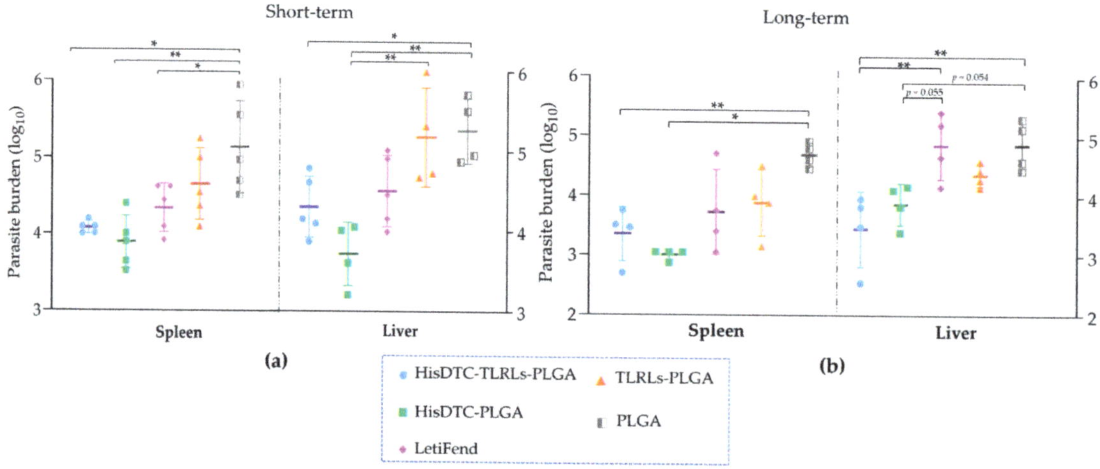

Figure 2. Splenic and liver parasite burden in immunized and infected BALB/c mice. BALB/c mice were infected with stationary phase *L. infantum* promastigotes four weeks after the last immunization in short-term vaccinated and evaluated mice (**a**) or fourteen weeks after the last immunization in long-term evaluated mice (**b**). Seven weeks post-infection, mice were sacrificed, and parasite burden was calculated by limiting the dilution assay in the spleen and liver. Data are represented as the mean ± SD. Statistical analysis was performed with two-way ANOVA followed by Tukey's test for multiple comparisons. Asterisks (*) indicate statistically significant differences (*; $p < 0.05$, **; $p < 0.01$) between groups.

Similarly, for the long-term evaluation (Figure 2b), both HisDTC-PLGA and HisDTC-TLRLs-PLGA groups were the only ones that maintained durable protection, exhibiting statistically significant ($p < 0.05$) lower parasite loads compared to the PLGA control group in both target organs and the LetiFend group in the liver. Surprisingly, the analysis revealed that the parasite burden displayed a similar load in LetiFend and both control groups (TLRLs-PLGA and PLGA) when long-term protection was evaluated.

3.3. *Multipeptide HisDTC Elicits a Robust and Durable Antigen-Specific T-Cell Immune Response*

The ability of PLGA-NP formulations to induce and activate a protective antigenic specific cellular immune response against *L. infantum* infection in vaccinated mice was assessed by quantifying Th1 (IFN-γ and TNF-α) and Th2 (IL-10) characteristic cytokine profiles. For this purpose, cytokine concentrations were measured using ELISA in co-culture supernatants of splenocytes from immunized and infected mice with SLA-stimulated naïve BMDCs for 96 h. For the short-term evaluation (Figure 3), all HisDTC-vaccinated mice exhibited a significant increase ($p < 0.01$) in pro-inflammatory IFN-γ and TNF-α cytokine production compared to other groups (Figure 3a). Notably, HisDTC-immunized mice showed a significant reduction ($p < 0.01$) in anti-inflammatory IL-10 cytokine production compared with the PLGA control group. In contrast, the LetiFend vaccinated group did not show significant differences in proinflammatory cytokine production when compared to the control groups (Figure 3a).

To assess the Th1/Th2 balance and overall immune response, cytokine ratios IFN-γ/IL-10 and TNF-α/IL-10 were calculated. HisDTC-immunized groups (HisDTC-TLRLs-PLGA and HisDTC-PLGA) exhibited the highest IFN-γ/IL-10 ratios ($p < 0.01$) compared to the rest of the groups (Figure 3b). Additionally, mice immunized with HisDTC without TLRs as adjuvants showed statistically significant differences ($p < 0.05$) in the TNF-α/IL-10 ratio compared to TLRLs-PLGA and PLGA control groups (Figure 3b). Despite the LetiFend and TLRLs-PLGA groups showing a statistically significant lower IL-10 production ($p < 0.01$) compared to the PLGA group (Figure 3a), no differences in cytokine ratios were observed among them (Figure 3b). These results pointed to the predominance of a Th1 immune response in the Th1/Th2 immunological balance, which is related to the better control of parasite multiplication in terms of the parasite load showed previously in all animals that were immunized with the HisDTC peptide encapsulated.

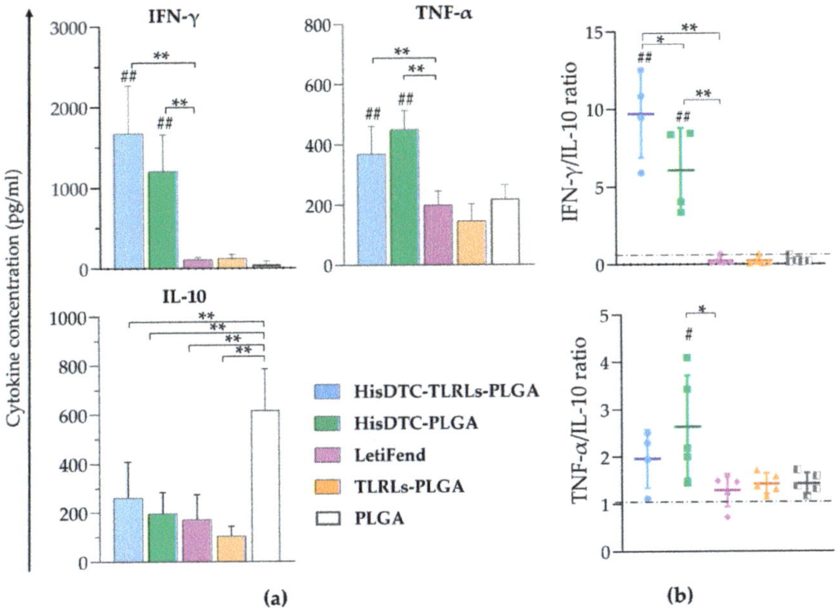

Figure 3. Cytokine production in splenocytes from immunized mice for the short-term evaluation. Spleen cells from short-term-evaluated vaccinated and infected animals were collected and co-cultured with naïve BMDCs stimulated with SLA. Levels of IFN-γ, TNF-α, and IL-10 were assessed using ELISA in culture supernatants. Panel (**a**) shows the mean ± SD of cytokine concentrations (pg/mL). Panel (**b**) shows the IFN-γ/IL-10 and TNF-α/IL-10 ratios. Statistical analysis was performed with one-way ANOVA followed by Tukey's test for multiple comparisons. Hashes indicate significant differences (#; $p < 0.05$; ##, $p < 0.01$) between immunized and control groups (TLRLs-PLGA and PLGA). Asterisks indicate statistically significant differences (*; $p < 0.05$, **; $p < 0.01$) between groups.

To investigate the underlying cause of the lasting protection observed in HisDTC-immunized mice for long-term evaluation, we examined their immunological responses. HisDTC-vaccinated groups consistently produced higher levels of IFN-γ and TNF-α cytokines and lower concentrations of the IL-10 cytokine in the co-cultured system compared to the other groups (Figure 4a). The IFN-γ/IL-10 and TNF-α/IL-10 ratios (Figure 4b), similar to the short-term evaluation, indicated a Th1-dominant response in mice vaccinated with the HisDTC peptide, which is associated with the effective control of *Leishmania* multiplication. These results highlight the capacity of the encapsulated multi-peptide HisDTC to promote a robust and durable Th1-dominant response, contributing to the sustained protection observed over time.

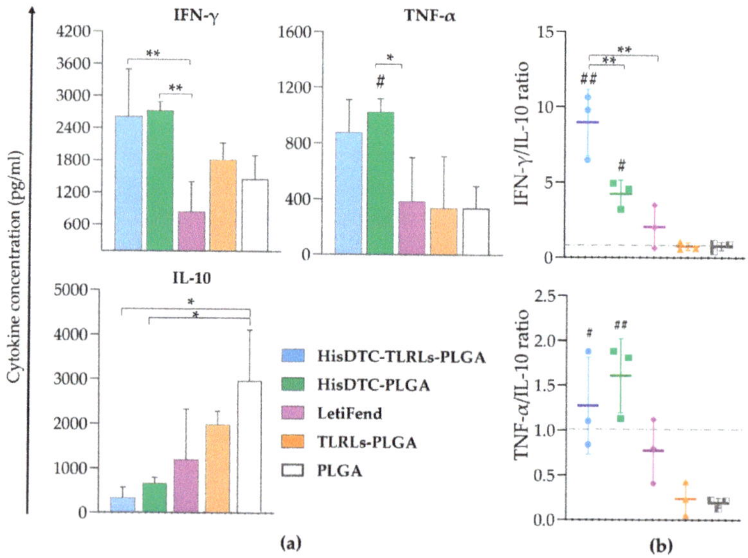

Figure 4. Cytokine production in splenocytes from immunized mice for long-term evaluation. Spleen cells from long-term-evaluated vaccinated and infected animals were collected and co-cultured with naïve BMDCs stimulated with SLA. Levels of IFN-γ, TNF-α, and IL-10 were assessed using ELISA in culture supernatants. Panel (**a**) shows the mean ± SD of cytokine concentrations (pg/mL). Panel (**b**) shows the IFN-γ/IL-10 and TNF-α/IL-10 ratios. Statistical analysis was performed with one-way ANOVA followed by Tukey's test for multiple comparisons. Hashes indicate significant differences (#; $p < 0.05$; ##; $p < 0.01$) between immunized and control groups (TLRLs-PLGA and PLGA). Asterisks indicate statistically significant differences (*; $p < 0.05$, **; $p < 0.01$) between groups.

3.4. HisDTC Nanoformulations Induce a Robust and Long-Term Th1 Memory Response Triggered by T CD8⁺ Lymphocytes

In some parasitic infectious diseases such as leishmaniosis or malaria, the efficacy of vaccines relies not only on the balance of cytokine production but also on the activation and induction of memory subsets of $CD4^+$ and $CD8^+$ T lymphocytes [32]. Thus, flow cytometry assays were conducted to assess the populations of T $CD4^+$ and $CD8^+$ cells producing IFN-γ and IL-10 as well as their respective memory lymphocyte subpopulations.

While no significant differences were observed in cytokine-producing $CD4^+$ T cells (data not shown), mice immunized with the chimera HisDTC-encapsulated w/o TLR adjuvants exhibited higher percentages of IFN-γ^+-producing $CD8^+$ T-lymphocytes compared to the rest of the groups, including the LetiFend group. Notably, statistically significant differences were observed only in the case of HisDTC nanoformulations ($p < 0.01$) for the short-term evaluation and in both HisDTC w/o adjuvant groups for the long-term evaluation (Figure 5). It is worth noting the increased presence of $CD8^+$ IFN-γ-producing T cells in HisDTC-vaccinated animals for the long-term evaluation was up to three-fold times higher compared to LetiFend and control groups.

An inverse relationship between the percentage of IFN-γ^+-producing T CD8 cells and IL-10^+-producing T $CD8^+$ cells was observed for both the short- and long-term evaluation (Figure 5). Consequently, HisDTC-vaccinated groups, characterized by the highest percentages of IFN-γ^+ T $CD8^+$ lymphocytes, exhibited fewer IL-10^+ T $CD8^+$ cells in contrast to the control and LetiFend groups. This difference was statistically significant during the long-term evaluation ($p < 0.01$). These findings highlight the ability of HisDTC encapsulated in PLGA w/o TLR as an adjuvant to promote a Th1-polarized response through the secretion of IFN-γ over IL-10, triggered by T $CD8^+$ lymphocytes against *L. infantum* infection in contrast to the control groups and even the LetiFend group.

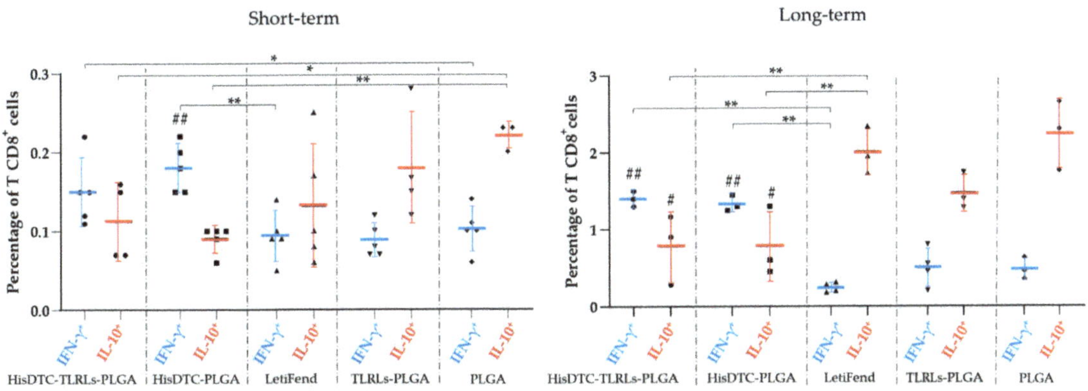

Figure 5. *L. infantum*-specific IFN-γ and IL-10-producing CD8⁺ T cells in immunized mice. BALB/c mice were immunized with two doses of different nanoformulations and the LetiFend® vaccine. Four weeks and fourteen weeks after the last immunization, animals were infected with *L. infantum* and euthanized seven weeks post-infection. Spleen cells were collected and stimulated with SLA. After 48 h, the frequency of $CD8^+$ T lymphocytes producing IFN-γ or IL-10 was assessed using the flow cytometry assay. Data show the mean ± SD of the percentage of Ag-specific $CD8^+$ T cells. Statistical analysis was performed with two-way ANOVA followed by Turkey's post hoc test. Hashes indicate significant differences (#; $p < 0.05$; ##; $p < 0.01$) between HisDTC-immunized and control groups (TLRLs-PLGA and PLGA). Asterisks indicate statistically significant differences (*; $p < 0.05$, **; $p < 0.01$) between groups.

In addition to cytokine-specific production, the induction of memory lymphocytes due to vaccination was also evaluated. The results revealed that, for the short-term evaluation, chimera HisDTC encapsulated w/o TRL as this adjuvant induced the highest percentages of central memory lymphocytes in both $CD4^+$ and $CD8^+$ cell subpopulations ($CD4^+$/$CD8^+$ $CD44^+$ $CD62L^+$) in infected mice when compared to the other groups. This was statistically significant ($p < 0.05$) when compared to the PLGA control group (Figure 6).

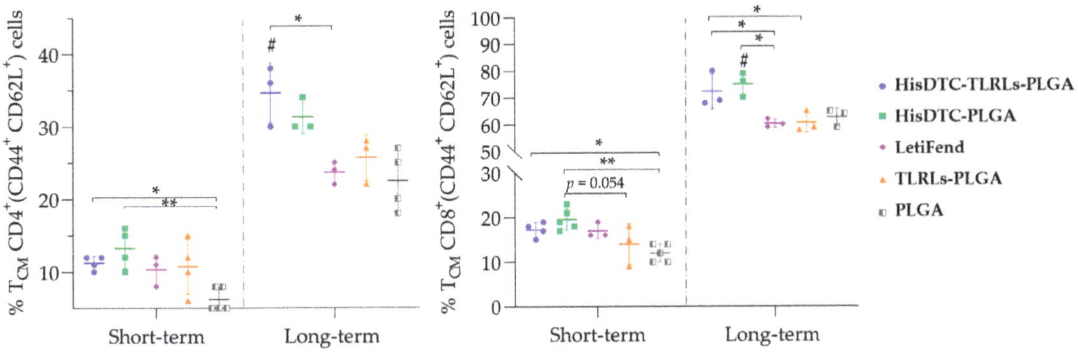

Figure 6. Frequencies of central memory T lymphocytes in immunized and infected mice. Spleen cells were previously stimulated for 12 h with SLA and analyzed via flow cytometry for immunophenotyping T cell memory based on CD62L and CD44 expression. The $CD62L^+$ $CD44^+$ central memory (T_{CM}) CD4 and CD8 T lymphocyte frequencies are depicted. Data represent the mean ± SD. Statistical analysis was performed with one-way ANOVA followed by Turkey's post hoc test. Hashes indicate significant differences (#; $p < 0.05$) between HisDTC-immunized and control groups (TLRLs-PLGA and PLGA). Asterisks indicate statistically significant differences (*; $p < 0.05$, **; $p < 0.01$) between groups.

To determine the long-lasting nature of these memory T cells and their implication in long-term protection against *L. infantum*, these memory cells were re-evaluated twenty-one weeks after the last immunization (long-term evaluation). Interestingly, HisDTC-TLRLs-PLGA and HisDTC-PLGA presented higher percentages of CD4$^+$ and CD8$^+$ memory T cells. It is noteworthy that control groups, as well as the LetiFend group, exhibited the same percentages in both CD4$^+$ and CD8$^+$ T cell subsets for the short- and long-term evaluation. These results may be correlated with the long-lasting capacity of HisDTC nanoformulations observed in vaccinated animals to inhibit *L. infantum* multiplications in target organs.

3.5. Vaccination with HisDTC Elicited Antigen-Specific Humoral Immune Responses after L. infantum Infection

After the successful control of parasite multiplication in vaccinated animals, we investigated the immune correlates of protection associated with immunoglobulin (Ig) production. Since the production of IgG2a antibodies is typically associated with a Th1-type immune response, whereas IgG1 is associated with a Th2-type immune response, we quantified the presence of anti-SLA antibodies for these two IgG isotypes in sera samples from immunized and infected mice seven weeks post-infection. As expected, for the short-term evaluation, the median anti-SLA IgG1 titer obtained by control groups (TLRs-PLGA and PLGA) was significantly higher ($p < 0.01$) than that obtained via immunization with encapsulated HisDTC (with and without adjuvants) (Figure 7). Interestingly, mice immunized with the LetiFend® vaccine exhibited a similar median IgG1 titer as the control groups and a statistically greater titer ($p < 0.01$) than mice immunized with HisDTC-encapsulated w/o adjuvants (Figure 7). Conversely, although no statistically significant differences were observed, the titer of IgG2a antibodies was higher in mice immunized with HisDTC and adjuvants (1:800), HisDTC alone (1:800), and LetiFend (1:800) compared to the control TLRLs-PLGA (1:400) and PLGA (1:200) groups.

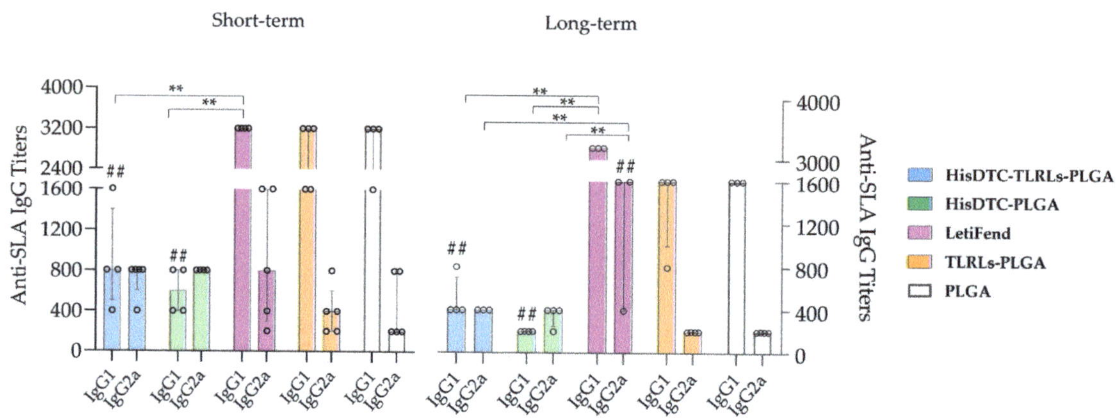

Figure 7. *Leishmania infantum*-specific IgG1 and IgG2a subclass antibody titers in immunized and infected mice. Serum samples were collected from all mice groups prior to sacrifice. Anti-SLA-specific IgG responses were measured via ELISA using an SLA-coated ELISA plate and HRP-conjugated antisera to detect IgG1 and IgG2a isotype levels in short- and long-term evaluations. Data are represented as the median and interquartile range. Hashes (##; $p < 0.01$; Kruskal–Wallis's test) denote a significant difference between median IgG levels in the immunized (HisDTC-TLRLs-PLGA, HisDTC-PLGA, and LetiFend) and control (TLRLs-PLGA and PLGA) groups. Asterisks denote significant differences between HisDTC- and LetiFend-immunized mice groups (**; $p < 0.01$; Kruskal–Wallis's test).

Consistent with previous results, antibody levels showed the same pattern for the long-term evaluation, with lower IgG1 titers ($p < 0.01$) in mice immunized with HisDTC compared to the LetiFend and control groups (Figure 7). The only notable difference observed was that LetiFend when evaluating long-term immunity, induced a statistically significant higher production ($p < 0.01$) of IgG2a antibodies compared to the rest of the groups. However, the only group that demonstrated the predominance of the IgG2a antibody over IgG1 was the group of mice vaccinated with HisDTC chimera without adjuvants, pointing to a greater capacity to effectively control the infection.

3.6. Immunization with Encapsulated HisDTC Induces a Protective L-Arginine Metabolic Pathway against Parasite Multiplication

The metabolism of the amino acid L-arginine, involving the enzymes nitric oxide synthetase (iNOS) and arginase, has significant implications for *Leishmania* infection in mice. iNOS generates NO from L-arginine, a potent leishmanicidal molecule that aids in parasite control. However, arginase synthesizes ornithine and polyamine products, promoting parasite survival [33]. Consequently, the balance of arginase activity and NO production is pivotal in determining resistance or susceptibility profiles to *Leishmania* infection. After 96 h of the co-culture of splenocytes from immunized animals and naïve BMDCs stimulated with SLA, these two enzymatic parameters were quantified.

The results obtained reveal that mice vaccinated with encapsulated peptide HisDTC, w/o TLRLs both in short- and long-term evaluations, exhibited higher NO production and lower arginase levels when compared to the control groups (TLRLs-PLGA and PLGA). Notably, this difference was statistically significant only in data obtained from the arginase activity measurement (Figure 8). These findings indicate that encapsulated HisDTC has the capability to induce, in vaccinated animals, L-arginine metabolism that favors the control of disease progression.

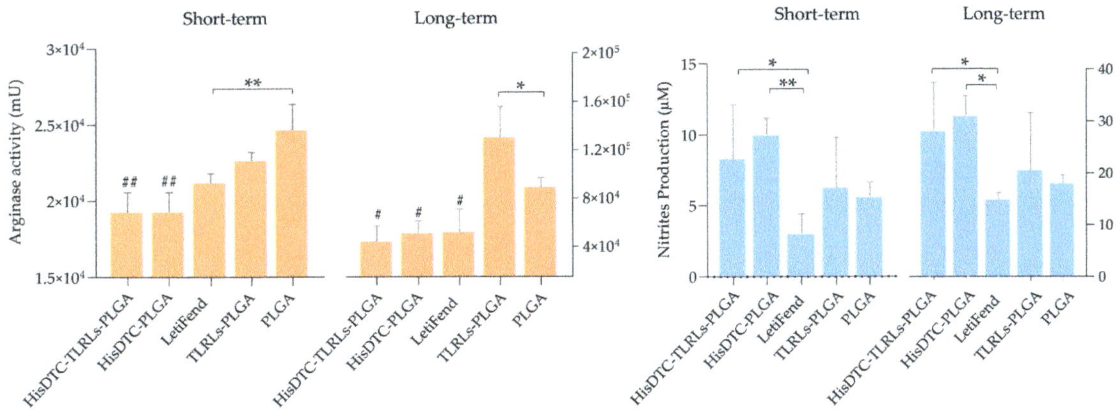

Figure 8. Arginine metabolism in naïve BMDCs co-cultured with splenocytes from vaccinated and infected mice. (**a**) The intracellular arginase activity (Mu) was estimated using the enzymatic-colorimetric assay method; (**b**) NO production was measured indirectly from the concentration of nitrites in supernatants using the Griess reaction. Data are represented as the mean ± SD. Hashes (one-way ANOVA: #; $p < 0.05$; ##; $p < 0.01$) denote a significant difference between mean levels in the immunized (HisDTC-TLRLs-PLGA, HisDTC-PLGA, and LetiFend) and control (TLRLs-PLGA and PLGA) groups. Asterisks (one-way ANOVA: *; $p < 0.05$; **; $p < 0.01$) denote significant differences between groups.

4. Discussion

CanL, caused by the protozoa *Leishmania infantum* species, poses significant clinical, epidemiological, and zoonotic challenges. The development of an effective preventive canine vaccine is crucial for its control [1]. Currently, commercially available vaccines such as Leishtec® (Hertape, Brazil) in Brazil and LetiFend® (LETI Pharma, S.L.U., Spain) in Europe exist but demonstrate limited efficacy against *L. infantum* infection in terms of disease progression and transmission [8]. These vaccines exhibit suboptimal efficacy against *L. infantum* infection, both in terms of disease progression and transmission, which is potentially attributed to the formulation of recombinant protein vaccines. Second-generation vaccines of this nature, while effective for other diseases, may encounter concerns related to low immunogenicity when administered without an adjuvant [10]. The successful development of an effective vaccine against leishmaniosis relies on the establishment of a specific and durable T-cell immune response, primarily driven by the activation of memory and effector T CD4$^+$ and CD8$^+$ lymphocytes [34,35]. To address this challenge, there is a growing exploration of peptide vaccines delivered through various platforms and adjuvants. [36,37]. Along this line, our group innovated by designing and developing a chimeric multi-epitope peptide named HisDTC. This peptide, whether administered alone or in combination with saponin, has demonstrated its capacity to induce a specific T-cell response, aiding immunized animals in controlling *L. infantum* multiplication in murine models [23,38]. Nevertheless, the optimization of this strategy is imperative to enhance the control of disease progression and ascertain the durability of the induced immunity.

In the present study, we conducted a comprehensive assessment of the short- and long-term immunoprophylactic efficacy of the multi-epitope peptide HisDTC. We explored its potential in the presence and absence of TLRL-2 and TLRL-3 and recognized immunomodulators co-encapsulated in PLGA polymeric nanoparticles in BALB/c mice. TLRL-2, a synthetic lipopeptide, has demonstrated its efficacy as a potent adjuvant for various vaccines, including those against leishmaniosis [18]. Similarly, TLRL-3, a synthetic analog of double-stranded RNA, has exhibited the ability to enhance a strong T-cell response in mice [16,17]. The HisDTC peptide, both with and without TLRLs, was encapsulated in widely used PLGA nanoparticles known for low cytotoxicity and efficient uptake by dendritic cells [39] as an essential aspect of promoting an effective cellular immune response against *Leishmania* spp. [40]. To assess the efficacy of these peptide-based nanovaccine strategies, we compared them with the LetiFend® vaccine, the only commercially available vaccine against CanL in Europe. Our goal was to evaluate which vaccine strategy elicited the most effective and long-lasting immune response under comparable conditions in the BALB/c model, providing valuable insights into their relative effectiveness.

Studying vaccine-induced immunogenicity prior to infection is a relevant aspect of assessing vaccine efficacy. The DTH response serves as an immunologic parameter, offering insights into cell-mediated immune responses associated with protection or sensibilization to *Leishmania* antigens [41]. To assess the immunogenicity of vaccine nanoformulations, immunized animals were exposed to *L. infantum*-soluble antigens before infection. The DTH reaction observed in mice immunized with HisDTC was significantly stronger and more prolonged compared to the control groups, indicating the development of a Th1-type immune response. In addition, the footpad depth swelling in the LetiFend group was lower than that of HisDTC-vaccinated animals but slightly higher than the control groups. Previous studies in the canine model of CanL revealed that, although the LetiFend® vaccine can induce a heightened DTH response in asymptomatic animals, this response is inconsistent and can vary among individuals [42]. These findings align with previously reported data, suggesting a correlation between the progression of VL and a diminished DTH response, indicating the absence or inadequate cell-mediated immunity [43].

To better analyze vaccine efficacy in controlling parasite multiplication, we examined the parasite burden in target organs, namely the spleen and liver, the primary organs affected by *L. infantum* infection [44]. For the short-term evaluation, subcutaneous immunization with encapsulated HisDTC without TLRLs demonstrated superior efficacy in reducing the parasite load significantly in both the liver and spleen, aligning with prior studies highlighting the protective efficacy of HisDTC against *L. infantum* without any adjuvant [38]. Notably, the HisDTC peptide exhibited better control than LetiFend®, a subunit vaccine lacking adjuvants and composed of protein Q, which is a polyprotein resulting from the combination of different antigenic fragments from four different *L. infantum* proteins (Lip2a, Lip2b, P0 and H2a) [43]. HisDTC, comprised specific T cell epitopes with high-MHC class-I- and class-II-binding affinity obtained from four *L. infantum* histones (H2a, H2b, H3, and H4) via bioinformatic analyses, leverages epitope prediction for peptide vaccine development, facilitating the selection of immunogenic epitopes that can promote a more specific and robust Th1 response against *Leishmania* [23,38,45]. The enhanced control capacity of HisDTC in controlling parasite multiplication may also be a consequence of the peptide delivery system. Both groups using the HisDTC peptide were encapsulated in PLGA NPs of around 250 nm in size. Previous research indicates that these polymeric nanoparticles are more efficiently recognized by antigen-presenting cells (APCs) when they are smaller than 500 nm [22]. Additionally, PLGA NPs not only shield the antigen from degradation and enzymatic activity but also offer sustained and prolonged antigen exposure via APCs, facilitating more efficient antigen uptake and presentation, resulting in stronger and more durable responses [46]. In accordance with this, our findings demonstrated the protective capacity in terms of the parasite load sustained over the long-term in HisDTC-immunized animals, a crucial aspect in *Leishmania* vaccine development. Even though parasite multiplication control is achieved in the vaccinated HisDTC group, total parasite clearance is not reached. Nevertheless, in parasitic vaccine development, rather than trying to achieve the complete clearance of the parasite, the goal should be minimizing or eliminating the pathological consequence of infections along with preventing disease transmission [47].

To further investigate the immunological parameters underlying the protective capacity of HisDTC against *Leishmania*, splenocytes and serum were collected from vaccinated and infected mice. The balance of Th1/Th2 immune responses, crucial for controlling *Leishmania* infections, was initially assessed. Cytokines, pivotal in T helper cells' function regulation, play a role in this balance, with increased production of Th1-associated cytokines, which are essential for infection control, while Th2-related cytokines, such as IL-10, are linked to susceptibility against *L. infantum* multiplication [48]. We focused on key pro-inflammatory cytokines, namely IFN-γ and TNF-α, and the anti-inflammatory cytokine IL-10 to comprehend differences in parasite burden in target organs. Cytokine production in the splenocyte/BMDC co-culture supernatant was analyzed for both short-term and long-term evaluations. Throughout the study, animals immunized with the encapsulated HisDTC peptide, either alone or with TLRLs, consistently induced higher amounts of both IFN-γ and TNF-α compared to other groups, alongside the lower production of the Th2-associated cytokine, IL-10. This pattern aligns with earlier findings for HisDTC, whether used alone or with saponin as an adjuvant in the murine model of VL [38]. No significant differences were observed between short- and long-term evaluation results or when comparing the HisDTC encapsulated peptide alone and in combination with TLRLs. TLRs are specialized in pathogen recognition, activating specific signaling pathways impacting pro-inflammatory and anti-inflammatory cytokine production, which is crucial for protection against *Leishmania* infection [49]. Despite the potential of TLR-based adjuvants to enhance vaccine efficacy [11,50], our results indicate that careful consideration is required due to potential complexities in their effects on the innate and adaptive immunity design [11,50]. Notably, both HisDTC-immunized groups exhibited a pattern that suggests a polarization toward the Th1-type immune response, correlating with the improved control of parasite multiplication in target organs. This Th1-type immune response bias was further supported by the Th1/Th2 characteristic cytokines ratio, highlighting the capacity of both vaccinated groups to induce Th1 predominance in both the short- and long-term evaluation.

In the context of cytokine production, the pivotal role of T $CD4^+$ Th1 cells in controlling *L. infantum* infections is well-established. These cells act as mediators by activating dendritic cells, subsequently inducing macrophage activation leading to parasite elimination [51–53]. Likewise, $CD8^+$-T cell responses have been recognized for their central role in conferring resistance against *Leishmania* spp. This immune response bias not only occurs during initial infection stages but also significantly contributes to resistance upon subsequent reinfection. $CD8^+$ T cells can exert cytotoxic activity, killing parasitized macrophages or synthesizing IFN-γ, activating macrophages to enhance their parasite-killing activities, or both [54–56]. To ascertain the involvement of T $CD4^+$ and $CD8^+$ cells in the observed protection against *L. infantum* in HisDTC-vaccinated groups, we evaluated IFN-γ and IL-10 intracellular cytokine production in these cell populations. Our results demonstrated a direct correlation between IFN-γ-producing T $CD8^+$ cells and the lowest parasite burden. This aligns with previous studies highlighting the importance of IFN-γ production by T $CD8^+$ lymphocytes in protection against *L. infantum* infection [57]. Regarding the percentage of T $CD8^+$ T cells producing IL-10 cytokine, several studies support the idea that the absence of IL-10 is correlated with a protective capacity against *L. infantum* infection in mice [58], humans [59], and canids [60]. In line with these findings, HisDTC-immunized animals showed a significantly lower percentage of IL-10^+-producing $CD8^+$ T cells and a preferential secretion of IFN-γ over IL-10 compared to the control and LetiFend groups. This suggests the crucial role of IL-10 in *L. infantum* protection.

One of the key aspects in assessing vaccine efficacy lies in its capacity to elicit and sustain a long-lasting immunological memory response. Effector T lymphocytes, responsible for the initial specific immune response against different pathogens, are complemented by memory T lymphocytes, offering a long-lasting immunity and a rapid response upon re-exposure [61]. In the case of *Leishmania* vaccine development, central memory T (T_{CM}) lymphocytes are especially noteworthy, playing a pivotal role in establishing and upholding durable immunity, even in the absence of parasites [62]. Nanoparticle delivery systems and TLRL adjuvants are instrumental in fostering the generation of memory T cells and long-lasting immunity against *Leishmania* [18,36,63]. Their ability to activate antigen-presenting cells in a controlled and sustained manner enhances their value in this context [64]. Our results underscore this, demonstrating that immunization with HisDTC-PLGA formulations led to an increase in the differentiation of both central memory T $CD4^+$ and $CD8^+$ cells in the spleens of immunized and infected mice. This is evident in the elevated percentage of the T memory cell subpopulation in the long-term evaluation compared to the short-term assessment. The sustained presence of these memory cell subpopulations over an extended period reinforced the notion that mice immunized with encapsulated HisDTC possess the capability to generate robust and enduring T-cell responses upon encountering the parasite with heightened protection. Contrasting the observed reduction in the parasite load in target organs of LetiFend-vaccinated animals, for the short-term evaluation, the LetiFend® vaccine fails to induce a robust and enduring memory T cell response. This is evident from the lack of impact on parasite loads and cytokine production during long-term vaccination (fourteen weeks after vaccination).

Regarding humoral responses, our results reveal that while encapsulated HisDTC did not induce significantly elevated IgG2a production in vaccinated animals, it did reduce IgG1 production compared to the control groups. Interestingly, immunization with HisDTC alone enhanced a higher IgG2a production over IgG1, highlighting the peptide's role in upregulating Th1-type responses, which correlated with a reduction in parasite burden. It is important to mention the discrepancy from previous results where the HisDTC peptide, administered alone or in combination with saponin as an adjuvant, induced a higher production of IgG2a subclass antibodies [38]. This variance may be explained by the use of PLGA nanoparticles, which, despite inducing a Th1 bias [65], did not result in a higher production of the IgG2a subclass immunoglobulin. Our findings concerning LetiFend groups align with previously described data, affirming the capacity of the LetiFend® vaccine to induce robust IgG2a subclass anti-SLA antibodies in immunized and infected animals [42].

Furthermore, the balance of Th1 immune cell responses associated with a resistant phenotype that allows the control of the disease has been associated with the activation of antigen-presenting cells and the modification of the arginine metabolism. This modification leads to a higher production of NO to the detriment of the activation of the arginase enzyme [66]. In this regard, it can be observed how vaccination strategies capable of inducing better control of parasite multiplication (HisDTC-based vaccines) were the ones capable of inducing the robust production of Th1-characteristic cytokines. IFN-γ can modulate L-arginine metabolism, leading to the control of disease progression. These data are in correlation with previously published findings where the HisDTC peptide, both alone and in combination with saponin, demonstrated the ability to induce parasite control in infected animals due to arginine metabolism modulation [38]. On the other hand, LetiFend and control groups produced higher amounts of IL-10 cytokine, leading to lower iNOS activity, thereby allowing parasite multiplication in target organs as evidenced by parasite loads.

Despite the considerable interest in TLR agonists, specifically TLRL2 and TLRL3, as potential adjuvants in vaccine development, our study revealed a lack of efficacy for these agonists. No significant differences were observed between the groups administered with HisDTC-TLRLs-PLGA and HisDTC-PLGA, indicating that the use of TLR2 (Pam3CSK4) and TLR3 (Poly(I:C)) agonists together did not enhance the synthesis of Th1-type cytokines, including IFN-γ and TNF-α. This finding is unexpected, given the interest in TLR agonists for their capacity to promote a Th1 profile immune response. The role of TLR agonists in *L. infantum* infection is relatively understudied, with most existing data based on *L. major* animal models [67–69]. Our results indicate that the combined administration of TLR2 and TLR3 agonists does not significantly impact Th1-type cytokine synthesis. However, there is some immune modulation, reflected in the lower concentration of the IL-10 cytokine both extracellularly and intracellularly in CD8$^+$ T cells, particularly during long-term evaluations. This modulation may account for the slight reduction in parasite loads observed in the TLRL-PLGA group compared to the PLGA control group for the long-term evaluation, highlighting the potential contribution of IL-10's reduction to the successful reduction in parasite burden, as previously described [48,59]. Collectively, our data suggest that the simultaneous administration of TLR2 and TLR3 agonists does not enhance protection against *L. infantum* infection or the immunogenicity induced by the HisDTC peptide. No significant differences were observed in parasite burden, memory T cells, or IgG titers between groups vaccinated with the peptide with or without TLR2/3 agonists. These results are in accordance with a previous study demonstrating that a TLR2 agonist is unable to contribute to protection against *L. infantum* infection in an in vivo murine model [70]. Further investigation of alternative TLR agonists or combinations is warranted in the search for suitable vaccine adjuvants against *L. infantum*.

The results presented in this study, combined with previously reported data, strongly indicate that the multi-epitope HisDTC is a promising tool in combating CanL. Moreover, the use of the HisDTC peptide encapsulated in PLGA nanoparticles, and w/o the presence of adjuvants, enhances a resistant and long-lasting T cell immune response characterized by robust DTH responses, high pro-inflammatory cytokines levels, and central memory T cells. These immune responses correlate with a reduction in the parasite load in the spleen and liver.

5. Conclusions

The present study shows the robust and enduring immunoprophylactic efficacy of the chimeric multi-epitope peptide HisDTC encapsulated in PLGA nanoparticles against *L. infantum* infection in the BALB/c murine model. This efficacy is characterized by the establishment of a Th1-type immune response in terms of the vaccine-induced delayed-type hypersensitivity response, the production of IFN-γ and TNF-α, and low IgG1 production. Notably, the activation of IFN-γ-producing T CD8$^+$ cells emerged as a pivotal factor in controlling parasite multiplication and promoting the expansion of central memory T cells, contributing to long-lasting immunity. Our study highlights the importance of conducting both short- and long-term evaluation experiments in order to analyze the establishment of a durable and specific immune response, a crucial requirement for the continued exploration of any vaccine candidate. Additionally, the novel use of HisDTC-encapsulated w/o TLR2 and TLR3 agonists demonstrated no significant differences among the vaccinated animals, prompting questions about the efficacy of TLR2 and TLR3 agonists as effective adjuvants for vaccine development against *L. infantum*. These findings open avenues for further research in the field of TLRLs used as adjuvants, considering their potential variability when employed with different delivery systems or against different pathogens, as observed in our study. This study underscores the promising prospects of the HisDTC nanovaccine as a preventive strategy against CanL, emphasizing the importance of PLGA nanoparticles as a promising choice for delivering antigens when strong and lasting T-cell responses. However, for practical application, further studies are imperative to validate the safety and efficacy of this vaccination strategy in dogs, the primary target for protection against this complex infectious disease. Consequently, this study establishes a robust foundation for future research in the ongoing quest to effectively combat CanL.

6. Patents

The HisDTC peptide is covered by a patent (EP4163303A1). This patent is the property of the Universidad Complutense de Madrid, Spain.

Author Contributions: Conceptualization, A.M.-R., A.M. and G.D.-B.; Formal analysis, C.H.-M., A.M.-R. and G.D.-B.; Funding acquisition, G.D.-B.; Investigation, C.H.-M., A.M. and G.D.-B.; Methodology, C.H.-M., A.M.-R. and L.d.U.-F.; Project administration, G.D.-B.; Supervision, A.M.-R., A.M. and G.D.-B.; Writing—original draft, C.H.-M. and A.M.-R.; Writing—review and editing, A.M.-R., J.A.O., A.M. and G.D.-B. All authors have read and agreed to the published version of the manuscript.

Funding: This research was partially supported by grants from the Spanish Government project PID2019-106487RB-I00/AEI/10.13039/501100011033, and from the Comunidad de Madrid (Spain) PLATESA2-CM (S2018/BAA-4370). Alicia Mas (A.M.) was supported by a postdoctoral fellowship (Margarita Salas) from European Union funds.

Institutional Review Board Statement: The study was approved by the Animal Welfare Committee of the Community of Madrid, Spain, (reference: PROEX 211/18) following Spanish and EU legislation (Law 32/2007, R.D. 53/2013, and Council Directive 2010/63/EU).

Informed Consent Statement: Not applicable.

Data Availability Statement: The raw data supporting the conclusions of this article will be made available by the authors on request.

Conflicts of Interest: Clara Hurtado-Morillas and Laura de Urbina-Fuentes declare no conflicts of interest. Abel Martínez-Rodrigo, Alicia Mas, José A. Orden, and Gustavo Domínguez-Bernal are co-inventors of the patent (EP4163303A1). The funders had no role in the design of the study; in the collection, analyses, or interpretation of data; in the writing of the manuscript; or in the decision to publish the results.

References

1. Morales-Yuste, M.; Martín-Sánchez, J.; Corpas-Lopez, V. Canine Leishmaniasis: Update on Epidemiology, Diagnosis, Treatment, and Prevention. *Vet. Sci.* **2022**, *9*, 387. [CrossRef]
2. Rocha, R.; Pereira, A.; Maia, C. A global perspective on non-autochthonous canine and feline Leishmania infection and leishmaniosis in the 21st century. *Acta Trop.* **2023**, *237*, 106710. [CrossRef]
3. Dantas-Torres, F.; Solano-Gallego, L.; Baneth, G.; Ribeiro, V.M.; de Paiva-Cavalcanti, M.; Otranto, D. Canine leishmaniosis in the Old and New Worlds: Unveiled similarities and differences. *Trends Parasitol.* **2012**, *28*, 531–538. [CrossRef]
4. WHO. Leishmaniasis. Available online: https://www.who.int/es/news-room/fact-sheets/detail/leishmaniasis (accessed on 19 June 2023).
5. Manna, L.; Reale, S.; Vitale, F.; Picillo, E.; Pavone, L.M.; Gravino, A.E. Real-time PCR assay in Leishmania-infected dogs treated with meglumine antimoniate and allopurinol. *Vet. J.* **2008**, *177*, 279–282. [CrossRef]
6. Segarra, S.; Miro, G.; Montoya, A.; Pardo-Marin, L.; Boque, N.; Ferrer, L.; Ceron, J. Randomized, allopurinol-controlled trial of the effects of dietary nucleotides and active hexose correlated compound in the treatment of canine leishmaniosis. *Vet. Parasitol.* **2017**, *239*, 50–56. [CrossRef]
7. Agency, E.M. Positive opinion of the Committee for veterinary Medicinal Products for Neoleish. Available online: https://www.ema.europa.eu/en/medicines/veterinary/summaries-opinion/neoleish (accessed on 19 June 2023).

8. Velez, R.; Gallego, M. Commercially approved vaccines for canine leishmaniosis: A review of available data on their safety and efficacy. *Trop. Med. Int. Health* **2020**, *25*, 540–557. [CrossRef] [PubMed]
9. Calzetta, L.; Pistocchini, E.; Ritondo, B.L.; Roncada, P.; Palma, E.; di Cave, D.; Mattei, M.; Britti, D. Immunoprophylaxis pharmacotherapy against canine leishmaniosis: A systematic review and meta-analysis on the efficacy of vaccines approved in European Union. *Vaccine* **2020**, *38*, 6695–6703. [CrossRef] [PubMed]
10. De Brito, R.C.F.; Cardoso, J.M.O.; Reis, L.E.S.; Vieira, J.F.; Mathias, F.A.S.; Roatt, B.M.; Aguiar-Soares, R.; Ruiz, J.C.; Resende, D.M.; Reis, A.B. Peptide Vaccines for Leishmaniasis. *Front. Immunol.* **2018**, *9*, 1043. [CrossRef] [PubMed]
11. Yang, J.X.; Tseng, J.C.; Yu, G.Y.; Luo, Y.; Huang, C.F.; Hong, Y.R.; Chuang, T.H. Recent Advances in the Development of Toll-like Receptor Agonist-Based Vaccine Adjuvants for Infectious Diseases. *Pharmaceutics* **2022**, *14*, 423. [CrossRef]
12. Schaut, R.G.; Grinnage-Pulley, T.L.; Esch, K.J.; Toepp, A.J.; Duthie, M.S.; Howard, R.F.; Reed, S.G.; Petersen, C.A. Recovery of antigen-specific T cell responses from dogs infected with Leishmania (L.) infantum by use of vaccine associated TLR-agonist adjuvant. *Vaccine* **2016**, *34*, 5225–5234. [CrossRef] [PubMed]
13. Garcon, N.; Morel, S.; Didierlaurent, A.; Descamps, D.; Wettendorff, M.; Van Mechelen, M. Development of an AS04-adjuvanted HPV vaccine with the adjuvant system approach. *BioDrugs* **2011**, *25*, 217–226. [CrossRef] [PubMed]
14. Garçon, N.; Van Mechelen, M. Recent clinical experience with vaccines using MPL- and QS-21-containing adjuvant systems. *Expert. Rev. Vaccines* **2011**, *10*, 471–486. [CrossRef] [PubMed]
15. Didierlaurent, A.M.; Laupèze, B.; Di Pasquale, A.; Hergli, N.; Collignon, C.; Garçon, N. Adjuvant system AS01: Helping to overcome the challenges of modern vaccines. *Expert. Rev. Vaccines* **2017**, *16*, 55–63. [CrossRef] [PubMed]
16. Celis, E. Toll-like receptor ligands energize peptide vaccines through multiple paths. *Cancer Res.* **2007**, *67*, 7945–7947. [CrossRef]
17. Fransen, F.; Boog, C.J.; van Putten, J.P.; van der Ley, P. Agonists of Toll-like receptors 3, 4, 7, and 9 are candidates for use as adjuvants in an outer membrane vaccine against Neisseria meningitidis serogroup B. *Infect. Immun.* **2007**, *75*, 5939–5946. [CrossRef]
18. Jayakumar, A.; Castilho, T.M.; Park, E.; Goldsmith-Pestana, K.; Blackwell, J.M.; McMahon-Pratt, D. TLR1/2 activation during heterologous prime-boost vaccination (DNA-MVA) enhances CD8+ T Cell responses providing protection against Leishmania (Viannia). *PLoS Negl. Trop. Dis.* **2011**, *5*, e1204. [CrossRef] [PubMed]
19. Thomas, S.J.; Moreira, E.D., Jr.; Kitchin, N.; Absalon, J.; Gurtman, A.; Lockhart, S.; Perez, J.L.; Pérez Marc, G.; Polack, F.P.; Zerbini, C.; et al. Safety and Efficacy of the BNT162b2 mRNA COVID-19 Vaccine through 6 Months. *N. Engl. J. Med.* **2021**, *385*, 1761–1773. [CrossRef]
20. Prasanna, P.; Kumar, P.; Kumar, S.; Rajana, V.K.; Kant, V.; Prasad, S.R.; Mohan, U.; Ravichandiran, V.; Mandal, D. Current status of nanoscale drug delivery and the future of nano-vaccine development for leishmaniasis—A review. *Biomed. Pharmacother.* **2021**, *141*, 111920. [CrossRef]
21. Arevalo, C.P.; Bolton, M.J.; Le Sage, V.; Ye, N.; Furey, C.; Muramatsu, H.; Alameh, M.G.; Pardi, N.; Drapeau, E.M.; Parkhouse, K.; et al. A multivalent nucleoside-modified mRNA vaccine against all known influenza virus subtypes. *Science* **2022**, *378*, 899–904. [CrossRef]
22. Silva, A.L.; Rosalia, R.A.; Varypataki, E.; Sibuea, S.; Ossendorp, F.; Jiskoot, W. Poly-(lactic-co-glycolic-acid)-based particulate vaccines: Particle uptake by dendritic cells is a key parameter for immune activation. *Vaccine* **2015**, *33*, 847–854. [CrossRef]
23. Martínez-Rodrigo, A.; Mas, A.; Álvarez-Campos, D.; Orden, J.A.; Domínguez-Bernal, G.; Carrión, J. Epitope Selection for Fighting Visceral Leishmaniosis: Not All Peptides Function the Same Way. *Vaccines* **2020**, *8*, 352. [CrossRef]
24. Carrion, J.; Abengozar, M.A.; Fernandez-Reyes, M.; Sanchez-Martin, C.; Rial, E.; Dominguez-Bernal, G.; Gonzalez-Barroso, M.M. UCP2 deficiency helps to restrict the pathogenesis of experimental cutaneous and visceral leishmaniosis in mice. *PLoS Negl. Trop. Dis.* **2013**, *7*, e2077. [CrossRef]
25. Moreira, D.; Santarém, N.; Loureiro, I.; Tavares, J.; Silva, A.M.; Amorim, A.M.; Ouaissi, A.; Cordeiro-da-Silva, A.; Silvestre, R. Impact of continuous axenic cultivation in Leishmania infantum virulence. *PLoS Negl. Trop. Dis.* **2012**, *6*, e1469. [CrossRef]
26. Scott, P.; Pearce, E.; Natovitz, P.; Sher, A. Vaccination against cutaneous leishmaniasis in a murine model. I. Induction of protective immunity with a soluble extract of promastigotes. *J. Immunol.* **1987**, *139*, 221–227. [CrossRef]
27. Buffet, P.A.; Sulahian, A.; Garin, Y.J.; Nassar, N.; Derouin, F. Culture microtitration: A sensitive method for quantifying Leishmania infantum in tissues of infected mice. *Antimicrob. Agents Chemother.* **1995**, *39*, 2167–2168. [CrossRef]
28. Lutz, M.B.; Kukutsch, N.; Ogilvie, A.L.; Rössner, S.; Koch, F.; Romani, N.; Schuler, G. An advanced culture method for generating large quantities of highly pure dendritic cells from mouse bone marrow. *J. Immunol. Methods* **1999**, *223*, 77–92. [CrossRef]
29. Ding, A.H.; Nathan, C.F.; Stuehr, D.J. Release of reactive nitrogen intermediates and reactive oxygen intermediates from mouse peritoneal macrophages. Comparison of activating cytokines and evidence for independent production. *J. Immunol.* **1988**, *141*, 2407–2412. [CrossRef]
30. Garrido, V.V.; Dulgerian, L.R.; Stempin, C.C.; Cerbán, F.M. The increase in mannose receptor recycling favors arginase induction and Trypanosoma cruzi survival in macrophages. *Int. J. Biol. Sci.* **2011**, *7*, 1257–1272. [CrossRef] [PubMed]
31. Dominguez-Bernal, G.; Martinez-Rodrigo, A.; Mas, A.; Blanco, M.M.; Orden, J.A.; De La Fuente, R.; Carrion, J. Alternative strategy for visceral leishmaniosis control: HisAK70-Salmonella Choleraesuis-pulsed dendritic cells. *Comp. Immunol. Microbiol. Infect. Dis.* **2017**, *54*, 13–19. [CrossRef] [PubMed]
32. Scott, P. Immunoparasitology. *Immunol. Rev.* **2004**, *201*, 5–8. [CrossRef] [PubMed]

33. Wanasen, N.; Soong, L. L-arginine metabolism and its impact on host immunity against Leishmania infection. *Immunol. Res.* **2008**, *41*, 15–25. [CrossRef]
34. Giunchetti, R.C.; Silveira, P.; Resende, L.A.; Leite, J.C.; Melo-Junior, O.A.O.; Rodrigues-Alves, M.L.; Costa, L.M.; Lair, D.F.; Chaves, V.R.; Soares, I.D.S.; et al. Canine visceral leishmaniasis biomarkers and their employment in vaccines. *Vet. Parasitol.* **2019**, *271*, 87–97. [CrossRef]
35. Sabur, A.; Bhowmick, S.; Chhajer, R.; Ejazi, S.A.; Didwania, N.; Asad, M.; Bhattacharyya, A.; Sinha, U.; Ali, N. Liposomal Elongation Factor-1α Triggers Effector CD4 and CD8 T Cells for Induction of Long-Lasting Protective Immunity against Visceral Leishmaniasis. *Front. Immunol.* **2018**, *9*, 18. [CrossRef]
36. Agallou, M.; Margaroni, M.; Kotsakis, S.D.; Karagouni, E. A Canine-Directed Chimeric Multi-Epitope Vaccine Induced Protective Immune Responses in BALB/c Mice Infected with Leishmania infantum. *Vaccines* **2020**, *8*, 350. [CrossRef]
37. Agallou, M.; Margaroni, M.; Tsanaktsidou, E.; Badounas, F.; Kammona, O.; Kiparissides, C.; Karagouni, E. A liposomal vaccine promotes strong adaptive immune responses via dendritic cell activation in draining lymph nodes. *J. Control. Release Off. J. Control. Release Soc.* **2023**, *356*, 386–401. [CrossRef] [PubMed]
38. Mas, A.; Hurtado-Morillas, C.; Martínez-Rodrigo, A.; Orden, J.A.; de la Fuente, R.; Domínguez-Bernal, G.; Carrión, J. A Tailored Approach to Leishmaniases Vaccination: Comparative Evaluation of the Efficacy and Cross-Protection Capacity of DNA vs. Peptide-Based Vaccines in a Murine Model. *Int. J. Mol. Sci.* **2023**, *24*, 12334. [CrossRef] [PubMed]
39. Margaroni, M.; Agallou, M.; Kontonikola, K.; Karidi, K.; Kammona, O.; Kiparissides, C.; Gaitanaki, C.; Karagouni, E. PLGA nanoparticles modified with a TNFα mimicking peptide, soluble Leishmania antigens and MPLA induce T cell priming in vitro via dendritic cell functional differentiation. *Eur. J. Pharm. Biopharm.* **2016**, *105*, 18–31. [CrossRef] [PubMed]
40. Soong, L. Modulation of dendritic cell function by Leishmania parasites. *J. Immunol.* **2008**, *180*, 4355–4360. [CrossRef] [PubMed]
41. De Rossell, R.A.; Bray, R.S.; Alexander, J. The correlation between delayed hypersensitivity, lymphocyte activation and protective immunity in experimental murine leishmaniasis. *Parasite Immunol.* **1987**, *9*, 105–115. [CrossRef] [PubMed]
42. Carcelén, J.; Iniesta, V.; Fernández-Cotrina, J.; Serrano, F.; Parejo, J.C.; Corraliza, I.; Gallardo-Soler, A.; Marañón, F.; Soto, M.; Alonso, C.; et al. The chimerical multi-component Q protein from Leishmania in the absence of adjuvant protects dogs against an experimental Leishmania infantum infection. *Vaccine* **2009**, *27*, 5964–5973. [CrossRef] [PubMed]
43. Pacheco-Fernandez, T.; Volpedo, G.; Gannavaram, S.; Bhattacharya, P.; Dey, R.; Satoskar, A.; Matlashewski, G.; Nakhasi, H.L. Revival of Leishmanization and Leishmanin. *Front. Cell Infect. Microbiol.* **2021**, *11*, 639801. [CrossRef]
44. Loeuillet, C.; Bañuls, A.L.; Hide, M. Study of Leishmania pathogenesis in mice: Experimental considerations. *Parasit. Vectors* **2016**, *9*, 144. [CrossRef]
45. Agallou, M.; Athanasiou, E.; Koutsoni, O.; Dotsika, E.; Karagouni, E. Experimental Validation of Multi-Epitope Peptides Including Promising MHC Class I- and II-Restricted Epitopes of Four Known Leishmania infantum Proteins. *Front. Immunol.* **2014**, *5*, 268. [CrossRef]
46. Cruz, L.J.; Tacken, P.J.; Eich, C.; Rueda, F.; Torensma, R.; Figdor, C.G. Controlled release of antigen and Toll-like receptor ligands from PLGA nanoparticles enhances immunogenicity. *Nanomedicine* **2017**, *12*, 491–510. [CrossRef]
47. Donnelly, S. The immunology of parasite infections: Grand challenges. *Front. Parasitol.* **2022**, *1*, 1069205. [CrossRef]
48. Kaye, P.M.; Cruz, I.; Picado, A.; Van Bocxlaer, K.; Croft, S.L. Leishmaniasis immunopathology-impact on design and use of vaccines, diagnostics and drugs. *Semin. Immunopathol.* **2020**, *42*, 247–264. [CrossRef]
49. Martinez-Orellana, P.; Quirola-Amores, P.; Montserrat-Sangra, S.; Ordeix, L.; Llull, J.; Alvarez-Fernandez, A.; Solano-Gallego, L. The inflammatory cytokine effect of Pam3CSK4 TLR2 agonist alone or in combination with Leishmania infantum antigen on ex-vivo whole blood from sick and resistant dogs. *Parasit. Vectors* **2017**, *10*, 123. [CrossRef] [PubMed]
50. Lu, B.L.; Williams, G.M.; Brimble, M.A. TLR2 agonists and their structure-activity relationships. *Org. Biomol. Chem.* **2020**, *18*, 5073–5094. [CrossRef] [PubMed]
51. Romano, A.; Doria, N.A.; Mendez, J.; Sacks, D.L.; Peters, N.C. Cutaneous Infection with Leishmania major Mediates Heterologous Protection against Visceral Infection with Leishmania infantum. *J. Immunol.* **2015**, *195*, 3816–3827. [CrossRef] [PubMed]
52. Coler, R.N.; Goto, Y.; Bogatzki, L.; Raman, V.; Reed, S.G. Leish-111f, a recombinant polyprotein vaccine that protects against visceral Leishmaniasis by elicitation of CD4+ T cells. *Infect. Immun.* **2007**, *75*, 4648–4654. [CrossRef] [PubMed]
53. Hohman, L.S.; Peters, N.C. CD4(+) T Cell-Mediated Immunity against the Phagosomal Pathogen Leishmania: Implications for Vaccination. *Trends Parasitol.* **2019**, *35*, 423–435. [CrossRef]
54. Stäger, S.; Rafati, S. CD8(+) T cells in leishmania infections: Friends or foes? *Front. Immunol.* **2012**, *3*, 5. [CrossRef] [PubMed]
55. Basu, R.; Bhaumik, S.; Haldar, A.K.; Naskar, K.; De, T.; Dana, S.K.; Walden, P.; Roy, S. Hybrid cell vaccination resolves Leishmania donovani infection by eliciting a strong CD8+ cytotoxic T-lymphocyte response with concomitant suppression of interleukin-10 (IL-10) but not IL-4 or IL-13. *Infect. Immun.* **2007**, *75*, 5956–5966. [CrossRef]
56. Polley, R.; Stager, S.; Prickett, S.; Maroof, A.; Zubairi, S.; Smith, D.F.; Kaye, P.M. Adoptive immunotherapy against experimental visceral leishmaniasis with CD8+ T cells requires the presence of cognate antigen. *Infect. Immun.* **2006**, *74*, 773–776. [CrossRef] [PubMed]
57. Athanasiou, E.; Agallou, M.; Tastsoglou, S.; Kammona, O.; Hatzigeorgiou, A.; Kiparissides, C.; Karagouni, E. A Poly(Lactic-co-Glycolic) Acid Nanovaccine Based on Chimeric Peptides from Different Leishmania infantum Proteins Induces Dendritic Cells Maturation and Promotes Peptide-Specific IFNγ-Producing CD8(+) T Cells Essential for the Protection against Experimental Visceral Leishmaniasis. *Front. Immunol.* **2017**, *8*, 684. [CrossRef] [PubMed]

58. Murphy, M.L.; Wille, U.; Villegas, E.N.; Hunter, C.A.; Farrell, J.P. IL-10 mediates susceptibility to Leishmania donovani infection. *Eur. J. Immunol.* **2001**, *31*, 2848–2856. [CrossRef] [PubMed]
59. Vouldoukis, I.; Bécherel, P.A.; Riveros-Moreno, V.; Arock, M.; da Silva, O.; Debré, P.; Mazier, D.; Mossalayi, M.D. Interleukin-10 and interleukin-4 inhibit intracellular killing of Leishmania infantum and Leishmania major by human macrophages by decreasing nitric oxide generation. *Eur. J. Immunol.* **1997**, *27*, 860–865. [CrossRef]
60. de Oliveira Cardoso, J.M.; de Brito, R.C.F.; Costa, A.F.P.; Siqueira Mathias, F.A.; Soares Reis, L.E.; Vieira, J.F.P.; de Oliveira Aguiar Soares, R.D.; Reis, A.B.; Roatt, B.M. IL-10 receptor blockade controls the in vitro infectivity of Leishmania infantum and promotes a Th1 activation in PBMC of dogs with visceral leishmaniasis. *Mol. Immunol.* **2021**, *137*, 20–27. [CrossRef]
61. Ojdana, D.; Safiejko, K.; Lipska, A.; Radziwon, P.; Dadan, J.; Tryniszewska, E. Effector and memory CD4+ and CD8+ T cells in the chronic infection process. *Folia Histochem. Cytobiol.* **2008**, *46*, 413–417. [CrossRef]
62. Zaph, C.; Uzonna, J.; Beverley, S.M.; Scott, P. Central memory T cells mediate long-term immunity to Leishmania major in the absence of persistent parasites. *Nat. Med.* **2004**, *10*, 1104–1110. [CrossRef]
63. Agallou, M.; Margaroni, M.; Karagouni, E. Intramuscular Immunization with a Liposomal Multi-Epitope Chimeric Protein Induces Strong Cellular Immune Responses against Visceral Leishmaniasis. *Vaccines* **2023**, *11*, 1384. [CrossRef]
64. Demento, S.L.; Cui, W.; Criscione, J.M.; Stern, E.; Tulipan, J.; Kaech, S.M.; Fahmy, T.M. Role of sustained antigen release from nanoparticle vaccines in shaping the T cell memory phenotype. *Biomaterials* **2012**, *33*, 4957–4964. [CrossRef]
65. Gu, P.; Wusiman, A.; Wang, S.; Zhang, Y.; Liu, Z.; Hu, Y.; Liu, J.; Wang, D. Polyethylenimine-coated PLGA nanoparticles-encapsulated Angelica sinensis polysaccharide as an adjuvant to enhance immune responses. *Carbohydr. Polym.* **2019**, *223*, 115128. [CrossRef]
66. Martinez-Rodrigo, A.; D, S.D.; Ribeiro, P.A.F.; Roatt, B.M.; Mas, A.; Carrion, J.; Coelho, E.A.F.; Dominguez-Bernal, G. Immunization with the HisAK70 DNA Vaccine Induces Resistance against Leishmania Amazonensis Infection in BALB/c Mice. *Vaccines* **2019**, *7*, 183. [CrossRef]
67. Huang, L.; Hinchman, M.; Mendez, S. Coinjection with TLR2 agonist Pam3CSK4 reduces the pathology of leishmanization in mice. *PLoS Negl. Trop. Dis.* **2015**, *9*, e0003546. [CrossRef] [PubMed]
68. Halliday, A.; Turner, J.D.; Guimarães, A.; Bates, P.A.; Taylor, M.J. The TLR2/6 ligand PAM2CSK4 is a Th2 polarizing adjuvant in Leishmania major and Brugia malayi murine vaccine models. *Parasit. Vectors* **2016**, *9*, 96. [CrossRef] [PubMed]
69. Pandey, S.P.; Chandel, H.S.; Srivastava, S.; Selvaraj, S.; Jha, M.K.; Shukla, D.; Ebensen, T.; Guzman, C.A.; Saha, B. Pegylated bisacycloxypropylcysteine, a diacylated lipopeptide ligand of TLR6, plays a host-protective role against experimental Leishmania major infection. *J. Immunol.* **2014**, *193*, 3632–3643. [CrossRef] [PubMed]
70. Salgado, C.L.; Dias, E.L.; Stringari, L.L.; Covre, L.P.; Dietze, R.; Lima Pereira, F.E.; de Matos Guedes, H.L.; Rossi-Bergmann, B.; Gomes, D.C.O. Pam3CSK4 adjuvant given intranasally boosts anti-Leishmania immunogenicity but not protective immune responses conferred by LaAg vaccine against visceral leishmaniasis. *Microbes Infect.* **2019**, *21*, 328–335. [CrossRef] [PubMed]

animals

MDPI

Article

An Epidemiological Assessment of *Cryptosporidium* and *Giardia* spp. Infection in Pet Animals from Taiwan

Chia-Hui Hsu [1], Chi Liang [2], Shi-Chien Chi [2], Kuan-Ju Lee [2], Chung-Hsi Chou [2], Chen-Si Lin [2,*] and Wen-Yuan Yang [2,*]

1. Center for Animal Health and Food Safety, College of Veterinary Medicine, University of Minnesota, Saint Paul, MN 55108, USA; hsu00124@umn.edu
2. Department of Veterinary Medicine, National Taiwan University, Taipei 10617, Taiwan; r10629031@ntu.edu.tw (C.L.); b07606003@ntu.edu.tw (S.-C.C.); b06609062@ntu.edu.tw (K.-J.L.); cchou@ntu.edu.tw (C.-H.C.)

* Correspondence: cslin100@ntu.edu.tw (C.-S.L.); wenyuanyang108@ntu.edu.tw (W.-Y.Y.)

Simple Summary: *Cryptosporidium* spp. and *Giardia duodenalis*, enteric protozoan pathogens affecting humans and animals, elicit substantial global public concern. This study conducted in Taiwan sought to determine the prevalence and co-infection rates of Cryptosporidium spp. and *G. duodenalis* in dogs and cats. The investigation encompassed an analysis of infection rates and associated risk factors within the surveyed population. Predominantly identified species were *C. canis* and *C. felis*, aligning with canine-specific assemblages C and D. In contrast, the infrequent presence of human-specific assemblage A was noted in Giardia-positive samples. Phylogenetic analysis alluded to the potential for zoonotic transmission originating from domesticated animals. This underscores the role of pets as possible reservoirs for the transmission of cryptosporidiosis and giardiasis to humans in Taiwan.

Abstract: *Cryptosporidium* spp. and *Giardia duodenalis* are enteric protozoan pathogens in humans. and animals. Companion animals infected with zoonotic species/assemblages are a matter of major public concern around the world. The objectives of the present study were to determine the prevalences of *Cryptosporidium* spp. and *G. duodenalis* infections and their co-infection statuses in dogs and cats living in Taiwan and to identify the species and assemblages. Fecal samples were collected from local animal shelters (n = 285) and a veterinary hospital (n = 108). Nested polymerase chain reaction (PCR) was performed using the SSU-rRNA, β-giardin, and glutamate dehydrogenase genes for *Cryptosporidium* spp. and *G. duodenalis*, respectively. Results showed that the overall prevalences of *Cryptosporidium* and *G. duodenalis* were 7.38% (29/393) and 10.69% (42/393). In addition, co-infection was detected in 1.02% (4/393) of all samples. Sample source, clinical sign, and breed may be risk factors that influence the infection rate. In *Cryptosporidium*-positive samples, *C. canis* and *C. felis* were detected most frequently. Although the canine-specific assemblages C and D (37/42) were dominant, the zoonotic human-specific assemblage A (1/42) was also found in *Giardia*-positive samples. Phylogenetic analysis revealed that most positive samples belonged to host-specific subtypes/assemblages, while some *Cryptosporidium* or *Giardia*-positive samples could be zoonotic. The findings suggested that pet animals could be a cause of zoonotic transmission, causing human cryptosporidiosis and giardiasis in Taiwan.

Keywords: *Cryptosporidium*; SSU-rRNA; β-giardin; glutamate dehydrogenase; *Giardia duodenalis*; nested polymerase chain reaction; zoonoses; molecular epidemiology; assemblages

1. Introduction

Cryptosporidiosis and giardiasis are emerging infectious diseases caused by *Cryptosporidium* spp. and *Giardia intestinalis* (syn. *G. duodenalis* or *G. lamblia*), protozoan intestinal parasites which are of significant public health concern worldwide because infections are transmitted primarily through zoonotic, water-borne, and foodborne routes [1,2]. Human beings, companion animals, birds, domestic livestock, and a wide range of vertebrates all potentially contribute *Cryptosporidium* or *Giardia* spp. to the environment [3]. The main clinical presentation of cryptosporidiosis and giardiasis in humans and animals manifests as abdominal pain, watery diarrhea, dehydration, malabsorption, and wasting [4]. Generally, *Cryptosporidium* infections in immunocompetent people are mild and self-limiting. However, the lives of young children, the elderly and the immunocompromised can be threatened by infections [1,5]. In human beings, although *Cryptosporidium* and Giardia infections are only occasionally found, such as in immunocompromised patients, they still represent a significant health concern.

Among the approximately 40 recognized *Cryptosporidium* species, *C. hominis*, *C. parvum*, *C. meleagridis*, *C. canis*, and *C. felis* are the most prevalent in humans [6]. Similarly, within the eight common genotypes (A to H) of *G. duodenalis*, only assemblages A and B pose significant human health risks [7]. *Cryptosporidium* spp. and *G. duodenalis* are frequently found in dogs and cats worldwide [8,9]. *Cryptosporidium canis* and *C. felis* are the primary *Cryptosporidium* species in dogs and cats, respectively. However, occasional detections of *C. hominis*, *C. parvum*, *C. muris*, and *C. ubiquitum* have been reported in these animals [10–14]. Similarly, dog-adapted assemblages C and D and the cat-adapted assemblage F are the dominant *G. duodenalis* genotypes in these animals. However, zoonotic assemblages A and B have been identified in some studies [7,15].

Companion animals may play an essential role in the transmission routes of *Cryptosporidiosis* and *Giardiasis* [16]. Animal shelters are environments that facilitate the spread of gastrointestinal protozoan parasitic disease to incoming animals and shelter staff due to overcrowding and the multiple stressors of the animals. In addition to the animals, shelter staff and visitors are also considered *vulnerable to* exposure to zoonotic diseases [17]. Since both *Cryptosporidiosis and Giardiasis* are transmitted to humans through the fecal–oral route or contaminated food and water, infections can be more prevalent in crowded conditions if the shelter settings are unsanitary. In this research, we focused on sources in local animal shelters and household pets and analyzed the associations with risk factors. Hence, in this study, we collected stool specimens from one veterinary teaching hospital and three local animal shelters in Taiwan.

Traditionally, the diagnosis of *Cryptosporidium* and *Giardia* relies on detecting specific oocyst/cyst characteristics in stool samples, typically using acid-fast staining [18] or immunofluorescence assays with antibodies. However, these traditional methods cannot distinguish between species based on morphology or host occurrence. To address this issue and enable quick and accurate identification of these protozoan parasites, there is a need for novel, rapid, and discriminatory analytical methods applicable in both developed and developing regions, suitable for clinical and environmental samples.

Diagnosing *Cryptosporidium* and *Giardia* with the traditional visual microscopic method is difficult due to the inability to differentiate between different species using morphology and/or host occurrence. Instead, a molecular technique such as polymerase chain reaction (PCR) should be used to characterize *Cryptosporidium* and *Giardia* in feces or environmental samples [19]. Compared to the traditional microscopic method, nested PCR is more sensitive and reliable, so it is recommended for such detection and genotyping [20].

Although some surveys of *Cryptosporidium* oocysts and *Giardia* cysts have been conducted in livestock drinking water in Taiwan over the past decade, references or reports pertaining to *Cryptosporidium* and *Giardia* infection in small animal shelters are relatively limited. Thus, in this study, we performed the first cross-sectional investigation from different sources to offer a molecular assessment for future shelter animal medicine research. The primary aim of this research was to determine and elucidate the prevalences of *Cryptosporidium* and *Giardia* infection and their co-infection status in companion animals in Taiwan. The resulting information could serve as baseline data here.

2. Materials and Methods

2.1. Study Design

In 2020, from February 1 to December 31, fresh canine and feline stool specimens were randomly collected from three local animal shelters (TW-TPE, TW-TYN, and TW-TTT; abbreviations of sampling sites were based on assigned country code) and one veterinary teaching hospital in Taiwan. The majority of the stool specimens were specifically recruited for the study, while the rest were obtained from routine diagnostics at the Section of Clinical Pathology, National Taiwan University Veterinary Hospital (NTUVH), Taipei. Among the 393 canine and feline specimens, 285 were from 3 local animal shelters, and 108 were from NTUVH. All the specimens were collected from live animals, and the collection processes were approved and followed local government regulations on animal protection. The study was approved by the Institutional Animal Care and Use Committee of National Taiwan University, Taipei, Taiwan (protocol code: IACUC No. NTU107-EL-00200).

2.2. Sample Collection

Each specimen (0.5–1 g) was collected from the ground immediately after animal defecation, placed into a 2 mL microcentrifuge tube, and then mixed with an appropriate amount of distilled deionized water (DDW). The properties of each specimen and the breed, age, gender, and past history of the animal were recorded in detail if possible. After several rounds of mechanical agitation, aliquots of the stool specimen were exposed to 5 freeze–thaw cycles, as described previously [21]. The processed specimens were stored at −20 °C, and DNA was extracted within 1 week after collection.

2.3. DNA Extraction

After the freeze–thaw cycles, total genomic DNA (gDNA) was isolated from the feces sample using the NautiaZ Stool DNA Extraction Mini Kit (Nautia Gene, Taipei, Taiwan) and eluted according to the manufacturer's instructions. DNA concentration and purity were measured with a spectrophotometer (Eppendorf AG, Hamburg, Germany). The isolated gDNA samples were used immediately or stored at −20 °C for up to 1 month prior to screening.

2.4. Nested Polymerase Chain Reaction (PCR) Amplification

Cryptosporidium spp. was detected via nested-PCR amplification of the small subunit (SSU) rRNA gene. The DNA of *G. intestinalis* was amplified, targeting β-giardin (BG) and glutamate dehydrogenase (GDH) genes. As seen in Table 1, nested-PCR primer pairs were adopted in PCR amplification reactions in the study. A positive control (*Cryptosporidium* or *Giardia* DNA) and a negative control (distilled water) were run with every PCR batch. After nested-PCR amplification, 10 µL of each PCR product was electrophoresed by 2% agarose gel at 100 V for 30 min to identify the size of the products. PCR-positive products were purified using the PCR Clean-Up and Gel Extraction Kit (GeneDirex, Las Vegas, NV, USA), following the manufacturer's instructions.

To calculate the estimated prevalence rate, we count the number of positive samples and divide it by the total number of samples tested using nested PCR. The overall prevalence rate is determined by dividing the number of positive samples by the total number of companion animal fecal samples that are being tested, irrespective of whether they are dogs or cats. We also calculated stratum-specific prevalence rates for dogs and cats separately.

Table 1. Primers and annealing conditions of nested PCR for *Cryptosporidium* spp. SSU gene, *Giardia* BG gene, and *Giardia* GDH gene.

Primer Name	Primer Sequence (5′→3′)	Annealing (°C/s)	Product Size (bp)	Reference
Nested-PCR primers for *Cryptosporidium* SSU genes				
SSU-F1 SSU-R1	TTCTAGAGCTAATACATGCG CCCATTTCCTTCGAAACAGGA	55/45	1325	(Xiao et al., 1999) [22]
SSU-F2 SSU-R2	GGAAGGGTTGTATTTATTAGATAAAG CTCATAAGGTGCTGAAGGAGTA	55/45	840	
Nested-PCR primers for *Giardia* BG genes				
BG-F1 BG-R1	AAGCCCGACGACCTCACCCGCAGTGC GAGGCCGCCCTGGATCTTCGAGACGAC	65/30	735	(MarcoLalle et al., 2005) [23]
BG-F2 BG-R2	GAACGAACGAGATCGAGGTCCG CTCGACGAGCTTCGTGTT	65/30	511	
Nested-PCR primers for *Giardia* GDH genes				
GDH-F1 GDH-R1	TTCCGTRTYCAGTACAACTC ACCTCGTTCTGRGTGGCGCA	57.8/30	754	(S.M.Cacciò et al., 2008) [24]
GDH-F2 GDH-R2	ATGACYGAGCTYCAGAGGCACGT GTGGCGCARGGCATGATGCA	57.8/30	530	

2.5. Phylogenetic Analysis

To examine the genetic relationships among *Cryptosporidium* spp. and *Giardia duodenalis* assemblages in canines and felines, all positive secondary PCR amplicons (~840 bp for *Cryptosporidium*; ~510 bp for BG in *Giardia*; ~530 bp for GDH in *Giardia*) were sent for nucleotide sequencing at Mission Biotech (Taipei, Taiwan). The results were compared with corresponding sequences on the National Center for Biotechnology Information (NCBI) website using the BLAST® program (http://blast.ncbi.nlm.nih.gov/; accessed during 1 March~30 September 2020). A phylogenetic tree was constructed using the MEGA-X program with neighbor-joining (NJ) analysis of the SSU rRNA and BG gene sequences.

2.6. Software and Statistical Analysis

Geological maps were illustrated in the software QGIS 3.10.14. Statistical analyses were performed in Python 3.7 scipy.stats modules. Pearson's chi-squared (χ^2) test and Fisher's exact tests were performed to test for any significant differences in location, gender, age, breed, and clinical signs between the two groups of canine and feline populations. Results were considered statistically significant at $p < 0.05$.

3. Results

3.1. Prevalence of Cryptosporidium and Giardia Infection and Co-Infection Pattern in Companion Animals

A total number of 393 individual stool samples were obtained for this study. Of them, 27.4% (108/393) of the specimens were collected from the veterinary hospital (TW-NTU) and 72.5% (285/393) from local shelters (TW-TPE, TW-TYN, and TW-TTT) (Figures 1 and 2A). Of the canine fecal samples, 34.8% were from the veterinary hospital, and 65.2% were from local shelters. Of the feline samples, the percentages collected from the hospital and shelters were 14.7% and 85.3%, respectively. The overall prevalence rates of *Cryptosporidium* spp. and *G. duodenalis* by nested PCR were 7.38% (29/393) and 10.69% (42/393) (Figure 2B). In this study, the estimated prevalence rates of *Cryptosporidium* infection in dogs and cats were 8.4% (21/250) and 5.6% (8/143), while those of *G. duodenalis* infection in dogs and cats were 12.4% (31/250) and 7.69% (11/143), respectively (Table 2). Moreover, co-infection with both pathogens was detected in 1.02% (4/393) of all companion animal samples (Figure 2C,D). For identifying the specific assemblages of *G. duodenalis*, for example,

16 canine samples tested positive in Source-1 for BG, but only 8 samples for C and 6 for D were identified by genotyping. The remaining samples were not identified with specific assemblages. (Table 3).

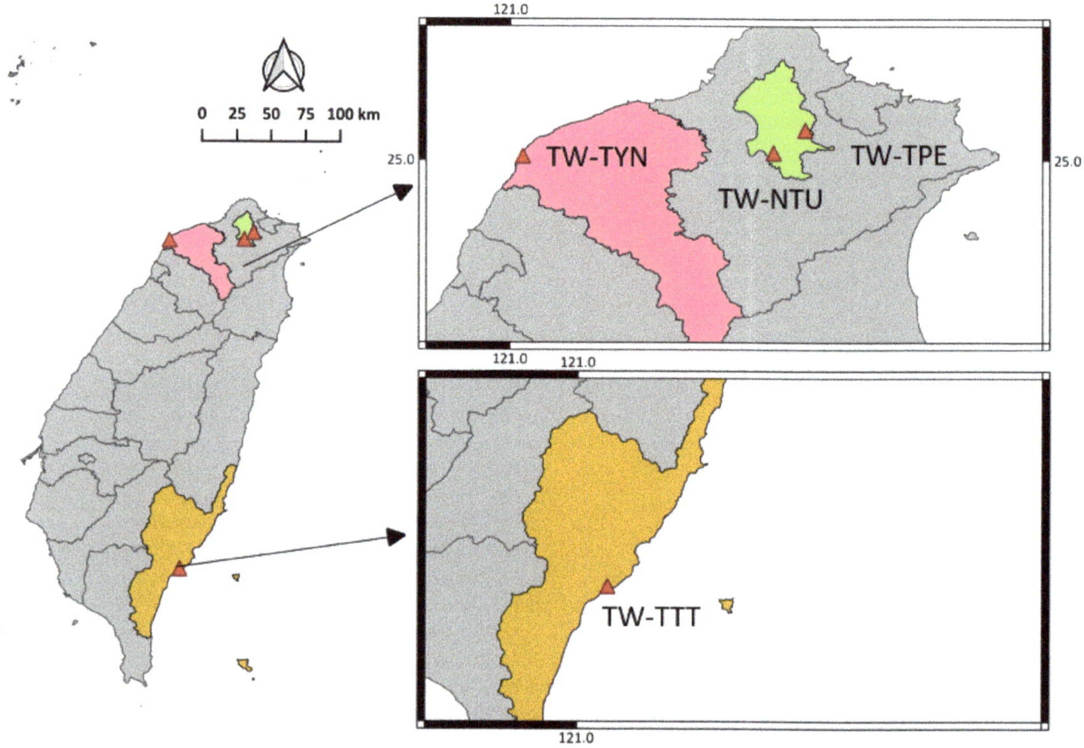

Figure 1. Map showing the sampling locations in Taiwan. Fresh canine and feline stool samples were collected in 2020 from three locations in Taiwan, including two urban areas (Taipei and Taoyuan) and a relatively rural area (Taitung). TW, Taiwan; TPE, Taipei; NTU, National Taiwan University; TYN, Taoyuan; TTT, Taitung.

Table 2. Percentages and numbers of nested-PCR-positive canine and feline samples.

Category	No. Examined	Canine				No. Examined	Feline			
Pathogen		*Cryptosporidium* (%)		Giardia (%)			*Cryptosporidium* (%)		Giardia (%)	
Source-1 TW-TPE	82	9	11.0%	19	23.2%	76	4	5.3%	6	7.9%
Source-2 TW-TYN	55	2	3.6%	7	12.7%	36	1	2.8%	3	8.3%
Source-3 TW-TTT	26	3	11.5%	2	7.7%	10	0	0.0%	1	10.0%
Subtotal	163	14	8.6%	28	17.2%	122	5	4.1%	10	8.2%
Source-4 TW-NTU	87	7	8.0%	3	3.4%	21	3	14.3%	1	4.8%
Total	250	21	8.4%	31	12.4%	143	8	5.6%	11	7.7%

Table 3. Percentages and numbers of *G. duodenalis* isolates of assemblage in canine and feline samples.

Category	No. Examined	Assemblages (n) of Canine					No. Examined	Assemblages (n) of Feline				
Pathogen		Overall	BG		GDH			Overall	BG		GDH	
Source-1 TW-TPE	82	19	16	C (8), D (6)	7	C (3), D (3)	76	6	6	A (1)	0	
Source-2 TW-TYN	55	7	6	C (5), D (1)	3	C (3)	36	3	3	C (2)	1	F (1)
Source-3 TW-TTT	26	2	1	D (1)	2	D (2)	10	1	0		1	D (1)
Source-4 TW-NTU	87	3	3	C (2)	0		21	1	1		0	
Total	250	31	26		12		143	11	10		2	

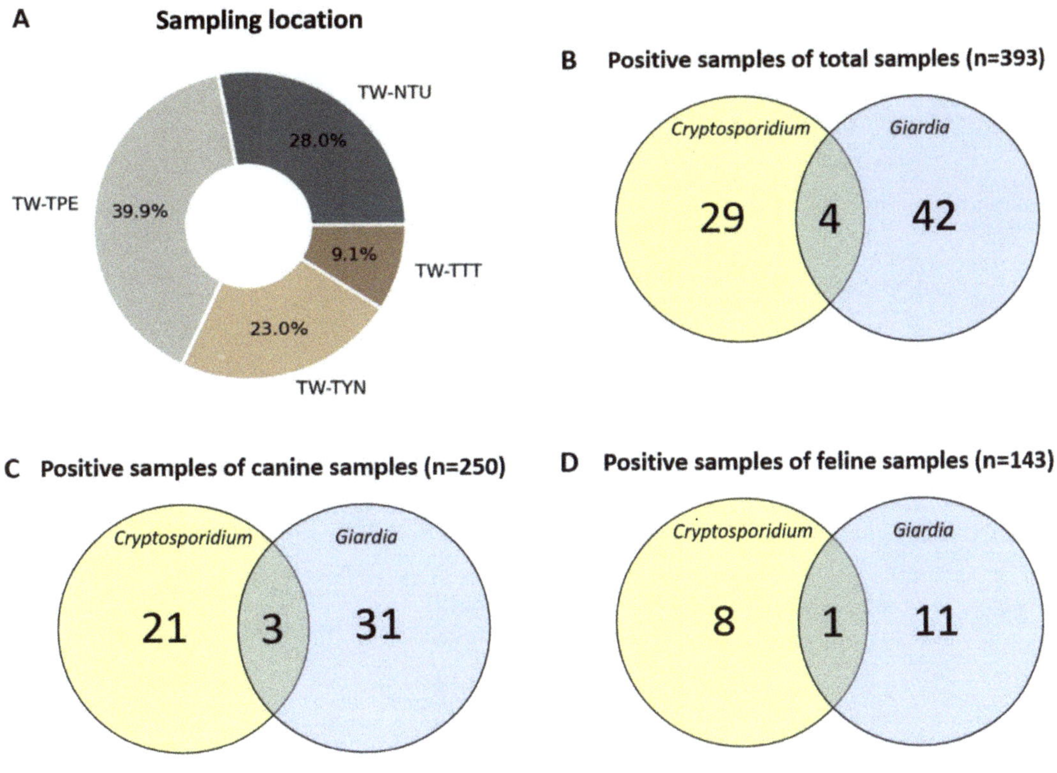

Figure 2. Percentages of the fecal samples from different locations, and the numbers of *Cryptosporidium* and *Giardia*-positive samples. (**A**) A pie chart of different sampling sources and sample sizes. (**B–D**) Venn diagrams showing the sample numbers of nested-PCR-positive results and co-infections of total, canine, and feline samples, respectively. TW, Taiwan; TPE, Taipei; NTU, National Taiwan University; TYN, Taoyuan; TTT, Taitung.

3.2. Risk Factor Analysis

The univariable analyses of risk factors associated with positive detection results are shown in Tables 4 and 5 for the two parasitic pathogens in canines and felines. Variables with p values < 0.05 were regarded as associated with infection. Animal gender, fecal source, breed, and clinical signs were evaluated as possible variables. Only gender was found to be not significantly related to both *Cryptosporidium* and *G. duodenalis* infections. The rates of *G. duodenalis* infection in canines and felines were higher in those from the local animal shelters than in those from the veterinary hospital. The probability of being infected by *G. duodenalis* was more than five times higher (OR = 5.8, 95% CI = 1.7–19.7, $p = 0.0048$) for the animal shelter source than for the veterinary hospital source. Dogs presenting diarrhea had a 2.9 times higher (OR = 2.9, 95% CI = 1.0–8.0, p = 0.0439) probability of *Cryptosporidium* infection. In contrast, cats, whether displaying noticeable clinical signs or not, were not linked to the prevalence of either disease. Interestingly, compared to mixed-breed felines, purebred cats were 9.5 times (OR = 9.5 95% CI = 1.9–47.1, p = 0.0058) more likely to be *Cryptosporidium*-positive. However, mixed-breed dogs were more likely to be *Giardia*-positive than purebred dogs (p = 0.0406).

Table 4. Risk factor analysis for canine cases.

Factor	*n*	*Cryptosporidium* spp.				*G. duodenalis*			
		No. Positive	OR	95%CI	*p*-Value	No. Positive	OR	95%CI	*p*-Value
Gender									
Female	113	11	1.4	(0.6–3.4)	0.49	12	0.7	(0.3–1.6)	0.44
Male	106	7	0.7	(0.3–1.7)	0.38	12	0.8	(0.4–1.8)	0.66
Unknown	31	3	1.2	(0.3–4.3)	0.78	7	2.4	(0.9–6.1)	0.07
Total	250	21				31			
Source									
Veterinary Hospital	87	7	1	(0.4–2.4)	0.88	3	0.2	(0.1–0.6)	0.0048 **
Animal Shelter	163	14	1.1	(0.4–2.8)		28	5.8	(1.7–19.7)	
Total	250	21				31			
Breed									
Purebred	64	4	0.7	(0.2–2.1)	0.47	3	0.3	(0.1–0.9)	0.0406 *
Mixed	186	17	1.5	(0.5–4.7)		28	3.6	(1.1–12.3)	
Total	250	21				31			
Clinical Sign									
Diarrhea	34	6	2.9	(1.0–8.0)	0.0439 *	7	2.1	(0.8–5.3)	0.13
No notable sign	216	15	0.3	(0.1–1.0)		24	0.5	(0.2–1.2)	
Total	250	21				31			

Note: * *p*-value < 0.05; ** *p*-value < 0.01.

Table 5. Risk factor analysis for feline cases.

Factor	*n*	*Cryptosporidium* spp.				*G. duodenalis*			
		No. Positive	OR	95% CI	*p*-Value	No. Positive	OR	95% CI	*p*-Value
Gender									
Female	50	3	1.1	(0.3–4.9)	0.88	1	0.17	(0.0–1.4)	0.10
Male	46	4	2.2	(0.5–9.3)	0.28	4	1.22	(0.3–4.4)	0.76
Unknown	47	1	0.3	(0.0–2.3)	0.24	6	2.66	(0.8–9.2)	0.12
Total	143	8				11			
Source									
Veterinary Hospital	21	3	3.9	(0.9–17.7)	0.08	1	0.61	(0.1–5.0)	0.65
Animal Shelter	122	5	0.3	(0.1–1.2)		10	1.6393	(0.2–13.5)	
Total	143	8				11			
Breed									
Purebred	11	3	9.5	(1.9–47.1)	0.0058 **	0	0.4594	(0.0–8.3)	0.60
Mixed	132	5	0.1	(0.0–0.5)		11	2.177	(0.1–39.3)	
Total	143	8				11			
Clinical Sign									
Diarrhea	20	0	0.3	(0.0–6.0)	0.45	1	0.5947	(0.1–4.9)	0.63
No notable sign	123	8	3	(0.2–54.3)		10	1.6814	(0.2–13.9)	
Total	143	8				11			

Note: ** *p*-value < 0.01.

3.3. Phylogenetic Analysis

Nested PCR and DNA sequencing results revealed that *Cryptosporidium* infection was detected in 7.38% of all fecal samples, whereas *Giardia* infection was detected in 10.69%. After DNA sequencing, the results showed that the species of *Cryptosporidium* were mainly *C. canis, C. felis, and C. meleagridis* (Figure 3A). Sequencing analysis of the SSU rRNA gene demonstrated that 72.4% (21/29) of the *Cryptosporidium*-positive samples shared 89–100% homology with the sequence of *C. canis (accession numbers:* AB210854.1, AF112576.1, KF516543.1, and MT329018.1) retrieved from the GenBank database. Within the nucleotide sequences of *Giardia*-positive samples isolated from companion animals, 61.9% (26/42) were successfully characterized as some specific assemblages, with most being assemblage C (17/26), followed by assemblage D (8/26) and assemblage A (1/26). Other parts of isolated *Giardia*-positive samples still presented as β-giardin-positive but matched as partial codons (Figure 3B). The most dominant species/assemblages of *Cryptosporidium-/Giardia*-positive samples were *C. canis* (*SSU-rRNA*; GenBank accession number AB210864.1, sequence homology 99–100%, $n = 17$) and *Giardia* canine-specific assemblage C (β-giardin; GenBank accession number LC437428.1, sequence homology 99–100%, $n = 13$).

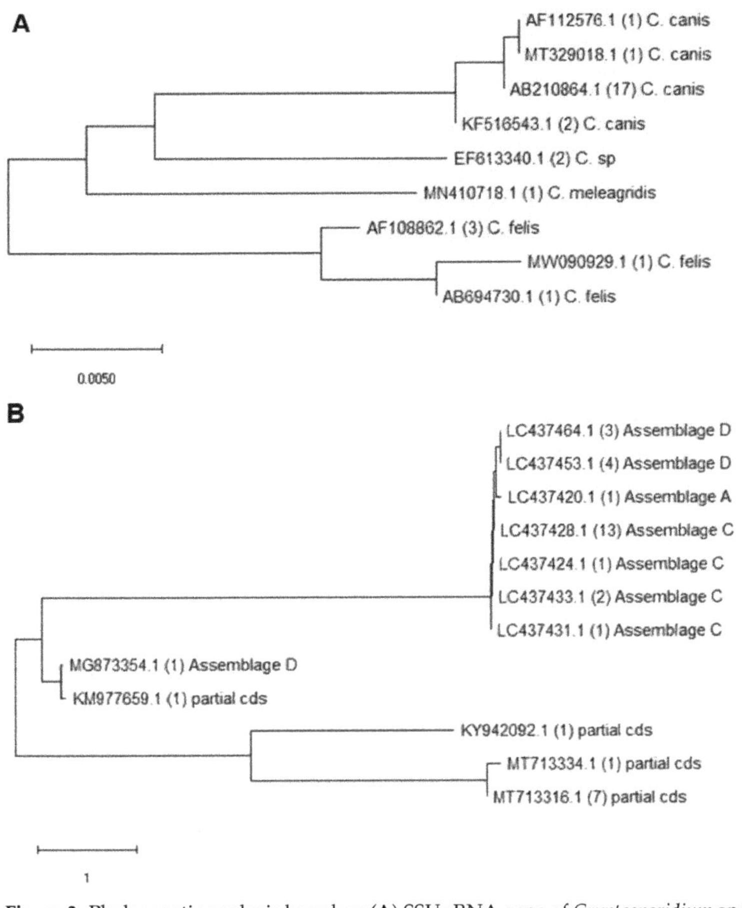

Figure 3. Phylogenetic analysis based on (**A**) SSU rRNA gene of *Cryptosporidium* and (**B**) BG gene of *Giardia*. The phylogenetic tree was constructed using MEGA software (1.0) by employing the neighbor-joining (NJ) method. The numbers in parentheses show the number of different species or assemblages. CDS, coding sequence.

4. Discussion

Our data showed that the overall *Cryptosporidium* infection rate was 7.38%. The prevalence of *Giardia* infection in companion animals identified at different genetic loci (*bg* and *gdh*) were 9.16% (36/393) and 3.56% (14/393), respectively. After the integration of positive samples of *Giardia* at the two loci, an overall 10.69% (42/393) prevalence was detected. A systemic review reported prevalence rates of *Giardia* of around 15.2% in dogs and 12% in cats worldwide [8]. In comparison, recent regional surveys to determine the prevalence of *Giardia* infection by ELISA antigen test or PCR have reported rates of 18.7% in Japan [25], 11.2% in South Korea [26], and 11% in China [27]. Our findings generally follow a pattern similar to those in surveys previously carried out in East Asia and worldwide.

The most noteworthy symptom in companion animals infected with protozoan parasites is diarrhea. According to our risk factor analysis, the risk of being positive for *Cryptosporidium* spp. was 2.9-fold higher in the diarrhea canine group than in the no-obvious-clinical-sign canine group. Meanwhile, our study also indicated that *Cryptosporidium* in felines and *Giardia* in both companion animals were not associated with diarrhea clinical signs. That is, although *Cryptosporidium* was confirmed in diarrhea cases, it was also commonly found in normal feces from patients without clinical signs. Both pathogens were detected regardless of companion animal diarrhea status. The detection of both endoparasites in normal feces suggests that the possibility of contamination with (oo)cysts in an entire population cannot be ruled out. This finding is important and has public health significance because pet owners and clinical veterinarians may neglect the possibility of infection of *Cryptosporidium* or *Giardia* due to their clinical presentation. These results should be considered by small-animal veterinarians in differential diagnosis.

Among the known *Cryptosporidium* species, human cryptosporidiosis is mostly caused by five species: *C. parvum*, *C. hominis*, *C. meleagridis*, *C. felis*, and *C. canis* [6,28,29]. In some countries, *C. meleagridis* may be as common as *C. parvum* [28,30,31]. In our findings, we detected the species *C. meleagridis* (n = 1) and *C. canis* (n = 18) in dogs, while *C. felis* (n = 5) and *C. canis* (n = 3) were identified in cats. For *Giardia*, 88.1% were classified as assemblage C or D. Moreover, one feline sample was successfully classified as *Giardia* assemblage A (β-giardin; GenBank accession number LC437420.1, sequence homology 100%). The three *Cryptosporidium* species and *Giardia* spp. genotype A recognized by DNA sequencing is known as zoonotic endoparasites.

Although human cryptosporidiosis infection through anthroponotic transmission is mainly caused by *C. parvum* and *C. hominis, which are* waterborne, *C. canis* and *C. felis* may infect humans through intrafamilial transmission by companion animals [32]. Thus, the horizontal infection of *Cryptosporidiosis* and *Giardiasis* in local animal shelters in Taiwan may present a public health threat to humans. As noted in our previous discussion, there is no significant difference between clinical signs in *Giardia*-positive pets. Therefore, we may neglect the potential risk when contacting clinically healthy companion animals.

The hosts of *Cryptosporidium* are virtually all vertebrate animal species. In Taiwan, the shelter facilities vary greatly by size, capacity, budget, and the number of governmental staff. These shelters offer temporary housing for not only stray dogs and feral cats but also a wide range of animals brought in by animal rescue services in local regions. Thus, wild birds and wildlife such as ferret badgers may be housed simultaneously in a high-density facility. The increased probability of exposure to contaminated water and food in animal shelters may explain the higher percentage of parasitic opportunistic infections. For example, birds are the natural host of *C. meleagridis* [33]. It is not surprising that the endoparasite may spread from birds to canines within the stressful environment of a local animal shelter.

Giardia cysts and *Cryptosporidium* oocysts can survive for months in environments with high humidity, low temperature, and low exposure to sunlight [34]. In Taiwan, the average monthly temperatures range from 17.9 °C in January to 30.9 °C in July, and the average annual relative humidity ranges from 73.0% to 80.6% [35]. Additionally, in the past two decades, *Cryptosporidium* oocysts and *Giardia* cysts have been documented in both drinking water for livestock and urban tap water in Taiwan [36,37]. One of the major features of giardiasis and cryptosporidiosis is their low infectious doses. Fewer than 10 cysts given orally could cause infection of *G. duodenalis*, while *Cryptosporidium* variably requires from 9 to 1024 oocysts [34,38]. The potential for contamination through drinking water should be investigated.

On the other hand, dogs and cats in animal shelters commonly come into contact with adopters and visitors. Although veterinarians regularly impose the disinfection process and exo-parasite chemical deworming program before the animals are admitted, the infection risk of endoparasites still exists. Moreover, once *Cryptosporidium* or *Giardia* (oo)cytes are transmitted in animal shelters, the horizontal transmission may spread dramatically because of oral–fecal infection within companion animals, owing to the narrow space and high-contact environment. Thus, to reduce the risk of zoonotic disease occurrence, fecal samples of pet animals should be routinely submitted for parasitic diagnostic tests, and owners should be informed about the public health issues related to pet fecal pollution. In local animal shelters, expansion of the disinfection checkpoints to maintain the biosecurity of the animal shelters is warranted.

There are some limitations to this research. Firstly, given the intermittent shedding pattern of (oo)cytes of *Cryptosporidium* and *Giardia*, molecular examinations should not only be based on fecal samples so as to prevent underestimation of the prevalence of each parasite. Secondly, increases in the levels of *Cryptosporidium* spp. and *Giardia* infection in young companion animals have been documented previously [39,40]. The age of the companion animals is commonly included in risk factor analysis. Nevertheless, in our sampling reference, the medical records on age are relatively vague and incomplete, complicating the statistical analysis. Third, viral disease infections should be considered necessary in enteropathogenic co-infection in kennels and shelters. The co-infection pattern is not limited to dual infection. In cats in the United Kingdom, co-infection of feline panleukopenia virus, *Cryptosporidium* spp., and *Giardia* spp. has been reported with double or triple infections [41]. The concurrent presence of canine parvovirus type 2, *Cryptosporidium* spp., and *Giardia* spp. has also been noted in Brazil [42]. However, no routine examinations of gastrointestinal viral pathogens are required in Taiwan. An analysis of the necessity of such requirements may be the next step in our research. Given these limitations, the zoonotic infection findings should be interpreted cautiously when generalized to a larger population of companion animals.

5. Conclusions

The present study demonstrated that the most dominant *Cryptosporidium* and *Giardia* species/assemblage characteristics in companion animals were *C. canis*, *C. felis*, and *Giardia* assemblages C and D, respectively. This is the first cross-sectional molecular evaluation of animals derived from local animal shelters and household pets in Taiwan. Notably, *C. canis*, *C. felis*, *C. meleagridis*, and *Giardia* assemblage A have the potential for zoonotic transmission. These molecular results can serve as a baseline to assess the zoonotic potential of companion animals in Taiwan. Further epidemiological studies are necessary to understand the transmission better.

Author Contributions: Investigation and project administration: C.-H.H., C.L., S.-C.C. and K.-J.L.; Writing—original draft preparation: C.-H.H.; Methodology and formal analysis: C.-H.C.; Conceptualization, supervision, and writing—review and editing: C.-S.L.; Formal analysis and writing—review and editing: W.-Y.Y. All authors have read and agreed to the published version of the manuscript.

Funding: This research received no external funding.

Institutional Review Board Statement: The study was approved by the Institutional Animal Care and Use Committee of National Taiwan University, Taipei, Taiwan (protocol code: IACUC No. NTU107-EL-00200).

Informed Consent Statement: Not applicable.

Data Availability Statement: The data in the present study are available from the corresponding author upon reasonable request.

Conflicts of Interest: The authors declare no conflict of interest.

References

1. Rossle, N.F.; Latif, B. Cryptosporidiosis as threatening health problem: A review. *Asian Pac. J. Trop. Biomed.* **2013**, *3*, 916–924. [CrossRef]
2. Fayer, R.; Speer, C.; Dubey, J. *The General Biology of Cryptosporidium*; CRC Press: Boca Raton, FL, USA, 1997.
3. Wang, J.; Liew, C. Prevalence of *Cryptosporidium* spp. in birds in Taiwan. *Taiwan J. Vet. Med. Anim. Husb.* **1990**, *56*, 45–57.
4. Bodley-Tickell, A.; Kitchen, S.; Sturdee, A. Occurrence of *Cryptosporidium* in agricultural surface waters during an annual farming cycle in lowland UK. *Water Res.* **2002**, *36*, 1880–1886. [CrossRef] [PubMed]
5. Gerace, E.; Presti, V.D.M.L.; Biondo, C. *Cryptosporidium* infection: Epidemiology, pathogenesis, and differential diagnosis. *Eur. J. Microbiol. Immunol.* **2019**, *9*, 119–123. [CrossRef] [PubMed]
6. Feng, Y.; Ryan, U.M.; Xiao, L. Genetic Diversity and Population Structure of *Cryptosporidium*. *Trends Parasitol.* **2018**, *34*, 997–1011. [CrossRef]
7. Feng, Y.; Xiao, L. Zoonotic potential and molecular epidemiology of *Giardia* species and giardiasis. *Clin. Microbiol. Rev.* **2011**, *24*, 110–140. [CrossRef]
8. Bouzid, M.; Halai, K.; Jeffreys, D.; Hunter, P.R. The prevalence of *Giardia* infection in dogs and cats, a systematic review and meta-analysis of prevalence studies from stool samples. *Vet. Parasitol.* **2015**, *207*, 181–202. [CrossRef]
9. Santin, M. Clinical and subclinical infections with *Cryptosporidium* in animals. *N. Z. Vet. J.* **2013**, *61*, 1–10. [CrossRef]
10. Alves, M.E.M.; Martins, F.D.C.; Braunig, P.; Pivoto, F.L.; Sangioni, L.A.; Vogel, F.S.F. Molecular detection of *Cryptosporidium* spp. and the occurrence of intestinal parasites in fecal samples of naturally infected dogs and cats. *Parasitol. Res.* **2018**, *117*, 3033–3038. [CrossRef]
11. Yang, R.; Ying, J.L.; Monis, P.; Ryan, U. Molecular characterisation of *Cryptosporidium* and *Giardia* in cats (*Felis catus*) in Western Australia. *Exp. Parasitol.* **2015**, *155*, 13–18. [CrossRef]
12. Gil, H.; Cano, L.; de Lucio, A.; Bailo, B.; de Mingo, M.H.; Cardona, G.A.; Fernandez-Basterra, J.A.; Aramburu-Aguirre, J.; Lopez-Molina, N.; Carmena, D. Detection and molecular diversity of *Giardia duodenalis* and *Cryptosporidium* spp. in sheltered dogs and cats in Northern Spain. *Infect. Genet. Evol.* **2017**, *50*, 62–69. [CrossRef]
13. Li, W.; Li, Y.; Song, M.; Lu, Y.; Yang, J.; Tao, W.; Jiang, Y.; Wan, Q.; Zhang, S.; Xiao, L. Prevalence and genetic characteristics of *Cryptosporidium*, *Enterocytozoon bieneusi* and *Giardia duodenalis* in cats and dogs in Heilongjiang province, China. *Vet. Parasitol.* **2015**, *208*, 125–134. [CrossRef]
14. Lucio-Forster, A.; Griffiths, J.K.; Cama, V.A.; Xiao, L.; Bowman, D.D. Minimal zoonotic risk of cryptosporidiosis from pet dogs and cats. *Trends Parasitol.* **2010**, *26*, 174–179. [CrossRef]
15. Xu, H.; Jin, Y.; Wu, W.; Li, P.; Wang, L.; Li, N.; Feng, Y.; Xiao, L. Genotypes of *Cryptosporidium* spp., *Enterocytozoon bieneusi* and *Giardia duodenalis* in dogs and cats in Shanghai, China. *Parasites Vectors* **2016**, *9*, 121. [CrossRef] [PubMed]
16. Thompson, R.A.; Palmer, C.S.; O'Handley, R. The public health and clinical significance of *Giardia* and *Cryptosporidium* in domestic animals. *Vet. J.* **2008**, *177*, 18–25. [CrossRef]
17. Steneroden, K.; Hill, A.E.; Salman, M. Zoonotic disease awareness in animal shelter workers and volunteers and the effect of training. *Zoonoses Public Health* **2011**, *58*, 449–453. [CrossRef] [PubMed]
18. Garcia, L.S.; Bruckner, D.A.; Brewer, T.C.; Shimizu, R.Y. Techniques for the Recovery and Identification of *Cryptosporidium* Oocysts from Stool Specimens. *J. Clin. Microbiol.* **1983**, *18*, 185–190. [CrossRef]
19. Hunter, P.R.; Thompson, R.A. The zoonotic transmission of *Giardia* and *Cryptosporidium*. *Int. J. Parasitol.* **2005**, *35*, 1181–1190. [CrossRef] [PubMed]
20. Coupe, S.; Sarfati, C.; Hamane, S.; Derouin, F. Detection of *Cryptosporidium* and identification to the species level by nested PCR and restriction fragment length polymorphism. *J. Clin. Microbiol.* **2005**, *43*, 1017–1023. [CrossRef] [PubMed]
21. Hawash, Y. DNA extraction from protozoan oocysts/cysts in feces for diagnostic PCR. *Korean J. Parasitol.* **2014**, *52*, 263. [CrossRef]
22. Xiao, L.; Escalante, L.; Yang, C.; Sulaiman, I.; Escalante, A.A.; Montali, R.J.; Fayer, R.; Lal, A.A. Phylogenetic analysis of *Cryptosporidium* parasites based on the small-subunit rRNA gene locus. *Appl. Environ. Microbiol.* **1999**, *65*, 1578–1583. [CrossRef] [PubMed]
23. Lalle, M.; Pozio, E.; Capelli, G.; Bruschi, F.; Crotti, D.; Cacciò, S.M. Genetic heterogeneity at the β-giardin locus among human and animal isolates of and identification of potentially zoonotic subgenotypes. *Int. J. Parasitol.* **2005**, *35*, 207–213. [CrossRef] [PubMed]
24. Cacciò, S.M.; Beck, R.; Lalle, M.; Marinculic, A.; Pozio, E. Multilocus genotyping of reveals striking differences between assemblages A and B. *Int. J. Parasitol.* **2008**, *38*, 1523–1531. [CrossRef] [PubMed]
25. Ito, Y.; Itoh, N.; Kimura, Y.; Kanai, K. Prevalence of intestinal parasites in breeding cattery cats in Japan. *J. Feline Med. Surg.* **2016**, *18*, 834–837. [CrossRef] [PubMed]
26. Liu, J.; Lee, S.; Song, K. Prevalence of canine giardiosis in South Korea. *Res. Vet. Sci.* **2008**, *84*, 416–418. [CrossRef] [PubMed]
27. Li, J.; Zhang, P.; Wang, P.; Alsarakibi, M.; Zhu, H.; Liu, Y.; Meng, X.; Li, J.; Guo, J.; Li, G. Genotype identification and prevalence of *Giardia duodenalis* in pet dogs of Guangzhou, Southern China. *Vet. Parasitol.* **2012**, *188*, 368–371. [CrossRef] [PubMed]

28. Xiao, L. Molecular epidemiology of cryptosporidiosis: An update. *Exp. Parasitol.* **2010**, *124*, 80–89. [CrossRef]
29. Beser, J.; Toresson, L.; Eitrem, R.; Troell, K.; Winiecka-Krusnell, J.; Lebbad, M. Possible zoonotic transmission of *Cryptosporidium* felis in a household. *Infect. Ecol. Epidemiol.* **2015**, *5*, 28463. [CrossRef]
30. Cama, V.A.; Ross, J.M.; Crawford, S.; Kawai, V.; Chavez-Valdez, R.; Vargas, D.; Vivar, A.; Ticona, E.; Navincopa, M.; Williamson, J. Differences in clinical manifestations among *Cryptosporidium* species and subtypes in HIV-infected persons. *J. Infect. Dis.* **2007**, *196*, 684–691. [CrossRef]
31. Gatei, W.; Suputtamongkol, Y.; Waywa, D.; Ashford, R.; Bailey, J.; Greensill, J.; Beeching, N.; Hart, C.A. Zoonotic species of *Cryptosporidium* are as prevalent as the anthroponotic in HIV-infected patients in Thailand. *Ann. Trop. Med. Parasitol.* **2002**, *96*, 797–802. [CrossRef]
32. Xiao, L.; Cama, V.A.; Cabrera, L.; Ortega, Y.; Pearson, J.; Gilman, R.H. Possible transmission of *Cryptosporidium* canis among children and a dog in a household. *J. Clin. Microbiol.* **2007**, *45*, 2014–2016. [CrossRef]
33. Koompapong, K.; Mori, H.; Thammasonthijarern, N.; Prasertbun, R.; Pintong, A.-r.; Popruk, S.; Rojekittikhun, W.; Chaisiri, K.; Sukthana, Y.; Mahittikorn, A. Molecular identification of *Cryptosporidium* spp. in seagulls, pigeons, dogs, and cats in Thailand. *Parasite* **2014**, *21*, 52. [CrossRef]
34. Erickson, M.C.; Ortega, Y.R. Inactivation of protozoan parasites in food, water, and environmental systems. *J. Food Prot.* **2006**, *69*, 2786–2808. [CrossRef] [PubMed]
35. Taiwan Open Weather Data Communications. Central Weather Administration. 2020. Available online: https://opendata.cwb.gov.tw/index (accessed on 15 October 2020).
36. Watanabe, Y.; Kimura, K.; Yang, C.-H.; Ooi, H.-K. Detection of *Cryptosporidium* sp. oocyst and *Giardia* sp. cyst in faucet water samples from cattle and goat farms in Taiwan. *J. Vet. Med. Sci.* **2005**, *67*, 1285–1287. [CrossRef] [PubMed]
37. Hsu, B.-M.; Huang, C.; Hsu, C.-L.L.; Hsu, Y.-F.; Yeh, J. Occurrence of *Giardia* and *Cryptosporidium* in the Kau-Ping River and its watershed in southern Taiwan. *Water Res.* **1999**, *33*, 2701–2707. [CrossRef]
38. Okhuysen, P.C.; Chappell, C.L.; Crabb, J.H.; Sterling, C.R.; DuPont, H.L. Virulence of three distinct *Cryptospovidium parvum* isolates for healthy adults. *J. Infect. Dis.* **1999**, *180*, 1275–1281. [CrossRef] [PubMed]
39. Hamnes, I.S.; Gjerde, B.K.; Robertson, L.J. A longitudinal study on the occurrence of *Cryptosporidium* and *Giardia* in dogs during their first year of life. *Acta Vet. Scand.* **2007**, *49*, 22. [CrossRef]
40. Uehlinger, F.D.; Greenwood, S.J.; McClure, J.T.; Conboy, G.; O'Handley, R.; Barkema, H.W. Zoonotic potential of *Giardia duodenalis* and *Cryptosporidium* spp. and prevalence of intestinal parasites in young dogs from different populations on Prince Edward Island, Canada. *Vet. Parasitol.* **2013**, *196*, 509–514. [CrossRef]
41. Paris, J.K.; Wills, S.; Balzer, H.-J.; Shaw, D.J.; Gunn-Moore, D.A. Enteropathogen co-infection in UK cats with diarrhoea. *BMC Vet. Res.* **2014**, *10*, 13. [CrossRef]
42. da Rocha Gizzi, A.B.; Oliveira, S.T.; Leutenegger, C.M.; Estrada, M.; Kozemjakin, D.A.; Stedile, R.; Marcondes, M.; Biondo, A.W. Presence of infectious agents and co-infections in diarrheic dogs determined with a real-time polymerase chain reaction-based panel. *BMC Vet. Res.* **2014**, *10*, 23.

animals

Article

Intron Regions as Genetic Markers for Population Genetic Investigations of *Opisthorchis viverrini* sensu lato and *Clonorchis sinensis*

Chairat Tantrawatpan [1], Wanchai Maleewong [2,3], Tongjit Thanchomnang [4], Warayutt Pilap [5,6], Takeshi Agatsuma [7], Ross H. Andrews [8,9], Paiboon Sithithaworn [2,9] and Weerachai Saijuntha [4,6,*]

1 Division of Cell Biology, Department of Preclinical Sciences, Faculty of Medicine, and Center of Excellence in Stem Cell Research, Thammasat University, Rangsit Campus, Khlong Nueng 12120, Thailand; talent3003@yahoo.com
2 Department of Parasitology, Faculty of Medicine, Khon Kaen University, Khon Kaen 40002, Thailand; wanch_ma@kku.ac.th (W.M.); paibsit@gmail.com (P.S.)
3 Mekong Health Science Research Institute, Khon Kaen University, Khon Kaen 40002, Thailand
4 Faculty of Medicine, Mahasarakham University, Kham Riang 44000, Thailand; tongjit.t@msu.ac.th
5 Walai Rukhavej Botanical Research Institute, Mahasarakham University, Kham Riang 44150, Thailand; warayuttpilap@gmail.com
6 Center of Excellence in Biodiversity Research, Mahasarakham University, Kham Riang 44150, Thailand
7 Department of Environmental Medicine, Kochi Medical School, Kochi University, Oko, Nankoku 783-8505, Kochi, Japan; agatsuma@kochi-u.ac.jp
8 Department of Surgery & Cancer, Faculty of Medicine, Imperial College, South Kensington Campus, London SW7 2AZ, UK; rhandrews@gmail.com
9 Cholangiocarcinoma Research Institute, Faculty of Medicine, Khon Kaen University, Khon Kaen 40002, Thailand
* Correspondence: weerachai.s@msu.ac.th

Simple Summary: The zoonotic liver flukes *Opisthorchis viverrini* and *Clonorchis sinensis* infect small mammals, such as cats, dogs, pigs, rodents, and rabbits, as well as humans. Human infection subsequently develops into bile duct malignancy, also referred to as cholangiocarcinoma (CCA). Understanding the molecular systematics and population genetics of these liver flukes has an important role in prevention and control, and is important in comprehending their roles in zoonotic transmission. Different molecular markers have varying evolution rates and contain different genetic information. Polymorphic genetic markers are necessary and more suitable for such investigations. Therefore, we screened seven intron regions of the taurocyamine kinase gene (TK) to determine their potential as genetic markers for population genetic investigations of the liver flukes *O. viverrini* and *C. sinensis* which were collected from a range of geographical isolates and animal hosts. We identified a suitable intron region of TK, i.e., intron 5 of domain 1 (TkD1Int5) as having the most potential as a polymorphic marker. Results showed that TkD1Int5 is effective in examining the genetic variation and heterozygosity of *O. viverrini* and *C. sinensis*, but further studies are required to better understand the role of different species of animals as reservoir hosts of these zoonotic liver flukes.

Abstract: Opisthorchiasis and clonorchiasis are prevalent in Southeast and Far-East Asia, which are caused by the group 1 carcinogenic liver flukes *Opisthorchis viverrini* sensu lato and *Clonorchis sinensis* infection. There have been comprehensive investigations of systematics and genetic variation of these liver flukes. Previous studies have shown that *O. viverrini* is a species complex, called "*O. viverrini* sensu lato". More comprehensive investigations of molecular systematics and population genetics of each of the species that make up the species complex are required. Thus, other polymorphic genetic markers need to be developed. Therefore, this study aimed to characterize the intron regions of taurocyamine kinase gene (TK) to examine the genetic variation and population genetics of *O. viverrini* and *C. sinensis* collected from different geographical isolates and from a range of animal hosts. We screened seven intron regions embedded in TK. Of these, we selected an intron 5 of domain 1 (TkD1Int5) region to investigate the genetic variation and population genetics of theses liver flukes. The high nucleotide and haplotype diversity of TkD1Int5 was detected in *O. viverrine*. Heterozygosity with several insertion/deletion (indel) regions were detected in TkD1Int5 of the *O. viverrine* samples, whereas only an indel nucleotide was detected in one *C. sinensis* sample. Several *O. viverrine* samples contained three different haplotypes within a particular heterozygous sample. There were no genetic differences between *C. sinensis* isolated from various animal host. Heterozygous patterns specifically detected in humans was observed in *C. sinensis*. Thus, TkD1Int5 is a high polymorphic genetic marker, which could be an alternative marker for further population genetic investigations of these carcinogenic liver flukes and other related species from a wide geographical distribution and variety of animal hosts.

Keywords: zoonoses; cat; dog; gastrointestinal parasite; intron; genetic variation; liver fluke

1. Introduction

Liver flukes of the family Opisthorchiidae are among the medically important food-borne trematodes. Currently, three principal species are recognized as pathogens causing human diseases: *Opisthorchis viverrine*, *Opisthorchis felineus*, and *Clonorchis sinensis* [1]. Up to 680 million people around the world are at risk of infection [2]. Of these, two species, namely *O. viverrini* and *C. sinensis*, have distributions in Asia. For instance, *O. viverrini* is currently found in Thailand, Cambodia, Lao People's Democratic Republic (PDR), Myanmar, and Vietnam, with approximately 10 million people having opisthorchiasis, whereas *C. sinensis* is found in parts of Russia, China, Korea, and Vietnam, with an estimated 35 million clonorchiasis cases [1]. Both are classified as group I carcinogens because they are the causative agents of bile duct cancer (cholangiocarcinoma; CCA) [1,3]. At present, about 20,000 people die of CCA every year in the northeast of Thailand alone [4].

The life cycles of all two species are very similar, beginning snails as the first intermediate host with a usually low prevalence of infection, fish as second intermediate hosts with substantially higher levels of infection, and usually a carnivorous mammal as final hosts [4]. Most recently, there is evidence of genetic similarity between liver flukes recovered from mammalian hosts, especially cats and *O. viverrini* in people [5,6]. Therefore, the complete elimination of infection may not be possible if domestic cat and dogs act as reservoir hosts, thus maintaining the source of flukes, therefore playing a critical role in maintaining their life cycle [4]. Thus, it is recommended that transdisciplinary research, including systematics and population genetics of the flukes and their life cycle hosts, should be undertaken to combat the liver flukes, therefore contributing to the reduction in CCA, particularly in endemic areas [4].

The comprehensive molecular systematics and population genetics investigations of *O. viverrini* have previously been undertaken, and cryptic species were discovered in 2007 [7]. Investigations of genetic variation of *C. sinensis* have also been reported from Russia, Vietnam, China, and Korea [6]. Some genetic markers used previously were unsuitable and unreliable for systematics and population genetic studies, such as markers in mitochondrial DNA genes, and some conserved regions in nuclear genes, due to observed low variation reported in some studies [7]. However, various molecular markers show different evolutionary rates and contain different genetic information. Compared to multi-copy sequences, single-copy genes may be more advantageous in identifying heterozygotes [8]. Thus, other polymorphic molecular markers, such as the sequence of intron regions of taurocyamine kinase gene (TK) need to be characterized and used to explore the genetic variation and population genetics of these liver flukes [7].

Phosphagen kinases (PKs) are phosphotransferases essential for cellular energy metabolism. In cells with high and fluctuating energy turnover rates, these highly conserved enzymes catalyze the reversible phosphate transfer between ATP and guanidine molecules [9]. Eight PKs are currently known, with creatine kinase (CK) as the sole PK in vertebrates. Arginine kinase (AK), glycocyamine kinase (GK), hypotaurocyamine kinase (HTK), lumbricine kinase (LK), opheline kinase (OK), thalessemine kinase (ThK), and taurocyamine kinase (TK) are PKs that are present in a wide range of invertebrate taxa [10]. The phosphagen kinase gene in trematodes refers to a specific gene, referred to as the taurocyamine kinase gene, which encodes an enzyme that plays a key role in the energy metabolism of trematode parasites [11]. Studying TK in trematodes involves characterizing its sequence, structure, expression patterns, and functional properties for identifying potential vulnerabilities that can be targeted for therapeutic interventions [11]. Not only functional genes (exon regions) but also intron regions have been successfully used as a genetic marker(s) to investigate the systematics and genetic variation of various organisms [12,13].

Non-coding introns are now commonly used in molecular phylogenetics and population genetics in an attempt to construct phylogenetic trees, and determine heterozygosity, DNA recombination, and genetic hybridization [14–17]. Compared to coding regions, introns predict the acquisition of a large number of independent parsimony-informative characters from most sites equally, associated with less homoplasy and lower transition: transversion ratios. Moreover, it must be acknowledged that diploid spliceosomal intron alleles have an average effective population size four times that of mitochondrial DNA (mtDNA) and in animals, introns mutate at approximately one quarter the rate of animal mtDNA [18,19].

Based on several previous publications, the intron regions of TK could offer an adequate resolution to examine genetic variation, genetic differentiation, genetic relationships, and heterozygosity of medically important parasitic trematodes [15–17]. For example, the bridge intron (TkBridgeInt) and intron 4 of domain 2 (TkD2Int4) regions have been applied to differentiate *Fasciola gigantica* and *F. hepatica*, as well as to explore genetic recombination in a *Fasciola* species (intermediate form), and in the case of using TkBridgeInt and TkD2Int4 regions, providing evidence of hybridization between *F. gigantica* and *F. hepatica* [15]. The second intron of domain 1 (TkD1Int2) was successfully used to differentiate species of the lung fluke genus *Paragonimus* [16], and evidence of the DNA recombination of *P. heterotremus* and *P. pseudoheterotremus* suggested incomplete lineage sorting [16]. Moreover, high polymorphism and a heterozygous TkD1Int5 region were detected in the intestinal flukes *Echinostoma* spp., which could potentially be used to differentiate *E. miyagawai* and *E. revolutum* [17]. All previous reports have shown that the intron region of TK has a potential role as a polymorphic genetic marker to explore the systematics and population genetics, especially heterozygosity, DNA recombination, and genetic hybridization, in these medically important trematodes [15–17]. Therefore, this study aimed to characterize and use the TK's intron region(s) as a molecular marker(s) for the population genetic investigations of *O. viverrini* sensu lato and *C. sinensis* isolated from various geographical isolates and animal hosts.

2. Materials and Methods

2.1. DNA Samples

The DNA of *O. viverrini* from Thailand and Lao PDR were provided by Prof. Paiboon Sithithaworn. The DNA of *C. sinensis* from Vietnam included samples from *Haplorchis taichui*, *Haplorchis pumilio*, and *Stellentchasmas fulcatus* which were provided by Prof. Wanchai Maleewong. The DNA of *C. sinensis* was collected from different hosts in China and was provided by Prof. Takeshi Agatsuma. A total of 147 *O. viverrini* and 45 *C. sinensis* DNA samples from different geographical localities and animal hosts in Thailand, Lao PDR, Vietnam, and China were used for comparative analyses in this study. The DNA samples of *H. pumilio*, *H. taichui*, and *S. fulcatus* were used to test cross-reactivity for TkD1Int5 amplification.

2.2. Primer Design

We designed the primer pairs to anneal the conserved flanking regions (exon) based on the full-length TK sequence of *C. sinensis* accession number JX435779 [11] for amplification by polymerase chain reaction (PCR) of seven intron regions. The resultant primers used to amplify three intron regions of domain 1 were intron 1 (TkD1Int1), intron 2 (TkD1intron2), and intron 5 (TkD1Int5), and three regions of domain 2, namely intron 2 (TkD2Int2), intron 3 (TkD2Int3), intron 4 (TkD2Int4), including the bridge intron (TkBridInt) (Figure 1). We also design a second primer pair (F2 and R2) to amplify some intron regions by a second PCR in cases where we were unable to amplify or there were low concentrations of PCR product amplified by the first primer pair (F1 and R1) (Figure 1).

Intron region	Primer name	Forward primer (primer sequence)	Reverse primer (5' to 3')	Tm
Domain 1 intron 1	TkD1Int1	OvD1Int1-F (5'-GAGGCCAATCGATCTTTCAC-3')	OvD1Int1-R (5'-ATAACGGCGTCAAAGAACTC-3')	50 °C
Domain 1 intron 2	TkD1Int2	OvD1Int2-F (5'-GTAATCCGCGGTGCATCCTT-3')	OvD1Int2-R (5'-GGAAATAGTGGGACAGAATCC-3')	55 °C
Domain 1 intron 5	TkD1Int5	OvD1Int5-F1 (5'-TTACATCGACTGGCCGAAT-3')	OvD1Int5-R1 (5'-GGGACACGAGCATGCACAC-3')	58 °C
		OvD1Int5-F2 (5'-AAGAAAATTTCCTGGTTTG-3')	OvD1Int5-R2 (5'-AAGGTGATAAAACCAAATC-3')	58 °C
Bridge intron	TkBridInt	OvBridInt-F (5'-GGTGGTGTCTACGATTTGAG-3')	OvBridInt-R (5'-ACGAACCATGTGTGCAAGAG-3')	55 °C
Domain 2 intron 2	TkD2Int2	OvD2Int2-F1 (5'-TGCAGTGATTTTGGACTATC-3')	OvD2Int2-R1 (5'-ATCCAAACGATCCTGCTTGC-3')	50 °C
		OvD2Int2-F2 (5'-AGCACCCACCGCCAACTTTC-3')	OvD2Int2-R2 (5'-GTTGAGAATAGAAACCCATC-3')	50 °C
Domain 2 intron 3	TkD2Int3	OvD2Int3-F (5'-GGATCGTTTGGATTTGGAAAC-3')	OvD2Int3-R (5'-ATCTTCTTCACACAACCAGAC-3')	50 °C
Domain 2 intron 4	TkD2Int4	OvD2Int4-F1 (5'-TAGCGTCCTACTCAAGATTC-3')	OvD2Int4-R1 (5'-TCCTTGGCAGCTTCCAACTC-3')	55 °C
		OvD2Int4-F2 (5'-GCCAAGCTGGATGAAGTATG-3')	OvD2Int4-R2 (5'-CGTTTGTTACTGATATCGTG-3')	55 °C

Figure 1. List of primers using for PCR amplification of seven intron regions of taurocyamine kinase gene. Tm is optimum annealing temperature for PCR of each primer pair.

2.3. PCR Analysis

To amplify each intron region, gradient PCR was used to optimize suitable conditions using an annealing temperature ranging between 50 and 60 °C. The PCR conditions were initial denaturation at 94 °C for 30 s, annealing at 50–60 °C for 30 s, extension at 72 °C for 1 min, and a final extension 72 °C for 5 min. In the case of low concentration (no/faint band), one microliter of the first PCR product was used as the DNA template for the second PCR using the same conditions as used with the first PCR. The PCR mixture contained 1× TaKaRa Ex PCR buffer, 0.2 mM dNTPs (each), 0.2 μM of each primer, and 1.0 U of TaKaRa Ex *Taq* polymerase (Takara Bio Inc., Shiga, Japan). Subsequently, the PCR product underwent electrophoresis in 1% agarose gel and was visualized with GelRed^{TM} Nucleic Acid Gel Stain (Biotium, Inc., Hayward, CA, USA). The PCR product (~1000 bp) was cut for gel purification using an E.Z.N.A.® Gel Extraction kit (Omega bio-tek, Norcross, GA, USA) and was subsequently used for DNA sequencing and cloning.

2.4. DNA Sequencing and Molecular Cloning

The purified PCR products were cycle-sequenced via a commercial service provider (Eurofin, Tokyo, Japan). Sequencing results were visualized in a Sequence Scanner ver. 1.0 program to check for heterozygous patterns (Figure 2). Thereafter, heterozygous samples were detected from chromatograms, which showed the existence of indel nucleotide (Figure 2). Then, the purified PCR product of a particular heterozygous sample was cloned into a pGEM-T Easy Vector (Promega, Madison, WI, USA) according to manufacturer's instructions. The recombinant plasmid was transformed and propagated in *Escherichia coli* JM109, and then four to ten inserted colonies were randomly picked and cultured in Luria-Bertani (LB) broth at 37 °C with continuous horizontal shaking for 12 h. Then, plasmid DNA was extracted using the FastGene® Plasmid Mini kit (Nippon Genetics Co., Ltd., Tokyo, Japan) following manufacture's protocol. The DNA from those plasmids was used as a template for sequencing in both directions using M13F and M13R as sequencing primers to detect the different haplotypes mixed in each particular heterozygous sample.

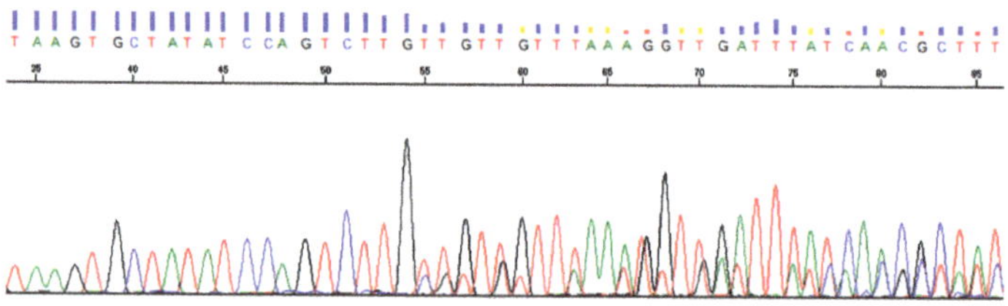

```
KBp7-C1   TAAGTGCTATATCCAGTCTTG-CGTTGTTTAAATGTTGATTTATCAACGCT
KBp7-C4   TAAGTGCTATATCCAGTCTTG-CGTTGTTTAAATGTTGATTTATCAACGCT
KBp7-C2   TAAGTGCTATATCCAGTCTTGTTGTTGTTTAAATGTTGATTTATCAACGCT
          *********************  ****************************
```

Figure 2. Heterozygous pattern of an *O. viverrini* sample KBp7 represents an insertion/deletion (indel) region as yellow highlighted. After the indel, the chromatogram of nucleotide sequence is unreadable. Three clones and plasmid sequencing revealed that different haplotypes of TkD1Int5 sequence were mixed in this sample. Identical nucleotide indicated by "*". Different line colors represent different nucleotide base, red: base T; blue: base C; greed: base A; black: base G.

2.5. Data Analysis

Sequences were manually checked and edited using the BioEdit program version 7.2.6 [20]. Exon regions were blasted in NCBI for checking, and the real intron of TK was amplified. Thereafter, those exon regions were trimmed out before being used in intron sequence analysis. The intron sequences were multiple aligned using the ClustalW program version 2.0 [21] to compare and search for the variable sites, including microsatellite and insertion/deletion (indel) regions. Diversity indices and haplotype data were calculated and generated using the DnaSp program version 5 [22]. A neighbor joining (NJ) tree was constructed by using the MEGA program version 10 [23].

3. Results

3.1. Intron Amplification

Primer pairs were designed to amplify the seven intron regions, of these, five intron regions (TkD1Int2, TkD1Int5, TkD2Int2, TkD2Int4, and TkBridInt) were successfully amplified. From our experience, the suitable intron region for use as a genetic marker for genetic investigation in the trematodes should amplify a PCR product of 800–1200 bp in length. Thus, we selected only TkD1Int5, which provided a PCR product in length ~1000 bp. The other intron regions produced a PCR product either less than 800 bp or larger than 3000 bp (Figure S1). We tested the cross reaction of TkD1Int5 primers with the other *O. viverrini*-like eggs flukes, namely *H. taichui*, *H. pumilio*, and *S. fulcatus*, using the same PCR condition with *O. viverrini* and *C. sinensis* amplification. We found no cross-reaction.

3.2. Nucleotide Sequence Analysis

After the exon sequences were trimmed, the length of TkD1Int5 used in this analysis ranged between 852 and 893 for *O. viverrini*, and was 871 bp for *C. sinensis*. All TkD1Int5 haplotype sequences of *O. viverrini* and *C. sinensis* examined in this study were deposited in GenBank under the accession numbers OR454257–OR454358 and OR440582–OR440599, respectively. Several insertion/deletion (indel) regions were observed in the TkD1Int5 region of *O. viverrini* (Table S1). While only one nucleotide insertion was found in the TkD1Int5 region of *C. sinensis* from Vietnam (Table S1). One repeated sequence/microsatellite "CTGTA" was embedded in the TkD1Int5 region of *O. viverrini*, and two large indel fragments, namely, fragments of nucleotide sites 356–372 and 423–450 were detected in *O. viverrini*. Based on variable sites observed in TkD1Int5 of 147 *O. viverrini* (185 sequences) and 53 *C. sinensis* sequences, 102 (Ov1–Ov102) and 18 (Cs1–Cs18) haplotypes were generated, respectively (Figure 3).

Parasites	Country	Codes	District/Province	n	het	seq	s	h	Uh	Hd±SD	Nd±SD
O. viverrini	Thailand	BR	Buri Ram	10	2	11	23	5	3	0.618±0.164	0.0057±0.0019
		KLp	Muang/Khon Khaen	9	4	12	37	12	9	1.000±0.034	0.0121±0.0024
		KBp	Ban Phai/Khon Khaen	10	2	13	28	11	8	0.974±0.039	0.0076±0.0020
		KPv	Phuviang/Khon Khaen	10	2	13	20	7	4	0.846±0.085	0.0052±0.0017
		CP	Chaiya Phum	6	3	10	27	10	6	1.000±0.045	0.0105±0.0022
		MD	Mukdahan	9	4	14	40	12	9	0.978±0.035	0.0104±0.0024
		MS	Maha Sarakham	10	0	10	9	7	4	0.911±0.077	0.0027±0.0013
		SKm	Muang/Sakon Nakhon	9	1	11	14	8	5	0.927±0.066	0.0037±0.0015
		SKp	Phang Khon/Sakon Nakhon	10	0	10	2	3	2	0.511±0.164	0.0008±0.0006
		LP	Lampang	9	2	11	28	8	6	0.891±0.092	0.0088±0.0022
	Lao PDR	SV	Savannakhet	8	1	9	27	6	4	0.833±0.127	0.0082±0.0022
		KM	Khammouan	9	3	10	25	10	6	1.000±0.045	0.0072±0.0021
		NG	Nam Ngum/Veintaine	10	2	11	18	6	3	0.800±0.114	0.0049±0.0017
		TH	Tha Haue/Veintaine	7	2	8	19	8	6	1.000±0.063	0.0069±0.0020
		VT	Kam Pang Nakhon/Veintaine	10	3	11	15	6	3	0.836±0.089	0.0039±0.0016
		VV	Vang Viang/Veintaine	11	8	21	40	21	18	1.000±0.015	0.0115±0.0021
			Total	147	39	185	130	102	96	0.943±0.010	0.0078±0.0024
C. sinensis	China	CS-N[1]	NanNing/Guangxi	9	0	9	0	1	0	0.000±0.000	0.0000±0.0000
		CS-G[2]	Guangzhou/Guangdong	8	0	8	1	2	0	0.250±0.180	0.0003±0.0004
		CS-J[3]	Jaimusi/Heilongjiang	22	15	22	9	12	9	0.879±0.055	0.0028±0.0013
		CS-D[4]	Daqing/Heilongjiang	11	0	11	5	6	5	0.727±0.144	0.0014±0.0019
	Vietnam	CS-V[1]	Thai Binh	3	0	3	5	2	1	0.677±0.098	0.0038±0.0017
			Total	53	15	53	14	18	15	0.741±0.060	0.0018±0.0013

Figure 3. Diversity indices of the sequences of domain 1 intron 5 of taurocyamine kinase gene in the *O. viverrini* and *C. sinensis* populations from various isolates and animal hosts. Isolates from naturally infected cats [1], dogs [2], humans [3], and experimental rabbits [4], whereas all *O. viverrini* samples were from experimental hamsters. n: Number of DNA sample; het: number of heterozygous samples; seq: number of sequences generated in a particular isolate; s: segregations site; h: number of haplotypes; Uh: number of unique haplotypes; Hd: haplotype diversity; Nd: nucleotide diversity; SD: standard deviation.

Seven haplotypes of *O. viverrini* were shared between isolates, namely haplotypes Ov15, Ov16, Ov20, Ov31, Ov48, Ov52, and Ov74. The most common haplotype was Ov48, which was found in 34 sequences from various isolates (Figure 4). The TkD1Int5 sequence analysis of *C. sinensis* revealed three haplotypes, Cs1, Cs2, and Cs16, shared between different isolates/animal hosts. The isolates from humans depicted a high number of specific haplotypes, while no specific haplotypes were observed in *C. sinensis* collected from cats and dogs (Figure 4).

3.3. Heterozygosity and Phylogenetic Tree Analysis

Several samples of *O. viverrini* contained indel and/or microsatellite regions which was the predominant heterozygous pattern observed in this study. The 37 *O. viverrini* samples exhibited heterozygous pattern(s) by direct DNA sequencing (Figure 2). All heterozygous samples were cloned and randomly picked inserted colonies for DNA sequencing. Of these, the 27 samples demonstrated the combination of at least two haplotypes within a particular heterozygous sample, while only one haplotype could be detected in the rest of the heterozygous samples. Interestingly, there were eight *O. viverrini* samples that contained three haplotypes within a particular heterozygous sample, namely CP8, KBp4, KLp5, KPv3, MD1, SK1, VV4, and VV6 (Figure 5). These haplotypes contained the indel regions and the repeated "CGCTA" region, which are the main evidence of heterozygosity (Figure 5).

We found a heterozygous pattern in *C. sinensis* with double peaks at one nucleotide site (Figure 6). This heterozygous pattern was found in the 15 *C. sinensis* recovered from humans, i.e., CSJ01–CSJ08, CSJ10–CSJ12, and CSJ15–CSJ18. The phylogenetic tree constructed by

TkD1Int5 sequences revealed that 102 *O. viverrini* haplotypes clustered as a monophyletic group (Figure S2), as also did *C. sinensis* (Figure S3).

Parasites	Codes	n	Shared haplotypes							Specific haplotypes
			Ov15	Ov16	Ov20	Ov31	Ov48	Ov52	Ov74	
O. viverrini	BR	11		1			7			Ov4, Ov39, Ov98
	KLp	12		1			1		1	Ov1, Ov3, Ov22, Ov26, Ov66, Ov72, Ov73, Ov78, Ov79
	KBp	13	2				1		2	Ov27, Ov29, Ov30, Ov32, Ov62, Ov83, Ov87, Ov99
	KPv	13	2				5		2	Ov19, Ov55, Ov88, Ov96
	CP	10		1	1		1		1	Ov7, Ov8, Ov23, Ov37, Ov86, Ov101
	MD	14		2				1	2	Ov5, Ov6, Ov13, Ov36, Ov42, Ov57, Ov58, Ov89, Ov100
	MS	10		1			3		2	Ov46, Ov51, Ov67, Ov71
	SKm	11		1			2		3	Ov35, Ov54, Ov77, Ov92, Ov95
	SKp	10	7			2				Ov33
	LP	11				1				Ov10, Ov47, Ov60, Ov69, Ov85, Ov97
	SV	9					4		1	Ov2, Ov17, Ov28, Ov70
	KM	10	1				1	1	1	Ov41, Ov44, Ov53, Ov61, Ov75, Ov84
	NG	11	2		1		5			Ov11, Ov34, Ov65
	TH	8		1	1					Ov9, Ov21, Ov40, Ov61, Ov68, Ov76
	VT	11	1				3		4	Ov12, Ov24, Ov49
	VV	21	1				1		1	Ov14, Ov18, Ov25, Ov38, Ov43, Ov45, Ov50, Ov56, Ov59, Ov63, Ov80, Ov81, Ov82, Ov90, Ov91, Ov93, Ov94, Ov102
	Total	185	16	8	3	3	34	2	20	99

Parasites	Codes	n	Shared haplotypes			Specific haplotypes
			Cs1	Cs2	Cs11	
C. sinensis	CS-N[1]	9		9		n/a
	CS-G[2]	8		6	1	n/a
	CS-J[3]	22	1	4	7	Cs6, Cs8, Cs9, Cs12(2), Cs13 – Cs17,
	CS-D[4]	11		6		Cs3, Cs4, Cs7, Cs10, Cs18
	CS-V[1]	3	2			Cs5
	Total	53	3	25	8	17

Figure 4. Frequency of shared and specific haplotypes of *O. viverrini* and *C. sinensis* populations examined by domain 1 intron 5 of taurocyamine kinase gene sequence. Isolates from naturally infected cats [1], dogs [2], humans [3], and experimental rabbits [4], whereas all *O. viverrini* samples were from experimental hamster. n: Number of sequences.

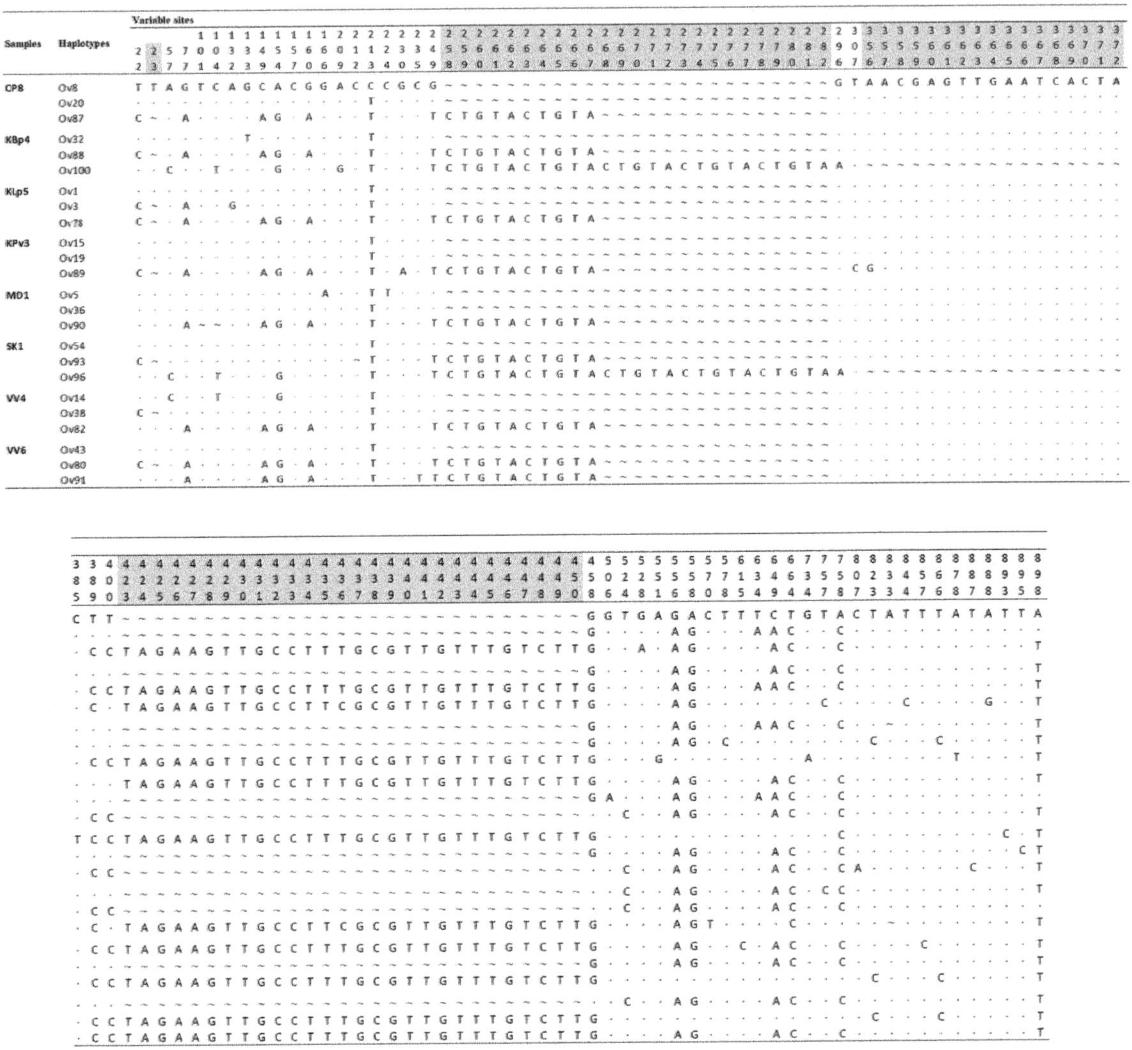

Figure 5. Variable nucleotide positions of TkD1Int5 of eight *O. viverrini* samples contained three haplotypes in a particular sample. Identical nucleotide indicated by "·"; insertion/deletion (indel) site indicated by "~"; indel site leading to heterozygosity observation indicated by gray shedding.

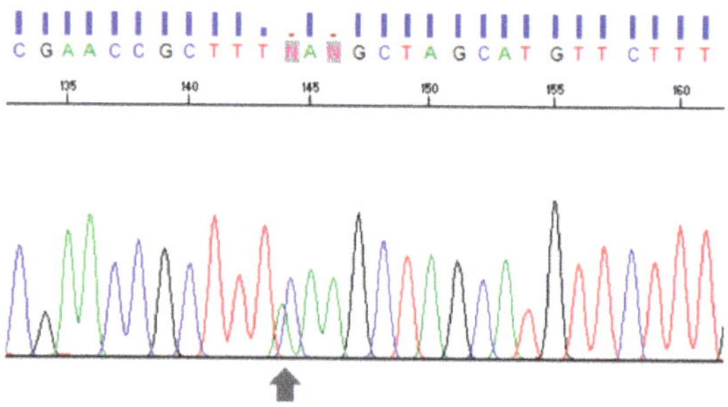

Figure 6. Heterozygous pattern of *C. sinensis* represents double peaks (A/C) at a particular nucleotide site as pointed by an arrow. Different line colors represent different nucleotide base, red: base T; blue: base C; greed: base A; black: base G.

4. Discussion

We successfully characterized a highly polymorphic intron region TkD1Int5, which can be used as a potential genetic marker for the investigation of the genetic variation and population genetics of *O. viverrini* sensu lato and *C. sinensis*. The TkD1Int5 region showed high polymorphism in *O. viverrini* but low variation in *C. sinensis*. This finding provided evidence that *C. sinensis* is more clonal than *O. viverrini*. In addition, the TkD1Int5 region of *O. viverrini* contained the repeated sequence and several indel regions, which caused the heterozygosity observed in this study. We unfortunately could not obtain *O. felineus*, another related species, to compare to *O. viverrini* and *C. sinensis*. The TkD1Int5 region could possibly be used for further genetic investigations of *O. felineus*. Interestingly, other intron and/or exon regions examined in this study could potentially be used to examine genetic differentiation between these liver flukes and other related species, especially co-endemic *O. viveriini*-like egg flukes. For instance, seven species of Southeast Asian *Paragonimus* could be differentiated by using variable nucleotide sites in the exon region of TK [16].

These liver flukes are diploid hermaphroditic trematodes. Therefore, reproduction can involve either self-fertilization or cross-fertilization [4]. Previous studies have commonly observed heterozygosity in *O. viverrini* by allozyme and microsatellite analyses, which indicates cross-fertilization as an optional mode of reproduction [7]. The results in our study found that 37 *O. viverrini* DNA samples (25.2%) depicted heterozygous patterns. Interestingly, seven *O. viverrini* DNA samples contained three different genotypes of TkD1Int5 within a particular sample. The samples of adult worms examined were from a hamster with an infected dose of 50–100 metacercariae, hence providing evidence that cross-breeding can occur in an infected hamster resulting in heterozygous fertilized eggs. In this case, DNA extraction was performed by using a whole adult worm. Therefore, DNA from fertilized eggs in their uterus can be extracted, which may have contaminated the total DNA sample. However, there is also the possibility that the triploid sample contained three haplotypes, as evidenced in *Paragonimus westermani* [16]. Only diploidy was observed in a previous chromosomal and karyotype analysis of *O. viverrini* [24], however, it was conducted on a small sample number of isolates of *O. viverrini*. A larger sample size from a wide variety of isolates and animal hosts needs to be investigated to provide a better understanding of ploidy.

Most recently, *O. viverrini* samples collected from Phang Khon District, Sakon Nakhon Province (SKp), have been previously classified as a cryptic genetic group by microsatellite, mitochondrial, and nuclear DNA markers [7]. We recently found that 10 *O. viverrini* samples from SKp could be classified into three haplotypes, i.e., Ov15, Ov31, and Ov33. Haplotype Ov15 was shared between the samples from several isolates, namely Lampang (LP), Khon Kaen (KPv and KBp), and SKp from Thailand; and Vientiane (VT), Khammouan (KM), Nam Ngum (NG), and Vang Viang (VV) from Lao PDR. Haplotype Ov31 was shared between LP and SKp, whereas Ov33 haplotypes were specifically detected only in SKp. However, in the phylogenetic tree, SKp was not clearly resolved as a genetically distinct group, as was seen when examining the TkD1Int5 region. Our observation raised an interesting question as to whether the SKp *O. viverrini* population consisted of two genetic groups.

Several closely related species of *O. viverrini* have recently been reported, i.e., *Opisthorchis lobatus* found in freshwater fish in Lao PDR, which may also cause zoonosis, but its role in humans is not known [25]. Similarly, the role of avian species is unknown; they include *O. cheelis*, *O. longissimus*, and *O. parageminus* which have been also reported from Southeast Asia [26,27]. Dao and colleagues [28] reported a sympatric distribution of duck and human genotypes of "*O. viverrini*". The discovery of several species in the genus *Opisthorchis* in addition to the species complex of *O. viverrini* reflects complicated host and parasite interactions and potential co-evolution. Thus, comprehensive molecular systematics investigations of *O. viverrini* and its sibling species are needed using TK intron sequences or the other polymorphic DNA sequences as genetic markers.

As these liver flukes can infect multiple hosts [7], they are potentially zoonotic; however, whether this is the case remains controversial. For instance, a recent report found that *O. viverrini* from infected cats and humans, despite being genetically similar, separated into two distinct genetic groups, suggestive of host specificity of 'human' and 'cat' genotype [5]. Unfortunately, we did not have access to *O. viverrini* samples collected from other animal hosts to use in comparative analyses in our study. Further investigations should be conducted using TkD1Int5 for the geno typing of *O. viverrini* collected from various natural species of animal hosts. It has been shown that genetic variation of *C. sinensis* is not related to its infected hosts [6], but significant differences have been found between different geographical isolates from Russia and Vietnam [29]. We also found that *C. sinensis* from infected cats, dogs, rabbits, and humans were not genetically distinct, which was shown when examining by TkD1Int5. These results provide compelling evidence that *C. sinensis* performs a significant role in zoonotic transmission. However, we found a heterozygous pattern of *C. sinensis* recovered from humans. This finding suggested that the cryptic genotype(s) of *C. sinensis* in their natural hosts may exist; hence, a larger sample size from wide variety of natural species of hosts needs to be further investigated to determine whether this is the case.

5. Conclusions

This study reported genetic variation of *O. viverrini* and *C. sinensis* collected from various geographical isolates and animal hosts using the intron region TkD1Int5 as a genetic marker. A high variation of the TkD1Int5 sequence was observed in *O. viverrini*, while a low variation was observed in *C. sinensis*, reflecting that *C. sinensis* is more clonal than *O. viverrini*. Heterozygosity was observed in several *O. viverrini* samples, which indicates that cross-fertilization is an optional mode of reproduction, and that TkD1Int5 was suitable for population genetic analysis of *O. viverrini*. However, we found no genetic differences between *C. sinensis* recovered from different animal hosts. This finding demonstrated that various animal hosts play significant roles in maintaining the life cycle of *C. sinensis*. Hence, it is highly possible that *C. sinensis* is circulated between human and animal hosts. In the case of *O. viverrini*, we need further investigations of the TkD1Int5 sequence from samples collected from a range of different species of animal hosts across a wide geographical range to better understand its role in zoonotic diseases. If these mammalian hosts are involved in the transmission cycle of these liver fluke circulated between humans, then control approaches must also include zoonotic cycles in carnivore reservoir hosts, which are significant as they will maintain the prevalence of infection and incidence of disease in humans.

Supplementary Materials: The following supporting information can be downloaded at: https://www.mdpi.com/article/10.3390/ani13203200/s1, Figure S1: PCR products of amplifiable five intron regions with an appropriate size ~800–1200 bp of TkD1Int5 region indicated in a square was selected in this study; Figure S2: Neighbor joining tree constructed based on TkD1Int5 sequences of *O. viverrini*; Figure S3: Neighbor joining tree constructed based on TkD1Int5 sequences of *C. sinensis*; Table S1: Variable nucleotide sites of *O. viverrini* (Ov1–Ov102) and *C. sinensis* (Cs1–Cs18) TkD1Int5 haplotypes.

Author Contributions: Conceptualization, C.T. and W.S.; methodology, C.T., W.S., T.T. and T.A.; formal analysis, C.T. and W.S.; investigation, C.T., W.S. and W.P.; resources, W.M., P.S. and T.A.; writing—original draft preparation, C.T. and W.S.; writing—review and editing, R.H.A., P.S., W.M. and T.A.; project administration, C.T. and W.S.; funding acquisition, W.S. and C.T. All authors have read and agreed to the published version of the manuscript.

Funding: This research project was financially supported by Mahasarakham University. This study was supported by the Thailand Science Research and Innovation Fundamental Fund (TUFF53/2566). We wish to acknowledge the support of the Japan Society for the Promotion of Science (JSPS) for research in Japan (grant no. S22093) awarded to W.S. hosted by T.A. at the Department of Environmental Medicine, Kochi Medical School, Japan.

Institutional Review Board Statement: The study was conducted in accordance with The Animal Care and Use Committee, Mahasarakham University with approval no. IACUC-MSU-14/2021, issued on 25 May 2021. Biosafety was approved by the Institute Biosafety Committee of Thammasat University, permission no. 035/2565, approved on 26 April 2022.

Informed Consent Statement: Not applicable.

Data Availability Statement: All data are available upon request.

Acknowledgments: We would like to thank Adrian R. Plant for English proofreading.

Conflicts of Interest: The authors declare no conflict of interest.

References

1. WHO. *Control of Foodborne Trematode Infections*; World Health Organization: Geneva, Switzerland, 1995.
2. Keiser, J.; Utzinger, J. Emerging foodborne trematodiasis. *Emerg. Infect. Dis.* **2005**, *11*, 1507–1514.
3. IARC. A review of carcinogen—Part B: Biological Agents. In *Monographs on the Evaluation of Carcinogenic Risks to Humans*; International Agency for Research on Cancer: Lyon, France, 2011.
4. Saijuntha, W.; Sithithaworn, P.; Kiatsopit, N.; Andrews, R.H.; Petney, T.N. Liver flukes: *Clonorchis* and *Opisthorchis*. In *Digenetic Trematodes*, 2nd ed.; Toledo, R., Fried, B., Eds.; Advances in Experimental Medicine and Biology; Springer: Cham, Switzerland, 2019; Volume 1154, pp. 139–180.
5. Sato, P.; Suttiprapa, S.; Tangkawattana, S.; Sripa, M.; Blair, D.; Sripa, B. Does *Opisthorchis viverrini* circulate between humans and domestic cats in an endemic area in Thailand? *Parasitology* **2022**, *149*, 1334–1338.
6. Wang, D.; Young, N.D.; Korhonen, P.K.; Gasser, R.B. *Clonorchis sinensis* and Clonorchiasis: The Relevance of Exploring Genetic Variation. *Adv. Parasitol.* **2018**, *100*, 155–208. [PubMed]
7. Saijuntha, W.; Sithithaworn, P.; Petney, T.N.; Andrews, R.H. Current assessment of the systematics and population genetics of *Opisthorchis viverrini* sensu lato (Trematoda: Opisthorchiidae) and its first intermediate host *Bithynia siamensis* sensu lato (Gastropoda: Bithyniidae) in Thailand and Southeast Asia. *Infect. Genet. Evol.* **2022**, *97*, 105182.
8. Tang, M.; Zhou, Y.; Liu, Y.; Cheng, N.; Zhang, J.; Xu, X. Molecular identification and genetic-polymorphism analysis of Fasciola flukes in Dali Prefecture, Yunnan Province, China. *Parasitol. Int.* **2021**, *85*, 102416.
9. Platzer, E.G.; Wang, W.; Thompson, S.N.; Borchardt, D.B. Arginine kinase and phosphoarginine, a functional phosphagen, in the rhabditoid nematode steinernema carpocapsae. *J. Parasitol.* **1999**, *85*, 603–607.
10. Uda, K.; Fujimoto, N.; Akiyama, Y.; Mizuta, K.; Tanaka, K.; Ellington, W.R.; Suzuki, T. Evolution of the arginine kinase gene family. *Comp. Biochem. Physiol. Part D Genom. Proteom.* **2006**, *1*, 209–218.
11. Xiao, J.Y.; Lee, J.Y.; Tokuhiro, S.; Ngataki, M.; Jarilla, B.R.; Nomura, H.; Kim, T.I.; Hong, S.J.; Agatsuma, T. Molecular cloning and characterization of taurocyamine kinase from *Clonorchis sinensis*: A candidate chemotherapeutic target. *PLoS Negl. Trop. Dis.* **2013**, *7*, e2548.
12. Creer, S. Choosing and using introns in molecular phylogenetics. *Evol. Bioinform.* **2007**, *3*, 99–108.
13. Doumen, C. Variable intron/exon structure in the oligochaete lumbricine kinase gene. *Gene* **2012**, *505*, 276–282.
14. Lynch, M. Intron evolution as a population-genetic process. *Proc. Natl. Acad. Sci. USA* **2002**, *99*, 6118–6123.
15. Saijuntha, W.; Tantrawatpan, C.; Wang, C.; Agatsuma, T.; Intapan, P.M.; Maleewong, W.; Petney, T.N. Revealing genetic hybridization and DNA recombination of *Fasciola hepatica* and *Fasciola gigantica* in nuclear introns of the hybrid *Fasciola* flukes. *Mol. Biochem. Parasitol.* **2018**, *223*, 31–36. [PubMed]

16. Tantrawatpan, C.; Tapdara, S.; Agatsuma, T.; Sanpool, O.; Intapan, P.M.; Maleewong, W.; Saijuntha, W. Genetic differentiation of Southeast Asian *Paragonimus* Braun 1899 (Digenea: Paragonimidae) and genetic variation in the *Paragonimus heterotremus* complex examined by nuclear DNA sequences. *Infect. Genet. Evol.* **2021**, *90*, 104761.
17. Saijuntha, W.; Tantrawatpan, C.; Agatsuma, T.; Duenngai, K.; Sithithaworn, P.; Andrews, R.H.; Petney, T.N. Intron sequence variation of the echinostomes (Trematoda; Echinostomatidae): Implications for genetic investigations of the 37 collar-spined, *Echinostoma miyagawai* Ischii, 1932 and *E. revolutum* (Fröelich, 1802). *Parasitol. Res.* **2020**, *199*, 2485–2494.
18. Prychitko, T.M.; Moore, W.S. Comparative evolution of the mitochondrial cytochrome b gene and nuclear ß-fi brinogen intron 7 in woodpeckers. *Mol. Biol. Evol.* **2000**, *17*, 1101–1111.
19. Prychitko, T.M.; Moore, W.S. Alignment and phylogenetic analysis of β-fi brinogen intron 7 sequences among avian orders reveal conserved regions within the intron. *Mol. Biol. Evol.* **2003**, *20*, 762–771. [PubMed]
20. Hall, T.A. BioEdit: A user-friendly biological sequence alignment editor and analysis program for Windows 95/98/NT. *Nucl. Acids Symp. Ser.* **1999**, *41*, 95–98.
21. Larkin, M.A.; Blackshields, G.; Brown, N.P.; Chenna, R.; McGettigan, P.A.; McWilliam, H.; Valentin, F.; Wallace, I.M.; Wilm, A.; Lopez, R.; et al. Clustal W and Clustal X version 2.0. *Bioinformatics* **2007**, *23*, 2947–2948.
22. Librado, P.; Rozas, J. DnaSP v5: A software for comprehensive analysis of DNA polymorphism data. *Bioinformatics* **2009**, *25*, 1451–1452.
23. Kumar, S.; Stecher, G.; Li, M.; Knyaz, C.; Tamura, K. MEGA X: Molecular Evolutionary Genetics Analysis across computing platforms. *Mol. Biol. Evol.* **2018**, *35*, 1547–1549.
24. Kaewkong, W.; Choochote, W.; Kanla, P.; Maleewong, W.; Intapan, P.M.; Wongkham, S.; Wongkham, C. Chromosomes and karyotype analysis of a liver fluke, *Opisthorchis viverrini*, by scanning electron microscopy. *Parasitol. Int.* **2012**, *61*, 504–507. [PubMed]
25. Thaenkham, U.; Nuamtanong, S.; Vonghachack, Y.; Yoonuan, T.; Sanguankiat, S.; Dekumyoy, P.; Prommasack, B.; Kobayashi, J.; Waikagul, J. Discovery of *Opisthorchis lobatus* (Trematoda: Opisthorchiidae): A new record of small liver flukes in the Greater Mekong Sub-region. *J. Parasitol.* **2011**, *97*, 1152–1158. [CrossRef] [PubMed]
26. Nawa, Y.; Doanh, P.N.; Thaenkham, U. Is *Opisthorchis viverrini* an avian liver fluke? *J. Helminthol.* **2015**, *89*, 255–256. [PubMed]
27. Doanh, P.N.; Nawa, Y. *Clonorchis sinensis* and *Opisthorchis* spp. in Vietnam: Current status and prospects. *Trans. R. Soc. Trop. Med. Hyg.* **2016**, *110*, 13–20.
28. Dao, T.T.H.; Nguyen, T.T.G.; Gabriel, S.; Bui, K.L.; Dorny, P.; Le, T.H. Updated molecular phylogenetic data for *Opisthorchis* spp. (Trematoda: Opisthorchioidea) from ducks in Vietnam. *Parasit. Vectors* **2017**, *10*, 575. [PubMed]
29. Chelomina, G.N.; Tatonova, Y.V.; Hung, N.M.; Ngo, H.D. Genetic diversity of the Chinese liver fluke *Clonorchis sinensis* from Russia and Vietnam. *Int. J. Parasitol.* **2014**, *44*, 795–810.

animals

MDPI

Brief Report

Efficacy of Sustained-Release Formulation of Moxidectin (Guardian SR) in Preventing Heartworm Infection over 18 Months in Dogs Living in a Hyperendemic Area

Agustina Isabel Quintana-Mayor [1], Elena Carretón [2,*] and José Alberto Montoya-Alonso [2]

1 "El Parque" Veterinary Center, C/ María Encarnación Navarro no. 42 Local Bajo, 35200 Telde, Spain; clinica_el_parque@hotmail.com

2 Internal Medicine, Veterinary Medicine and Therapeutic Research Group, Faculty of Veterinary Medicine, Research Institute of Biomedical and Health Sciences (IUIBS), Universidad de Las Palmas de Gran Canaria (ULPGC), 35016 Las Palmas de Gran Canaria, Spain; alberto.montoya@ulpgc.es

* Correspondence: elena.carreton@ulpgc.es

Simple Summary: This study evaluates the efficacy of a sustained-release (SR) moxidectin microsphere formulation in preventing canine heartworm infection over an 18-month period in Canary Hound dogs. Conducted in a hyperendemic region, the study included 109 outdoor dogs from 11 kennels, with heartworm prevalence rates ranging from 11.1% to 57.1%. Twenty healthy, heartworm-negative dogs received a single subcutaneous injection of moxidectin SR. Antigen and Knott's tests were conducted at 6, 12, 18, and 24 months, and no infections were detected. All treated dogs remained healthy and heartworm-negative throughout the study. The results suggest that a single dose of moxidectin SR may prevent heartworm infection for 18 months; however, further research is needed to confirm these findings. Extending the efficacy interval of moxidectin may enhance owner compliance, especially in populations with lower awareness of animal health.

Abstract: This study investigates the efficacy of a sustained-release (SR) moxidectin microsphere formulation in preventing canine heartworm infection over 18 months in Canary Hound dogs, a hunting breed common in the Canary Islands, which is a hyperendemic region. These dogs typically do not receive preventive treatments and act as reservoirs for the disease. This field study was conducted across 11 hunting kennels with 109 dogs living outdoors, none of whom were receiving heartworm prophylaxis, with *Dirofilaria immitis* prevalence ranging from 11.1% to 57.1% (average 36.7%). Among these, 20 clinically healthy, heartworm-negative dogs were randomly selected to receive a single subcutaneous injection of moxidectin SR (0.17 mg/kg body weight). Antigen and Knott's tests were performed at 6, 12, 18, and 24 months. All dogs completed the study without adverse reactions and remained heartworm-negative throughout. By the end of the study, kennel heartworm prevalence ranged from 14.3% to 46.7% (average 35.4%). A single subcutaneous dose of moxidectin SR at the recommended dosage may prevent patent heartworm infection in dogs for up to 18 months in hyperendemic regions. Further studies are required to confirm these findings. Extending the efficacy period of moxidectin could improve owner compliance, particularly among those with lower animal health awareness.

Keywords: *Dirofilaria immitis*; heartworm; chemoprophylaxis; canine; macrocyclic lactones; moxidectin; microspheres

1. Introduction

Heartworm disease (*Dirofilaria immitis*) is a parasitic zoonosis with global distribution, particularly prevalent in temperate, tropical, and subtropical regions due to the climatic requirements of its mosquito vectors [1]. However, areas with warm summers, large rivers, lakes, or extensive irrigation systems also provide ideal mosquito habitats, leading to high prevalence rates [2]. The incidence of heartworm has dramatically increased in recent years, spreading to colder regions, likely influenced by climate change and global warming [3]. This has led to longer mosquito activity periods, accelerated larval development stages, and expanded transmission across multiple regions. Other anthropogenic factors, such as urban heat islands, the introduction of new mosquito species capable of acting as vectors, and increased movement of reservoir dogs, further facilitate this spread [4,5].

The Canary Islands have long been considered a hyperendemic area for heartworm [6–12]. Veterinarians often prioritize heartworm in the differential diagnosis of cardiopulmonary symptoms in dogs not receiving regular chemoprophylaxis. In Gran Canaria, the prevalence of heartworm has significantly declined since the publication of the first epidemiological studies, from 67.02% in 1994 to 16.03% in the most recent reports, largely due to the continued awareness efforts of veterinarians [6,12]. However, despite this reduction, the prevalence of heartworm has remained stagnant at 16–20.7% for over 15 years. This suggests that further reducing the disease's incidence on the islands, particularly in Gran Canaria, may be difficult [7–10,12].

A key reason for the persistently high prevalence of heartworm, despite educational efforts, may be the handling of dogs by hunters and rural populations. These dogs, primarily of the Canary Hound breed, have shown heartworm prevalence rates twice as high as those of the general dog population in Gran Canaria, remaining slightly above 40% in all published studies [8–10]. This elevated prevalence is likely due to factors such as the absence of chemoprophylaxis, poor sanitary conditions, outdoor housing on farms or in kennels, and exposure to mosquito bites [10]. Consequently, this breed acts as a reservoir for heartworm, complicating control efforts in the Canary Islands.

Previous studies have assessed the efficacy of sustained-release (SR) moxidectin microsphere formulations, which are used to provide 12 months of protection against heartworm [13,14]. These studies have also indicated that injectable moxidectin results in higher compliance with heartworm prevention [15,16], although the duration of protection remains uncertain, with some suggesting it could last longer [13]. Therefore, this study aims to evaluate the efficacy of a moxidectin SR formulation over 18 months in a hyperendemic area using the highly exposed Canary Hound breed.

2. Materials and Methods

2.1. Location and Climate of the Canary Islands

Gran Canaria is part of the Canary Islands, a volcanic archipelago located 97 km off the coast of the Sahara. The island features a variety of climates based on altitude and geographical position, including desert (BW), steppe (BS), temperate with hot and dry summers (Csa), and temperate with dry and warm summers (Csb) climates, as per the Köppen climate classification system [17,18]. Gran Canaria is considered a hyperendemic area for heartworm, with prevalence variations depending on the island's isoclimatic zones [10–12].

2.2. Studied Animals

This observational field study was conducted across 11 kennels housing Canary Hounds (Figure 1). Inclusion criteria required that the dogs had not received any chemoprophylaxis for heartworm or mosquitoes and lived outdoors continuously.

At the study's outset, each kennel housed between 6 and 14 dogs, with a total of 109 dogs (Table 1). Blood samples were drawn from the cephalic vein of each dog to determine heartworm prevalence through antigen detection using a commercial test (Uranotest® Dirofilaria, Uranovet, Barcelona, Spain) following the manufacturer's instructions. Additionally, microfilariae detection was performed using the modified Knott's test.

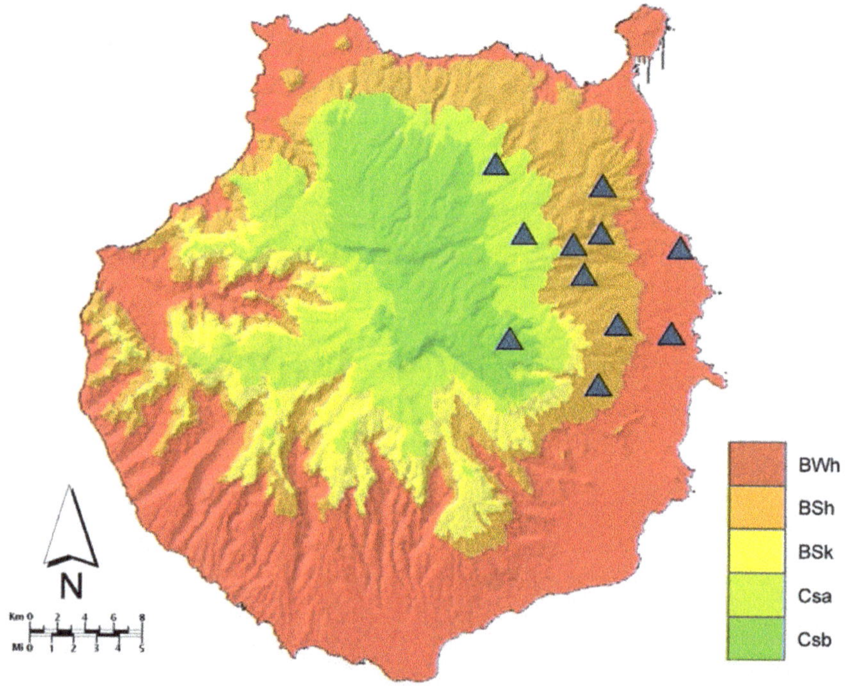

Figure 1. Map of Gran Canaria with the geographical distribution of the sampled kennels. Kennels are marked as blue triangles. Legend: BWh (hot desert climate), BSh (hot steppe), BSk (cold steppe), Csa (temperate with hot and dry summers), Csb (temperate with dry and warm summers). Map of Gran Canaria with the Köppen–Geiger climate classification extracted and modified from the Climate Atlas of the Archipelagos of the Canary Islands, Madeira and the Azores, with permission [18].

Twenty dogs were randomly selected from the 11 participating kennels (1–3 dogs per kennel). The selected animals were clinically healthy and negative for circulating *D. immitis* antigens and microfilariae. Each dog (13 females and 7 males, aged between 7 months and 8 years) received a subcutaneous injection of moxidectin SR (Guardian® SR, Elanco, Spain) at 0.17 mg/kg body weight (0.05 mL/kg body weight). None of the dogs experienced adverse reactions following moxidectin SR administration. All dogs continued their regular activities, including hunting.

Table 1. Details regarding the selected kennels, the climate, the number of dogs for each kennel, and the prevalence of heartworm.

			Beginning of Study			End of Study		
Kennel	**Climate**	**Studied Dogs**	*N*	**+ *D. immitis***	**%–CI (95%)**	*N*	**+ *D. immitis***	**%–CI (95%)**
1	BS	3	12	5	41.7–(14.8–70.7)	11	5	45.4–(17.9–70.5)
2	BS	2	9	3	33.3–(12–64.5)	9	4	44.4–(8.7–64.7)
3	BS	3	8	3	37.5–(7.3–73.8)	9	4	44.4–(8.7–64.7)
4	BS	1	9	2	22.2–(0.1–56.9)	9	4	44.4–(8.7–64.7)
5	BW	2	9	2	22.2–(0.1–56.9)	10	3	30–(4.8–66.6)

Table 1. *Cont.*

Kennel	Climate	Studied Dogs	Beginning of Study *N*	+ *D. immitis*	%–CI (95%)	End of Study *N*	+ *D. immitis*	%–CI (95%)
6	BS	2	10	4	40–(10.5–72.1)	10	3	30–(4.8–66.6)
7	BW	1	6	1	11.1–(0.1–65.9)	7	1	14.3–(0.3–45.5)
8	Csa	2	12	5	41.7–(14.8–70.7)	13	4	30.8–(11.9–57.2)
9	Csa	1	14	8	57.1–(37.2–75.4)	15	7	46.7–(26.3–68.8)
10	Csb	1	9	3	33.3–(4.1–66.8)	10	2	20–(0–55.8)
11	BS	2	11	4	36.4–(10.3–67.2)	10	3	30–(5.2–60.4)
Total		20	109	40	36.7–(27.6–45.8)	113	40	35.4–(26.6–44.2)

Legend: BW: desert climate; BS: steppe climate; Csa: temperate with hot and dry summers climate; Csb: temperate with dry and warm summers climate; (*N*): number of dogs that participated in the study in each kennel; (+ *D. immitis*): number of dogs positive to the antigen test; (%): prevalence of heartworm; (CI 95%): 95% confidence interval.

Antigen detection and Knott's tests were repeated at 6, 12, and 18 months (in the 20 treated dogs) and at 24 months (in all dogs). A second dose of moxidectin SR was administered at 24 months in the 20 study dogs. No additional curative or preventive products were given to any of the dogs. All animals were vaccinated annually for rabies. Owners were informed and provided consent for the participation of their dogs in the study.

3. Results

At the start of the study, 109 dogs were sampled, revealing an overall heartworm prevalence of 36.7% (40/109). Prevalence varied by kennel, ranging from 11.1% to 57.1% (Table 1).

All 20 dogs finished the study and remained healthy and negative to the antigens and Knott's tests throughout the study in months 6, 12, 18, and 24. The number of dogs per kennel fluctuated due to deaths ($n = 13$), losses ($n = 8$), sales/leases ($n = 7$), and new acquisitions ($n = 32$). By the end of the study, kennel populations ranged from 7 to 15 dogs, with a total of 113 dogs. The overall prevalence dropped slightly to 35.4% (40/113), varying by kennel from 14.3% to 46.7% (Table 1).

4. Discussion

The heartworm prevalence observed in this study aligns with previous findings, indicating that Canary Hound dogs exhibit a higher prevalence compared to the general dog population in Gran Canaria. The prevalences varied in each of the kennels, in some cases increasing and in other cases decreasing, which was probably due to the movement of animals during the study (deaths, losses, sales, leases, and new acquisitions). In a 2016 study, the heartworm prevalence in Canary Hounds was approximately 40%, compared to 20.7% in the general dog population [10]. Similar results were reported in studies conducted between 2000 and 2011, where heartworm prevalence in Canary Hounds ranged from 40.42% (2000) to 43% (2011), while the general dog population showed a decline from 30.19% to 19% during the same period [8,9]. As discussed by previous authors, this is probably due to the fact that these dogs live in unhygienic conditions and do not receive any vaccination, deworming, or chemoprophylaxis against heartworm, in addition to a higher exposure to the vector and the fact that they mainly inhabit the climatic zones with the highest heartworm prevalences, which are the Csa, Csb, and BS climates [10,19].

In these circumstances, Canary Hounds act as a natural reservoir for the disease and act as a barrier to the overall prevalence of the infection on the islands, which remain hyperendemic despite general awareness and the efforts of veterinarians and dog owners. Therefore, it is imperative to seek chemoprophylactic methods that will lead to greater compliance by the owners of these animals in order to help control heartworm in this reservoir animal sector. In this regard, other studies have shown that moxidectin SR offers very satisfactory results in terms of compliance, as observed in the present study, where all dogs completed the study and owners agreed to a second dose of moxidectin SR [15,16].

Contrary to what has been observed in other studies, no dogs experienced adverse reactions following administration of moxidectin SR. These results are in line with those published in another large study, which defined the incidence of adverse reactions to a similar moxidectin SR preparation as 14.3/10,000 doses (0.143%) [20]. However, in the study published by Vercelli et al. (2022) [21], it was observed that 12.7% (53/418) of dogs receiving moxidectin SR for the first time experienced adverse reactions (such as mild and temporary discomfort at the injection site, as well as local swelling in the muzzle, paws, eyelids, and lips or general allergic responses such as hives and itching) [21]. The reasons for these differences in the prevalence of adverse reactions are unclear but may be due to the different commercial preparations of moxidectin SR used in the studies. Both studies agreed that adverse reactions were more common in young dogs [20,21]; however, this did not appear to be reflected in the animals in this study.

As adult parasites are not detectable by antigen testing until 5–6 months after infection [1,2,22], testing at 6 months excluded infection prior to moxidectin SR administration as the dogs were not on preventive treatment. Similarly, the subsequent tests excluded infection during this study. Finally, the test performed at 24 months excluded the possibility of infection at least until 18 months after the administration of moxidectin SR. This methodology was similar to that developed by other authors [13], who reported the efficacy of moxidectin SR for the prevention of heartworm in dogs for 12 months.

Moxidectin is a macrocyclic lactone whose potent chemoprophylactic activity against heartworm has been widely demonstrated in different formats [16,23], including against some resistant strains [24,25]. The development of the injectable moxidectin SR formulation provided an interesting alternative to monthly drugs, initially demonstrating that a single subcutaneous injection was effective in protecting dogs against patent heartworm infection for at least 180 days after treatment [26], while subsequent studies have established the efficacy of injectable moxidectin, suggesting efficacy for up to 12 months [13,14]. However, efficacy over a longer period has never been demonstrated. The results of this study showed, for the first time, that the administration of moxidectin SR at the recommended dose could have a chemoprophylactic effect against *D. immitis* infection for 18 months. The possibility of prolonging the efficacy interval of moxidectin SR allows for a higher compliance among pet owners, especially in those groups with less awareness of animal health, which could contribute to reducing the overall heartworm prevalence in dogs in Gran Canaria.

This study is limited by the fact that the dogs studied could not have been bitten by infected mosquitoes during the duration of the study. More specifically, kennels were selected that were located in the most endemic areas of the island, so that dogs living outdoors 24 h a day would have a very high risk of being infected, which is reflected in the high prevalences found in these kennels. Another limitation of the study is that the sensitivity of the antigen detection test used in the study is 94.4% (compared to necropsy), so, although unlikely, false negatives could be possible in some of the dogs studied. The test used in this study has the advantage of detecting antigens that are not associated with the female genital tract of the parasite, i.e., it detects both males and females, although studies have shown that in hyperendemic areas, all infected dogs most likely have female infections [27]. Finally, the limited number of animals studied makes it necessary to extend this research to a larger number of dogs in order to confirm the duration of the chemoprotective effect.

5. Conclusions

Based on these results, it can be concluded that a single dose of subcutaneous, injectable moxidectin SR at the package insert-recommended dose (0.17 mg/kg; 0.05 mL/kg body weight) may be effective in preventing patent heartworm infection for at least 18 months in dogs living constantly exposed to mosquito vectors in hyperendemic areas. However, further research in larger numbers of animals is needed to confirm these results. Furthermore, if these results are confirmed, the use of this chemopreventive protocol could contribute to a reduction in heartworm prevalence in hyperendemic areas.

Author Contributions: Conceptualization, J.A.M.-A. and E.C.; methodology and resources, A.I.Q.-M.; investigation, J.A.M.-A. and A.I.Q.-M.; resources, A.I.Q.-M., E.C. and J.A.M.-A.; data curation, E.C.; writing—original draft preparation, E.C. and A.I.Q.-M.; writing—review and editing; visualization, all authors; supervision and funding acquisition, J.A.M.-A. All authors have read and agreed to the published version of the manuscript.

Funding: This study has been funded by ELANCO Valquimica SA (Project I.D. Sd-08/00-240/030/0026).

Institutional Review Board Statement: Ethical review and approval were not required for the animals in this study, as it involved the clinical application of a licensed drug and in all cases involved owned dogs. All of the dog owners were informed about the present study and consented to participate. The study was carried out in accordance with the current Spanish and European legislation on animal protection (Spanish Royal Decree 53/2013 and 2010/63/UE Directive).

Informed Consent Statement: Informed consent was obtained from all owners of dogs included in the study.

Data Availability Statement: All data generated or analyzed during this study are included in this article. The datasets used and/or analyzed during the present study are available from the corresponding author upon reasonable request.

Acknowledgments: The authors would like to thank Uranovet for kindly providing the Urano test Dirofilaria®. Also, they would like to thank the hunters who agreed to participate in this study.

Conflicts of Interest: The authors declare no conflicts of interest.

References

1. Simón, F.; Siles-Lucas, M.; Morchón, R.; González-Miguel, J.; Mellado, I.; Carretón, E.; Montoya-Alonso, J.A. Human and animal dirofilariasis: The emergence of a zoonotic mosaic. *Clin. Microbiol. Rev.* **2012**, *25*, 507–544. [CrossRef]
2. McCall, J.W.; Genchi, C.; Kramer, L.H.; Guerrero, J.; Venco, L. Heartworm disease in animals and humans. *Adv. Parasitol.* **2008**, *66*, 193–285.
3. Morchón, R.; Montoya-Alonso, J.A.; Rodríguez-Escolar, I.; Carretón, E. What Has Happened to Heartworm Disease in Europe in the Last 10 Years? *Pathogens* **2022**, *11*, 1042. [CrossRef]
4. Morchón, R.; Carretón, E.; González-Miguel, J.; Mellado-Hernández, I. Heartworm Disease (*Dirofilaria immitis*) and Their Vectors in Europe—New Distribution Trends. *Front. Physiol.* **2012**, *12*, 196. [CrossRef]
5. Montoya-Alonso, J.A.; Morchón, R.; Falcón-Cordón, Y.; Falcón-Cordón, S.; Simón, F.; Carretón, E. Prevalence of heartworm in dogs and cats of Madrid, Spain. *Parasit. Vectors* **2017**, *10*, 354. [CrossRef]
6. Montoya, J.A.; Morales, M.; Ferrer, O.; Molina, J.M.; Corbera, J.A. The prevalence of *Dirofilaria immitis* in Gran Canaria, Canary Islands, Spain (1994–1996). *Vet. Parasitol.* **1998**, *75*, 221–226. [CrossRef]
7. Montoya-Alonso, J.A.; Mellado, I.; Carretón, E.; Cabrera-Pedrero, E.D.; Morchón, R.; Simón, F. Canine dirofilariosis caused by *Dirofilaria immitis* is a risk factor for the human population on the island of Gran Canaria, Canary Islands, Spain. *Parasitol. Res.* **2010**, *107*, 1265–1269. [CrossRef]
8. Montoya-Alonso, J.A.; Carretón, E.; Juste, M.C.; Mellado, I.; Morchón, R.; Simón, F. Epidemiological survey of canine heartworm disease on the island of Gran Canaria (Canary Islands–Spain) between 2000 and 2008. *Vet. Parasitol.* **2010**, *173*, 165–168. [CrossRef]
9. Montoya-Alonso, J.A.; Carretón, E.; Corbera, J.A.; Juste, M.C.; Mellado, I.; Morchón, R.; Simón, F. Current prevalence of *Dirofilaria immitis* in dogs, cats and humans from the island of Gran Canaria, Spain. *Vet. Parasitol.* **2011**, *176*, 291–294. [CrossRef]
10. Montoya-Alonso, J.A.; Carretón, E.; Morchón, R.; Silveira-Viera, L.; Falcón, Y.; Simón, F. The impact of the climate on the epidemiology of *Dirofilaria immitis* in the pet population of the Canary Islands. *Vet. Parasitol.* **2016**, *216*, 66–71. [CrossRef]
11. Montoya-Alonso, J.A.; Morchón, R.; Costa-Rodríguez, N.; Matos, J.I.; Falcón-Cordón, Y.; Carretón, E. Current Distribution of Selected Vector-Borne Diseases in Dogs in Spain. *Front. Vet. Sci.* **2020**, *7*, 564429. [CrossRef]
12. Montoya-Alonso, J.A.; Morchón, R.; García-Rodríguez, S.N.; Falcón-Cordón, Y.; Costa-Rodríguez, N.; Matos, J.I.; Rodríguez Escolar, I.; Carretón, E. Expansion of Canine Heartworm in Spain. *Animals* **2022**, *12*, 1268. [CrossRef]
13. Genchi, C.; Rossi, L.; Cardini, G.; Kramer, L.H.; Venco, L.; Casiraghi, M.; Genchi, M.; Agostini, A. Full season efficacy of moxidectin microsphere sustained release formulation for the prevention of heartworm (*Dirofilaria immitis*) infection in dogs. *Vet. Parasitol.* **2002**, *110*, 85–91. [CrossRef]

14. Lok, J.B.; Knight, D.H.; Nolan, T.J.; Grubbs, S.T.; Cleale, R.M.; Heaney, K. Efficacy of an injectable, sustained-release formulation of moxidectin in preventing experimental heartworm infection in mongrel dogs challenged 12 months after administration. *Vet. Parasitol.* **2005**, *128*, 129–135. [CrossRef]
15. Mwacalimba, K.; Amodie, D.; Swisher, L.; Moldavchuk, M.; Brennan, C.; Walther, C.; Bowman, K. Pharmacoeconomic Analysis of Heartworm Preventive Compliance and Revenue in Veterinary Practices in the United States. *Front. Vet. Sci.* **2021**, *8*, 602622. [CrossRef]
16. Mwacalimba, K.; Sheehy, J.; Adolph, C.; Savadelis, M.; Kryda, K.; Poulsen Nautrup, B. A review of moxidectin vs. other macrocyclic lactones for prevention of heartworm disease in dogs with an appraisal of two commercial formulations. *Front. Vet. Sci.* **2024**, *11*, 1377718. [CrossRef]
17. Söllheim, A.L.L.; Suarez, P.M.; Hernández, F.G. The digital climate atlas of the Canary Islands: A tool to improve knowledge of climate and temperature and precipitation trends in the Atlantic islands. *Clim. Serv.* **2024**, *34*, 100487. [CrossRef]
18. Production Department of the State Meteorological Agency of Spain, Department of Meteorology and Climatology of the Institute of Meteorology of Portugal. Climate atlas of the archipelagos of the Canary Islands, Madeira and the Azores, Agencia Estatal de Meteorología, Ministerio de Agricultura, Alimentación y Medio Ambiente, Spain. 2012. Available online: https://www.aemet.es/documentos/es/conocermas/recursos_en_linea/publicaciones_y_estudios/publicaciones/2Atlas_climatologico/Atlas_Clima_Macaronesia___Baja.pdf (accessed on 17 July 2014).
19. Montoya-Alonso, J.A.; García-Rodríguez, S.N.; Matos, J.I.; Costa-Rodríguez, N.; Falcón-Cordón, Y.; Carretón, E.; Morchón, R. Change in the Distribution Pattern of *Dirofilaria immitis* in Gran Canaria (Hyperendemic Island) between 1994 and 2020. *Animals* **2024**, *14*, 2037. [CrossRef]
20. Saito, E.K.; Moore, G.E.; Morrison, J.; Spofford, N.; Yang, M. Younger dogs of specific breeds are more likely to experience a rare adverse event after administration of extended-release injectable moxidectin heartworm preventive. *J. Am. Vet. Med. Assoc.* **2023**, *261*, 1666–1672. [CrossRef]
21. Vercelli, C.; Bertolotti, L.; Gelsi, E.; Gazza, C.; Re, G. Evaluation of Side Effects and Long-Term Protection of a Sustained-Release Injectable Moxidectin Formulation against *Dirofilaria immitis* Infection in Dogs: An Observational-In Field Multicentric Study. *Vet. Sci.* **2022**, *9*, 408. [CrossRef]
22. American Heartworm Society. Canine Guidelines for the Prevention, Diagnosis, and Management of Heartworm Infection (*Dirofilaria immitis*) in Dogs. 2024. Available online: https://www.heartwormsociety.org (accessed on 17 July 2024).
23. Nolan, T.J.; Lok, J.B. Macrocyclic lactones in the treatment and control of parasitism in small companion animals. *Curr. Pharm. Biotechnol.* **2012**, *13*, 1078–1094. [CrossRef]
24. Savadelis, M.D.; McTier, T.L.; Kryda, K.; Maeder, S.J.; Woods, D.J. Moxidectin: Heartworm disease prevention in dogs in the face of emerging macrocyclic lactone resistance. *Parasit. Vectors* **2022**, *15*, 82. [CrossRef]
25. Bowman, D.D.; McTier, T.L.; Adams, E.L.; Mahabir, S.P.; Login, J.A.; Bidgood, T.; Woods, D.J. Evaluation of the efficacy of ProHeart® 6 (moxidectin) against a resistant isolate of *Dirofilaria immitis* (JYD-34) in dogs. *Parasit. Vectors* **2017**, *10*, 502. [CrossRef]
26. Lok, J.B.; Knight, D.H.; Wang, G.T.; Doscher, M.E.; Nolan, T.J.; Hendrick, M.J.; Steber, W.; Heaney, K. Activity of an injectable, sustained-release formulation of moxidectin administered prophylactically to mixed-breed dogs to prevent infection with *Dirofilaria immitis*. *Am. J. Vet. Res.* **2001**, *62*, 1721–1726. [CrossRef]
27. Rishniw, M.; Schukken, Y.; Greiner, E. Sex ratios of *Dirofilaria immitis* in naturally infected dogs show female bias at low worm intensities. *Res. Vet. Sci.* **2012**, *93*, 1324–1328. [CrossRef]

animals

Systematic Review

Toxocara cati Infection in Cats (*Felis catus*): A Systematic Review and Meta-Analysis

Jorge Luis Bonilla-Aldana [1], Alba Cristina Espinosa-Nuñez [1], D. Katterine Bonilla-Aldana [2,*] and Alfonso J. Rodriguez-Morales [3,4]

1 School of Veterinary Medicine and Zootehcnics, Faculty of Agricultural Sciences, Universidad de la Amazonia, Florencia 111321, Caquetá, Colombia; jorg.bonilla@udla.edu.co (J.L.B.-A.); a.espinosa@udla.edu.co (A.C.E.-N.)
2 Research Unit, Universidad Continental, Huancayo 12001, Peru
3 Masters of Climate Change and Clinical Epidemiology and Biostatistics Program, Universidad Cientifica del Sur, Lima 15307, Peru; arodriguezmo@cientifica.edu.pe or alphonso.morales@lau.edu.lb
4 Gilbert and Rose-Marie Chagoury School of Medicine, Lebanese American University, Beirut P.O. Box 36-5053, Lebanon
* Correspondence: dbonilla@continental.edu.pe

Simple Summary: Toxocariasis, caused by species of *Toxocara*, affects canines, felines, humans, and other vertebrates. The primary mode of infection is by ingesting embryonated eggs. It poses environmental, human, and animal health risks, especially in park soils. This study aimed to assess the global prevalence of *Toxocara cati* in cats (*Felis catus*), a neglected species compared to *T. canis*, via a systematic literature review across six databases. Significant prevalence was observed using coproparasitological methods, with Nepal displaying the highest rates. The findings highlight the imperative of preventive measures against toxocariasis due to its widespread occurrence. Recognising the interconnectedness of animal, environmental, and human health underscores the importance of deworming cats, promoting hygiene, and educating the public to mitigate the risks of this zoonotic condition. Protecting feline health benefits cats and reduces the likelihood of human transmission, creating a positive outcome for both.

Abstract: Introduction: Toxocariasis is an infection caused in canines, felines, humans, and other vertebrates by species of the genus *Toxocara*, such as *T. canis* and *T. cati*. The embryonated eggs of these parasites are the primary means of acquiring the infection for both definitive hosts, dogs and cats, respectively, and for intermediates, such as humans and other vertebrates. When deposited on park soils, environmental contamination becomes a risk to environmental, human, and animal health. **Objective:** To determine the global prevalence of *Toxocara cati* in cats (*Felis catus*). **Methods:** A systematic review of the literature was carried out in six databases (Scopus, PubMed, ScienceDirect, SciELO and Google Scholar) to evaluate the global prevalence of *Toxocara cati* in cats, defined by coproparasitological, histological, and molecular techniques. A meta-analysis was performed using a random effects model to calculate pooled prevalence and 95% confidence intervals (95% CI). A two-tailed 5% alpha level was used for hypothesis testing. **Results:** Two hundred and eighty-nine studies were included. The global pooled prevalence of *Toxocara cati* in cats using coproparasitological methods was 17.0% (95.0% CI: 16.2–17.8%). In the subgroup analysis according to country, Nepal had the highest prevalence of *T. cati* infection (94.4%; 95% CI 89.7–99.2%). The pooled prevalence of *T. cati* infection by PCR in four studies was 4.9% (95.0% CI: 1.9–7.9%). **Conclusions:** This systematic review underscores the need for preventive action against toxocariasis due to its widespread prevalence. The interplay between animal and human health should be emphasised, necessitating measures like deworming cats, hygiene practices, and public education to mitigate risks. Safeguarding feline health can also reduce human transmission, benefiting both species.

Keywords: *Toxocara*; prevalence; cats; systematic review; meta-analysis

1. Introduction

Zoonoses are a group of infectious diseases transmissible between animals and humans [1], including conditions where the human is not a definitive host of the etiological agent [2]. Pets such as dogs and cats are considered opportune hosts of various pathogenic agents of zoonotic incidence, such as gastrointestinal helminths of the *Toxocara* genus [3]. Toxocariasis is a parasitic disease with worldwide distribution, and its etiological agents in dogs and cats are *Toxocara canis* and *Toxocara cati*, respectively [4]. The parasite is transmitted vertically (transplacental and transmammary) and horizontally, addressing the ingestion of embryonated eggs in infected animals' soil and fur and through consuming contaminated food [5–8]. The adult nematodes of *T. canis* and *T. cati* complete their reproductive cycle in the intestine of their definitive host (dogs and cats), reproducing and eliminating about 200,000 eggs per day, excreted in faeces into the environment [9].

In dogs and cats, *T. canis* and *T. cati* mainly affect young animals from birth, presenting clinical signs such as cachexia, emaciation, body weakness, rough coat, growth dekay, vomiting, cough, diarrhoea, and distended abdomen; the cough is due to larval migration to the lungs [10–12]. The disease can affect adult cats and dogs, but they do not usually present clinical signs [13]. *Toxocara* infection occurs accidentally in humans due to the ingestion of eggs in soil or contaminated food, including paratenic hosts, such as poultry, pork, and beef [14,15]. Based on the clinical manifestations observed in humans, the disease can be classified into four primary syndromes: visceral larva migrans (VLM), ocular toxocariasis (OT), covert toxocariasis (TC), and neuro toxocariasis (NT) [16].

At a global level, both stray and domestic cats contribute to the dispersion and contamination of embryonated *Toxocara* eggs into the environment [17,18]. The presence of eggs in public places represents a risk for animal health and even for humans, given that approximately 21% of public spaces worldwide are contaminated with *Toxocara* eggs [19]. In Latin America, it is estimated that the prevalence of *Toxocara* in public parks is 50%, which means it can be considered a transmission route for people who attend these places, mainly children who might play on the ground [20]. On the other hand, studies have reported that direct contact with the fur of cats infected with *T. cati* is a route of transmission since potentially infective embryonated eggs have been identified in perianal areas, extremities, and the lower part of the tail of cats [8,21]. The global prevalence of *T. cati* in cats starts from 17%, with an average of 134 million cats worldwide contributing to the dispersal of eggs in the environment, generating a public health problem [11].

Diagnostic tests for *Toxocara* spp. in pets are fundamentally based on the microscopic examination of faeces to find eggs and analyse their morphology, using different copro-diagnostic techniques such as direct smear, Kato–Katz, MacMaster, and Faust (flotation–sedimentation), among others [22]. However, other diagnostic tests have greater sensitivity and specificity, such as serological tests, ELISA to detect anti-*Toxocara* IgG antibodies, and molecular techniques, such as polymerase chain reaction (PCR) and Western blot, that determine the larval TES antigen [23–25].

2. Methods

Protocol: The protocol followed the recommendations established by the PRISMA statement.

Inclusion criteria: Peer-reviewed published articles were included in which infection with coproparasitological, histological, or molecular confirmation of *Toxocara catis* in cats (*Felis catus*) was reported. For parasitological tests, we considered egg detection for tests based on molecular biology and PCR. The article language limit was not set, and we included publications from 1 January 1950 to the date the search ended, 31 January 2024. Review articles, opinion articles, and letters that do not present original data were excluded from the study, as were studies that reported cases with incomplete information.

Information sources and search strategy: A systematic review was conducted using Medline/PubMed, Scopus, ScienceDirect, SciELO, and Google Scholar. The search terms used were the following: "Prevalence", "*Toxocara*", "*Toxocara cati*", and "cats".

Study selection: Results from the initial search strategy were first selected by title and abstract. The full texts of relevant articles were examined for the inclusion and exclusion criteria. When an article provided duplicate information from the same subjects, the information from both reports was combined to obtain complementary data, counting as only one study. Observational studies reporting the prevalence of *Toxocara cati* in cats were included for quantitative synthesis (meta-analysis).

Data collection process and data elements: Two researchers independently completed data extraction forms, including information on publication type, publishing institution, country, year and date of publication, and number of infected animals evaluated by serological or molecular tests. A third researcher verified the list of articles and data extractions to ensure no duplicate articles or information from the same study were presented, and resolved any discrepancies regarding study inclusion.

Assessment of methodological quality and risk of bias: We used the IHE case series study quality assessment checklist and the critical appraisal tool to assess the quality of cross-sectional studies (AXIS) [26]. Publication bias was assessed using a funnel plot. Given the varying degrees of data heterogeneity and the heterogeneity inherent in any systematic review of published literature studies, a random effects model was used to calculate the pooled prevalence and 95% confidence interval (95%CI).

Statistical approach: Unit discordance for variables was resolved by converting all units to a standard measurement for that variable. Percentages and means ± standard deviation (SD) were calculated to describe the distributions of categorical and continuous variables, respectively. Since individual case information was unavailable, we will report weighted means and SDs. Baseline data were analysed using Stata version 14.0, licensed. Meta-analyses were performed with Stata, the licensed Open Meta [Analyst], and Comprehensive Meta-Analysis ve.3.3® software. The pooled prevalences and their 95% confidence intervals (95% CI) were used to summarise the weighted effect size for each study pooling variable using the binary random effects model of the individual studies (weighting took into account sample sizes), except for median age, where a continuous random effects model was applied (DerSimonian–Laird procedure). A random effects meta-analysis model will imply the assumption that the effects estimated in the different studies are not identical but rather follow a particular distribution. For random effects analyses, the pooled estimate and 95% CIs refer to the centre of the pooled prevalence distribution but do not describe the width of the distribution. Often, the pooled estimate and its 95% CI are cited in isolation as an alternative estimate of the quantity evaluated in a fixed-effects meta-analysis, which is inappropriate. The 95% CI of a random effects meta-analysis describes the uncertainty in the location of the systematically different mean prevalence in different studies. Measures of heterogeneity, including Cochran's Q statistic, I^2 index, and squared tau test, were estimated and reported. We performed subgroup analyses using techniques, countries, subregions, and meta-analyses for each variable of interest. Publication bias was assessed using a funnel plot. A random effects model was used to calculate pooled prevalence and 95% CI, given the varying degrees of data heterogeneity and inherent heterogeneity in any systematic review of published literature studies.

3. Results

3.1. Selection of Studies

Our search strategy yielded 16,266 records in the databases combined. After removing duplicates and screening for titles and abstracts, 329 articles underwent full-text review. Finally, 289 articles were included in the systemic review and meta-analysis [27–310] (Table 1). Figure 1 shows the PRISMA flow chart.

Table 1. Studies included.

Code	Study Title	Publication	Study	Location	Country	N	Ref.
RSM-1	Prevalence Of Intestinal Canine And Feline Parasites In Saitama Prefecture, Japan	2009	1999	Saitama	Japan	1079	[27]
RSM-2	Prevalencia De Helmintos Intestinales En Gatos Domésticos Del Departamento Del Quindío, Colombia	2012	2008	Quindío	Colombia	121	[28]
RSM-3	10-Year Parasitological Examination Results (2003 To 2012) Of Faecal Samples From Horses, Ruminants, Pigs, Dogs, Cats, Rabbits And Hedgehogs	2017	2003		Germany	903	[29]
RSM-4	A Comparative Study Of Some Intestinal Parasites In Fecal Samples Of Domestic And Stray Cats In Baghdad, Iraq	2022	2020	Baghdad	Iraq	121	[30]
RSM-5	A Cross-Sectional Study Of Tritrichomonas Foetus Infection In Feral And Shelter Cats In Prince Edward Island, Canada	2016	2011	Prince Edward Island	Canada	100	[31]
RSM-6	A Retrospective Investigation Of Feline Gastrointestinal Parasites In Western Canada	2013	1998		Canada	635	[32]
RSM-7	A Survey Of Gastrointestinal Helminths In Cats Of The Metropolitan Region Of Rio De Janeiro, Brazil	2004	2004	Rio de Janeiro	Brazil	135	[33]
RSM-8	A Survey Of Helminth Parasites Of Cats From Saskatoon	1999	1999	Saskatoon	Canada	52	[34]
RSM-9	A Survey Of Helminths In Domestic Cats In The Pretoria Area Of Transvaal, Republic Of South Africa, Part 1: The Prevalence And Comparison Of Burdens Of Helminths In Adult And Juvenile Cats	1989	1980		South Africa	1502	[35]
RSM-10	A Survey Of Helminths In Stray Cats From Copenhagen With Ecological Aspects	1984	1980	Copenhagen	Denmark	230	[36]
RSM-11	A Survey Of *Toxocara* Infections In Cat Breeding Colonies In The Netherlands	1998	1995		Netherlands	337	[37]
RSM-12	A Survey On Endoparasites And Ectoparasites In Domestic Dogs And Cats In Vladivostok, Russia 2014	2016	2013	Vladivostok,	Russia	54	[38]
RSM-13	A Survey On Endoparasites And Ectoparasites Of Stray Cats From Mashhad (Iran) And Association With Risk Factors	2011	2009	Mashhad	Iran	52	[39]
RSM-14	A Survey On The Prevalence Of *Toxocara cati*, *Toxocara* Canis And Toxascaris Leonina Eggs In Stray Dogs And Cats' Faeces In Northwest Of Iran: A Potential Risk For Human Health	2019	2017	Azarshahr, Marand	Iran	100	[40]
RSM-15	A Survey On *Toxocara cati* Eggs On The Hair Of Stray Cats: A Potential Risk Factor For Human Toxocariasis In Northeastern Iran	2019	2016	Mashhad	Iran	167	[8]
RSM-16	A Survey Study On Gastrointestinal Parasites Of Stray Cats In Azarshahr, (East Azerbaijan Province, Iran)	2016	2013	Azarshahr	Iran	50	[41]
RSM-17	Abundance, Zoonotic Potential And Risk Factors Of Intestinal Parasitism Amongst Dog And Cat Populations: The Scenario Of Crete, Greece	2017	2011	Crete	Greece	264	[42]
RSM-18	An Investigation Of The Potential For Spread Of *Sarcocystis* spp. And Other Parasites By Feral Cats	1990	1984		New Zealand	63	[43]
RSM-19	Aporte Al Conocimiento De Los Metazoos Parásitos Del Gato Doméstico En El Departamento De Montevideo, Uruguay.	2013	2002	Montevideo	Uruguay	22	[44]
RSM-20	Avaliação Das Endoparasitoses Intestinais Que Acometem Cães E Gatos Mantidos Em Um Abrigo	2021	2019	Zona da Mata Mineira	Brazil	26	[45]
RSM-21	Frequência De Parasitoses Com Potencial Zoonótico Em Cães E Gatos Naturalmente Infectados Na Cidade De Maringá-PR	2022	2021	Maringá	Brazil	19	[46]
RSM-22	Canine And Feline Helminth And Protozoan Infections In Belgium	1973	1973		Belgium	500	[47]
RSM-23	Caracterização Da Ocorrência De Parasitas Gastrointestinais De Gatos Na Zona De Pesca Da Ilha De Faro	2018	2014	Faro	Portugal	123	[48]
RSM-24	Caracterização Molecular De *Cryptosporidium* Spp. E Ocorrência Dos Principais Parasitas Gastrointestinais Em Amostras Fecais De Cães E Gatos Naturalmente Infectados	2018	2017		Brazil	49	[49]
RSM-25	Caracterización De La Infestación Parasitológica Gastrointestinal Y Respiratoria En Gatos Ferales (Felis Silvestris Catus) De La Ciudad De Córdoba, Comunidad Autónoma De Andalucía, España.	2022	2022	Cordova	Spain	33	[50]

Table 1. *Cont.*

Code	Study Title	Publication	Study	Location	Country	N	Ref.
RSM-26	Cardiorespiratory Nematodes And Co-Infections With Gastrointestinal Parasites In New Arrivals At Dog And Cat Shelters In North-Western Spain	2022	2019	Galicia	Spain	65	[51]
RSM-27	Challenging The Dogma Of The 'Island Syndrome': A Study Of Helminth Parasites Of Feral Cats And Black Rats On Christmas Island	2018	2018	Christmas Island	Australia	66	[52]
RSM-28	Characterisation Of Ecto- And Endoparasites In Domestic Cats From Tirana, Albania	2014	2008	Tirana	Albania	252	[53]
RSM-29	Co-Infection Of Intestinal Helminths In Humans And Animals In The Philippines	2022	2019		Philippines	27	[54]
RSM-30	Comparação Da Prevalência De Parasitos Entéricos Em Gatos Errantes E Domiciliados Em Goiânia-Goiás, Análise Da Acurácia De Técnicas Parasitológicas E Avaliação Da Copro-Pcr Para O Diagnóstico De Toxoplasma Gondii	2016	2015	Goiânia	Brazil	149	[55]
RSM-31	Comparative Clinical Epidemiology Of Toxocariosis In Dogs And Cats	2010	2007	Lahore	Pakistan	671	[56]
RSM-32	Comparison Of *Toxocara* Eggs In Hair And Faecal Samples From Owned Dogs And Cats Collected In Ankara, Turkey	2014	2014	Ankara	Turkey	100	[21]
RSM-33	Coprological Detection Of Toxocariosis In Domicile And Stray Dogs And Cats In Sulaimani Province, Iraq	2022	2020	Sulaymaniyah	Iraq	78	[57]
RSM-34	Cross-Sectional Survey Of Toxoplasma Gondii Infection In Colony Cats From Urban Florence (Italy)	2010	2010	Florence	Italy	50	[58]
RSM-35	Cross-Sectional Survey On Tritrichomonas Foetus Infection In Italian Cats	2016	2012		Italy	267	[59]
RSM-36	*Cryptosporidium* Spp. And Other Zoonotic Enteric Parasites In A Sample Of Domestic Dogs And Cats In The Niagara Region Of Ontario	2006	2001	Ontario	Canada	41	[60]
RSM-37	Current Status Of L. Infantum Infection In Stray Cats In The Madrid Region (Spain): Implications For The Recent Outbreak Of Human Leishmaniosis?	2014	2014	Madrid	Spain	287	[61]
RSM-38	Descriptive Epidemiology Of Intestinal Helminth Parasites From Stray Cat Populations In Qatar	2008	2006	Doha	Qatar	488	[62]
RSM-39	Detection Of Helminth Eggs And Identification Of Hookworm Species In Stray Cats, Dogs And Soil From Klang Valley, Malaysia	2015	2013	Klang Valley	Malaysia	152	[63]
RSM-40	Determinación De La Presencia De Enteroparásitos En Gatos Clínicamente Sanos En Cuatro Comunas De Santiago, Mediante Los Métodos De Teuscher Y Teleman	2018	2018	Santiago de Chile	Chile	40	[64]
RSM-41	Determinación De La Presencia De Helmintos Gastro Intestinales En Gatos En Las Parroquias Urbano Marginales De La Ciudad De Babahoyo	2023	2023	Babahoyo	Ecuador	60	[65]
RSM-42	Determinación De Prevalencia De Parásitos Intestinales Y Externos En Gatos Domésticos (Felis Catus) En Determinadas Zonas Del Ecuador	2012	2012	Quito, Manta	Ecuador	40	[66]
RSM-43	Diagnosis Of Feline Whipworm Infection Using A Coproantigen ELISA And The Prevalence In Feral Cats In Southern Florida	2018	2014	Miami	USA	35	[67]
RSM-44	Diagnóstico De Parásitos Gastrointestinales En Caninos Y Felinos: Estudio Retrospectivo En Dos Laboratorios Veterinarios	2021	2005		Costa Rica	57	[68]
RSM-45	Ectoparasitos E Helmintos Intestinais Em Felis Catus Domesticus, Da Cidade De Lages, SC, Brasil E Aspectos Sócioeconômicos E Culturais Das Famílias Dos Proprietários Dos Animais	2009	2005	Lages	Brazil	111	[69]
RSM-46	Endoparasite Prevalence And Infection Risk Factors Among Cats In An Animal Shelter In Estonia	2021	2015	Tartu	Estonia	290	[70]
RSM-47	Endoparasite Prevalence And Recurrence Across Different Age Groups Of Dogs And Cats	2009	1997	Pennsylvania	USA	1566	[71]
RSM-48	Endoparasites Detected In Faecal Samples From Dogs And Cats Referred For Routine Clinical Visit In Sardinia, Italy	2017	2011	Sacer	Italy	343	[72]
RSM-49	Endoparasites In Dogs And Cats Diagnosed At The Veterinary Teaching Hospital (VTH) Of The University Of Prince Edward Island Between 2000 And 2017. A Large-Scale Retrospective Study	2020	2000	Prince Edward Island	Canada	2391	[73]

Table 1. *Cont.*

Code	Study Title	Publication	Study	Location	Country	N	Ref.
RSM-50	Endoparasites In Dogs And Cats In Germany 1999–2002	2003	1999	Freiburg	Germany	3167	[74]
RSM-51	Endoparasites In Domestic Cats (Felis Catus) In The Semiarid Region Of Northeast Brazil	2023	2023	Sousa	Brazil	207	[75]
RSM-52	Endoparasites Of Cats From The Tirana Area And The First Report On Aelurostrongylus Abstrusus (Railliet, 1898) In Albania	2011	2008	Tirana	Albania	18	[76]
RSM-53	Endoparasites Of Household And Shelter Cats In The City Of Rio De Janeiro, Brazil	2020	2020	Rio de Janeiro	Brazil	393	[77]
RSM-54	Enteric Parasites Of Free-Roaming, Owned, And Rural Cats In Prairie Regions Of Canada	2015	2015		Canada	219	[78]
RSM-55	ENTEROPARÁSITOS EN PERROS (Canis Familiaris) Y GATOS (Felis Catus) DE LA PROVINCIA DE PUNO	2013	2013	Puno	Peru	96	[79]
RSM-56	Enteroparasitos Encontrados Em Cães E Gatos Atendidos Em Duas Clínicas Veterinárias Nacidade De Manaus, AM	2012	2010	Manaus	Brazil	13	[80]
RSM-57	Epidemiological Survey Of Zoonotic Helminths In Feral Cats I2016n Gran Canaria Island (Macaronesian Archipelago-Spain)	2016	2016	Gran Canaria island	Spain	48	[81]
RSM-58	Epidemiological Survey On Gastrointestinal And Pulmonary Parasites In Cats Around Toulouse (France)	2022	2015		France	498	[82]
RSM-59	Epidemiology Of *Toxocara* Spp. In Stray Dogs And Cats In Dublin, Ireland	1994	1990	Dublin	Ireland	181	[83]
RSM-60	Estudio Coprológico De Parasitosis En Gatos Del Área Periurbana De La Ciudad De Murcia Y Sus Implicaciones Zoonósicas	2017	2017	Murcia	Spain	61	[84]
RSM-61	Estudio De La Prevalencia De Parásitos Gastrointestinales Zoonosicos En Perros Y Gatos En El Barrio Carapungo De La Ciudad De Quito.	2010	2010	Quito	Ecuador	32	[85]
RSM-62	Estudio Retrospectivo De Casos De Parasitosis Gastrointestinales Presentados En Caninos Y Felinos En La Clínica Veterinaria Zooluciones Versátiles En La Ciudad De Bogotá	2021	2010	Bogotá	Colombia	38	[86]
RSM-63	Estudo Das Parasitoses Gastrintestinais De Cães E Gatos Domésticos No Município De São José Dos Campos—Sp.	2005	2004	São José dos Campos	Brazil	20	[87]
RSM-64	Fecal Survey Of Parasites In Free-Roaming Cats In Northcentral Oklahoma, United States	2018	2015	Oklahoma	USA	846	[88]
RSM-65	Feline Gastrointestinal Parasitism In Greece: Emergent Zoonotic Species And Associated Risk Factors	2018	2016	Macedonia, Islands, Central Greece, Epirus, Thrace, Peloponnesus, Thessaly	Greece	1150	[89]
RSM-66	Feline Immunodeficiency Virus, Feline Leukaemia Virus, Toxoplasma Gondii, And Intestinal Parasitic Infections In Taiwanese Cats	1990	1990		Taiwan	95	[90]
RSM-67	Feline Intestinal Parasites In Finland: Prevalence, Risk Factors And Anthelmintic Treatment Practices	2012	2009		Finland	411	[91]
RSM-68	Feline Parasites And The Emergence Of Feline Lungworm In The Portland Metropolitan Area, Oregon, USA 2016–2017	2021	2016	Portland	USA	126	[92]
RSM-69	First Report Of Echinococcus Multilocularis In Cats In Poland: A Monitoring Study In Cats And Dogs From A Rural Area And Animal Shelter In A Highly Endemic Region	2019	2017	Province Podkarpackie	Poland	67	[93]
RSM-70	Freqüência De Helmintos Em Gatos De Uberlândia, Minas Gerais	2004	2000	Uberlândia	Brazil	50	[94]
RSM-71	Frecuencia De Parásitos Gastrointestinales En Animales Domésticos Diagnosticados En Yucatán, México.	2001	1984	Yucatan	Mexico	46	[95]
RSM-72	Frecuencia De Parásitos Gastrointestinales En Felinos Domésticos (Felis Catus) En El Distrito De Jesús María—Lima	2022	2021	Lima	Peru	87	[96]
RSM-73	Frequência De Endoparasitas Em Gatos Internados Em Quatro Clínicas De Cascavel, Paraná	2018	2018	Cascavel	Brazil	23	[97]
RSM-74	Freqüência De Enteroparasitas Em Amostras Fecais De Cães E Gatos Dos Municípios Do Rio De Janeiro E Niterói	2005	1999		Brazil	40	[98]

Table 1. *Cont.*

Code	Study Title	Publication	Study	Location	Country	N	Ref.
RSM-75	Frequência De Helmintos Diagnosticados Em Cães E Gatos No Laboratório De Doenças Parasitárias Da Faculdade De Veterinária/Ufpel	2022	2019	Pelotas	Brazil	69	[99]
RSM-76	Frequência De Parasitas Gastrointestinais Em Cães E Gatos Domunicípio De Londrina, PR, Com Enfoque Em Saúde Pública	2013	2000	Londrina	Brazil	378	[100]
RSM-77	Freqüência De Parasitas Intestinais Em Cães (Canis Familiaris) E Gatos (Felis Catus Domestica) Em Araçatuba, São Paulo	1995	1992	Aracatuba	Brazil	32	[101]
RSM-78	Frequência De Parasitos Gastrintestinais, Presentes Em Fezes De Cães E Gatos, Analisadas No Laboratório De Doenças Parasitárias Da Ufpel, Durante O Ano De 2017	2019	2017	Pelotas	Brazil	25	[102]
RSM-79	Frequency Of Gastrointestinal Parasites In Cats Seen At The University Of São Paulo Veterinary Hospital, Brazil	2016	2005	Sao Paulo	Brazil	502	[103]
RSM-80	Freqüência De Parasitos Gastrintestinais Em Cães E Gatos Atendidos Em Hospital-Escola Veterinário Da Cidade De São Paulo	2007	2000	Sao Paulo	Brazil	327	[104]
RSM-81	Gastrointestinal Helminth Parasites In Stray Cats From The Mid-Ebro Valley, Spain	1998	1989	Zaragoza	Spain	58	[105]
RSM-82	Gastrointestinal Helminth Parasites Of Pets: Retrospective Study At The Veterinary Teaching Hospital, IPB University, Bogor, Indonesia	2023	2014	Bogor	Indonesia	171	[106]
RSM-83	Gastrointestinal Helminthes In Stray Cats (Felis Catus) From Aizawl, Mizoram, India	2011	2005	Aizawl	India	27	[107]
RSM-84	Gastrointestinal Helminthic Parasites Of Stray Cats (Felis Catus) In Northwest Iran	2021	2014	Meshkin-Shahr	Iran	104	[108]
RSM-85	Gastrointestinal Helminths And Ectoparasites In The Stray Cats (Felidae: Felis Catus) Of Ahar Municipality, Northwestern Iran	2017	2013	Ahar	Iran	51	[109]
RSM-86	Gastrointestinal Helminths Of Cat (Felis Catus) In Kashmir Valley, India.	2020	2017	Kashmir Valley	India	887	[110]
RSM-87	Gastrointestinal Parasite Infection In Cats In Daegu, Republic Of Korea, And Efficacy Of Treatment Using Topical Emodepside/Praziquantel Formulation	2019	2012	Daegu	Republic of Korea	407	[111]
RSM-88	Gastrointestinal Parasites In Dogs And Cats In Line With The One Health' Approach	2022	2018	Pernambuco	Brazil	105	[112]
RSM-89	Gastrointestinal Parasites In Feral Cats And Rodents From The Fernando De Noronha Archipelago, Brazil	2017	2016	Fernando de Noronha	Brazil	37	[113]
RSM-90	Gastrointestinal Parasites In Rural Dogs And Cats In Selangor And Pahang States In Peninsular Malaysia	2014	2011		Malaysia	28	[114]
RSM-91	Gastrointestinal Parasites In Shelter Cats Of Central Italy	2019	2011	Latium, Tuscany	Italy	132	[115]
RSM-92	Gastrointestinal Parasites In Stray And Shelter Cats In The Municipality Of Rio De Janeiro, Brazil	2017	2014	Rio de Janeiro	Brazil	263	[116]
RSM-93	Gastrointestinal Parasites Of Cats In Brazil: Frequency And Zoonotic Risk	2016	2016	Pernambuco	Brazil	173	[117]
RSM-94	Gastrointestinal Parasites Of Cats In Denmark Assessed By Necropsyand Concentration Mcmaster Technique	2015	2014		Denmark	99	[118]
RSM-95	Gastrointestinal Parasites Of Dogs And Cats In A Refuge In Nakhon Nayok, Thailand	2014	2014	Nakhon Nayok	Thailand	300	[119]
RSM-96	Gastrointestinal Parasites Of Domestic Cats In Perth, Western Australia	2003	2001	Pert	Australia	418	[120]
RSM-97	Gastrointestinal Parasites Of Feral Cats From Christmas Island	2008	2008	Christmas Island	Australia	28	[121]
RSM-98	Gastro-Intestinal Parasites Of Feral Cats In New South Wales	1976	1969	New South Wales	Australia	146	[122]
RSM-99	Gastrointestinal Parasites Of Stray Cats In Kashan, Iran	2009	2004	Kashan	Iran	113	[123]
RSM-100	Gastrointestinal Parasites Of Cats In Egypt: High Prevalence High Zoonotic Risk	2022	2021	Gharbia	Egypt	143	[124]
RSM-101	Giardia Is The Most Prevalent Parasitic Infection In Dogs And Cats With Diarrhea In The City Of Medellín, Colombia	2019	2018	Medellin	Colombia	203	[125]

Table 1. *Cont.*

Code	Study Title	Publication	Study	Location	Country	N	Ref.
RSM-102	Helminth And Protozoan Parasites In Dogs And Cats In Belgium	1991	1980		Belgium	30	[126]
RSM-103	Helminth Burden In Stray Cats From Thessaloniki, Greece	2014	2010	Thessaloniki	Greece	2015	[127]
RSM-104	Helminth Parasites And Arthropods Of Feral Cats	1981	1981		Australia	327	[128]
RSM-105	Helminth Parasites Of Cats From The Vientiane Province, Laos, As Indicators Of The Occurrence Of Causative Agents Of Human Parasitoses	2003	1989		Laos	55	[129]
RSM-106	Helminth Parasites Of Dogs And Cats And Toxoplasmosis Antibodies In Cats In Swansea, South Wales	1978	1977	Swansea	United Kingdom	46	[130]
RSM-107	Helminth Parasites Of The House Cat, Felis Catus, In Connecticut, U.S.A	2003	2003	Connecticut	USA	450	[131]
RSM-108	Helmintofauna De Gatos (Felis Silvestres Catus, Linnaeus, 1758) Da Região Metropolitana De Cuiabá	2012	2010		Brazil	146	[132]
RSM-109	Helmintofauna Parasitária Em Gatos Errantes De Lages, Santa Catarina, Brasil	2021	2012	Lages	Brazil	97	[133]
RSM-110	High Prevalence Of Covert Infection With Gastrointestinal Helminths In Cats	2015	2010	Oklahoma	USA	116	[134]
RSM-111	High Prevalence Of Helminth Parasites In Feral Cats In Majorca Island (Spain)	2009	2008	Majorca Island	Spain	58	[135]
RSM-112	Implications Of Zoonotic And Vector-Borne Parasites To Free-Roaming Cats In Central Spain	2018	2014		Spain	459	[136]
RSM-113	Importation Of Cats And Risk Of Parasite Spread: A Caribbean Perspective And Case Study From St Kitts	2020	2018		Saint Kitts and Nevis	74	[137]
RSM-114	Incidencia De Parásitos Gastrointestinales En Gatos En La Ciudad De Guayaquil	2013	2013	Guayaquil	Ecuador	1200	[138]
RSM-115	Incidencia De Parásitos En Gatos (Felis Silvestris Catus) En El Centro De Bienestar Animal Tecámac Municipio De Tecámac	2020	2020	Tecamac	Mexico	60	[139]
RSM-116	Incidencia De *Toxocara cati* En Felinos Domésticos De La Parroquia Veracruz, Cantón Pastaza, Provincia De Pastaza	2023	2023	Puyo	Ecuador	55	[140]
RSM-117	Infecções Por Parasitos Gastrintestinais Em Gatos Domésticos De Araguaína, Tocantins	2017	2017	Araguaina	Brazil	53	[141]
RSM-118	Infection Status With Helminthes In Feral Cats Purchased From A Market In Busan, Republic Of Korea	2005	1996	Busan	Republic of Korea	438	[142]
RSM-119	Infestação Por Ancilostomídeos E Toxoçarídeos Em Cães E Gatos Apreendidos Em Vias Públicas, São Paulo (Brasil)	1988	1980	Sao Paulo	Brazil	940	[143]
RSM-120	Insights To Helminth Infections In Food And Companion Animals In Bangladesh: Occurrence And Risk Profiling	2022	2020		Bangladesh	10	[144]
RSM-121	Internal Parasites Of Feral Cats From The Tasmanian Midlands And King Island	1976	1973	Midlands	Australia	107	[145]
RSM-122	Intestinal And Lung Parasites In Owned Dogs And Cats From Central Italy	2012	2008	Pisa	Italy	81	[146]
RSM-123	Intestinal Helminthic Infections Of Cats In Calabar, Nigeria	1988	1988	Calabar	Nigeria	52	[147]
RSM-124	Intestinal Helminths Of Cats In The Kainji Lake Area, Nigeria	1986	1986		Nigeria	83	[148]
RSM-125	Intestinal Helminths Of Feral Cat Populations From Urban And Suburban Districts Of Qatar	2010	2006	Doha	Qatar	658	[149]
RSM-126	Intestinal Parasites And Fecal Cortisol Metabolites In Multi-Unowned-Cat Environments: The Impact Of Housing Conditions	2021	2015		Spain	368	[150]
RSM-127	Intestinal Parasites And Lungworms In Stray, Shelter And Privately Owned Cats Of Switzerland	2019	2012		Switzerland	664	[151]
RSM-128	Intestinal Parasites And Risk Factors In Dogs And Cats From Rio De Janeiro, Brazil	2021	2017	Rio de Janeiro	Brazil	208	[152]
RSM-129	Intestinal Parasites In Dogs And Cats From The District Of Évora, Portugal	2011	2007	Evora	Portugal	20	[153]
RSM-130	Parásitos Intestinales En Caninos Y Felinos Con Cuadros Digestivos En Santiago, Chile. Consideraciones En Salud Pública	2006	1996	Santiago de Chile	Chile	230	[154]

Table 1. *Cont.*

Code	Study Title	Publication	Study	Location	Country	N	Ref.
RSM-131	Intestinal Parasites Of Cats Purchased In Seoul	1993	1993	Seoul	Republic of Korea	41	[155]
RSM-132	Intestinal Parasites Of Owned Dogs And Cats From Metropolitan And Micropolitan Areas: Prevalence, Zoonotic Risks, And Pet Owner Awareness In Northern Italy	2014	2010		Italy	127	[156]
RSM-133	Estudo De Endoparasitos E Ectoparasitos Em Gatos Domésticos De Área Urbana	2023	2023	Aracatuba	Brazil	61	[157]
RSM-134	Intestinal Parasites Of Pets And Other House-Kept Animals In Moscow	2019	2012	Moscow	Russia	1261	[158]
RSM-135	Intestinal Parasitic Infection In Multi-Cat Shelters In Catalonia	2017	2012	Catalonia	Spain	160	[159]
RSM-136	Investigations On The Endoparasite Fauna Of The Domestic Cat In Eastern Brandenburg [Untersuchungen Zur Endoparasitenfauna Der Hauskatze In Ostbrandenburg]	1997	1993	Brandenburg	Germany	155	[160]
RSM-137	Is There Any Change In The Prevalence Of Intestinal Or Cardiopulmonary Parasite Infections In Companion Animals (Dogs And Cats) In Germany Between 2004–2006 And 2015–2017? An Assessment Of The Impact Of The First ESCCAP Guidelines	2022	2015		Germany	72,200	[161]
RSM-138	Levels Of *Toxocara* Infections In Dogs And Cats From Urban Vietnam Together With Associated Risk Factors For Transmission	2016	2014	Hanoi	Vietnam	253	[162]
RSM-139	Macroparasite Communities In Stray Cat Populations From Urban Cities In Peninsular Malaysia	2013	2007	Kuala Lumpur, Georgetown, Kuantan, Malaca	Malaysia	543	[163]
RSM-140	Molecular Detection Of *Cryptosporidium* Spp. And The Occurrence Of Intestinal Parasites In Fecal Samples Of Naturally Infected Dogs And Cats	2018	2017	Santa Maria	Brazil	49	[164]
RSM-141	Molecular Evaluation Of *Toxocara* Species In Stray Cats Using Loop-Mediated Isothermal Amplification (Lamp) Technique As A Rapid, Sensitive And Simple Screening Assay	2021	2018	Khorramabad	Iran	95	[165]
RSM-142	National Study Of The Gastrointestinal Parasites Of Dogs And Cats In Australia	2008	2004		Australia	1063	[166]
RSM-143	Occurrence And Clinical Significance Of Aelurostrongylus Abstrusus And Other Endoparasites In Danish Cats	2016	2015		Denmark	259	[167]
RSM-144	Occurrence And Zoonotic Potential Of Endoparasites In Cats Of Cyprus And A New Distribution Area For Troglostrongylus Brevior	2017	2017		Republic of Cyprus	185	[168]
RSM-145	Occurrence Of Canine And Feline Extra-Intestinal Nematodes In Key Endemic Regions Of Italy	2019	2015	Abruzzo, Lazio, Molise, Marche, SanPietro Island, Piedmont, Veneto, Friuli-Venezia Giulia	Italy	1000	[169]
RSM-146	Ocorrência De Parasitos Gastrointestinais E Fatores De Risco De Parasitismo Em Gatos Domésticos Urbanos De Santa Maria, RS, Brasil	2013	2011	Santa Maria	Brazil	191	[170]
RSM-147	Seroprevalences Of Antibodies Against Bartonella Henselae And Toxoplasma Gondii And Fecal Shedding Of Cryptosporidium Spp, Giardia Spp, And *Toxocara cati* In Feral And Pet Domestic Cats	2004	2004	North Carolina	USA	153	[171]
RSM-148	Stray Dogs And Cats As Potential Sources Of Soil Contamination With Zoonotic Parasites	2017	2011	Lodz	Poland	68	[17]
RSM-149	Enquête Sur Le Parasitisme Digestif Des Chiens Et Des Chats De Particuliers De La Région Parisienne	2000	1998	Paris	France	34	[172]
RSM-150	Survey Of Helminth Parasites Of Cats From The Metropolitan Area Of Cuiabá, Mato Grosso, Brazil	2013	2010	Cuiaba	Brazil	146	[173]
RSM-151	Ocorrência De Parasitos Gastrintestinais Em Amostras Fecais De Felinos No Município De Andradina, São Paulo	2009	2009	Andradina	Brazil	51	[174]
RSM-152	Ocorrência De Parasitos Gastrintestinais Em Fezes De Gatos Das Cidades De São Paulo E Guarulhos	2002	2002		Brazil	138	[175]
RSM-153	Ocorrência De Protozoários E Helmintos Em Amostras De Fezes De Cães E Gatos Da Cidade De São Paulo	1999	1991	Sao Paulo	Brazil	187	[176]

Table 1. *Cont.*

Code	Study Title	Publication	Study	Location	Country	N	Ref.
RSM-154	Occurrence Of Toxoplasma Gondii And Other Gastrointestinal Parasites In Free-Roaming Cats From The Rio De Janeiro Zoo	2023	2023	Rio de Janeiro	Brazil	51	[177]
RSM-155	Ocorrência De Endoparasitas Com Potencial Zoonótico De Transmissão Em Fezes De Gatos (Felis Catus Domesticus Linnaeus, 1758) Domiciliados Na Área Urbana E Região Metropolitana De Castro—Paraná—Brasil	2012	2012	Castro	Brazil	38	[178]
RSM-156	Ocorrência De Endoparasitos Em Gatos De Cuiabá, Mato Grosso, Brasil.	2011	2009	Cuiaba	Brazil	50	[179]
RSM-157	Ocorrência De Nematódeos E Protozoários Em Gatos Com Tutores Da Cidade De Porto Alegre, RS, Brasil.	2020	2018	Porto Alegre	Brazil	266	[180]
RSM-158	Ocorrência De Parasitas Gastrintestinais De Gatos (Felis Catus) Domiciliados Nos Municípios De Patos-PB E Parelhas-RN.	2017	2008	Patos, Parelhas	Brazil	30	[181]
RSM-159	Ocorrência De Parasitas Gastrintestinais Em Fezes De Cães E Gatos, Curitiba-Pr	2005	2005	Curitiba	Brazil	30	[182]
RSM-160	Ocorrência De Parasitos Em Gatos (Felis Catus Domesticu5) E Pombos (Columba Livia) Procedentes De Algumas Localidades De Minas Gerais	1973	1973	Minas Gerais	Brazil	15	[183]
RSM-161	Ocorrência De Parasitas Gastrintestinais Em Cães E Gatos Na Rotina Do Laboratório De Enfermidades Parasitárias Da Fmvz/Unesp-Botucatu, Sp.	2008	2002	Botucatu	Brazil	140	[184]
RSM-162	One-Year Parasitological Screening Of Stray Dogs And Cats In County Dublin, Ireland	2018	2016	Dublin	Ireland	289	[185]
RSM-163	Parasitas Respiratórios, Gastrointestinais E Auriculares Em Gatos De Colónia, Na Casa Dos Animais De Lisboa	2020	2020	Lisbon	Portugal	47	[186]
RSM-164	Parasite Communities In Stray Cat Populations From Lisbon, Portugal	2014	2009	Lisbon	Portugal	120	[187]
RSM-165	Parasite Prevalence In Fecal Samples From Shelter Dogs And Cats Across The Canadian Provinces	2015	2015		Canada	636	[188]
RSM-166	Parasite Prevalence In Free-Ranging Farm Cats, Felis Silvestris Catus	1996	1989		United Kingdom	11	[189]
RSM-167	Parasite Prevalence Survey In Shelter Cats In Citrus County, Florida	2017	2017	Florida	USA	76	[190]
RSM-168	Parasite Richness And Abundance In Insular And Mainland Feral Cats: Insularity Or Density?	2001	1997	Lyon, Kerguelen	France	133	[191]
RSM-169	Parasites And Zoonotic Bacteria In The Feces Of Cats And Dogs From Animal Shelters In Carinthia, Austria	2023	2023	Carinthia	Austria	130	[192]
RSM-170	Parasites Of Domestic Owned Cats In Europe: Co-Infestations And Risk Factors	2014	2012	Budapest	Hungry, Italy, Romania, France, Austria, Spain, Belgium	300	[193]
RSM-171	Parasites Of Feral Cats From Southern Tasmania And Their Potential Significance	1997	1997		Australia	39	[194]
RSM-172	Parasites Of Stray Cats (Felis Domesticus L., 1758) On St. Kitts, West Indies	2010	2005	Basseterre	Saint Kitts and Nevis	100	[195]
RSM-173	Parasitic Infections Of Domestic Cats, Felis Catus, In Western Hungary	2013	2013		Hungry	235	[196]
RSM-174	Exame Parasitológico De Fezes De Gatos (Felis Catus Domesticus) Domiciliados E Errantes Da Região Metropolitana Do Rio De Janeiro, Brasil	2003	2003	Rio de Janeiro	Brazil	131	[197]
RSM-175	Survey Of Infectious And Parasitic Diseases In Stray Cats At The Lisbon Metropolitan Area, Portugal	2010	2003	Lisbon	Portugal	74	[198]
RSM-176	Survey On The Prevalence Of Intestinal Parasites In Domestic Cats (Felis Catus Linnaeus, 1758) In Central Nepal	2023	2020	Ratnanagar	Nepal	90	[199]
RSM-177	Parasitos De Interesse Zoonótico Em Felinos (Felis Catus Domesticus), Campo Grande, Mato Grosso Do Sul	2016	2014	Campo Grande	Brazil	210	[200]
RSM-178	Técnica De Centrífugo-Flutuação Com Sulfato De Zinco No Diagnóstico De Helmintos Gastrintestinais De Gatos Domésticos	2007	2004	Rio de Janeiro	Brazil	13	[201]

Table 1. *Cont.*

Code	Study Title	Publication	Study	Location	Country	N	Ref.
RSM-179	The First Study On The Prevalence Of Gastrointestinal Parasites In Owned And Sheltered Cats In Yangon, Myanmar	2023	2022	Yangon	Myanmar	230	[202]
RSM-180	Ocena Zależności Zarażenia Pasożytami Wewnętrznymi Psów I Kotów Od Przygotowania Hodowlano–Weterynaryjnego Wła–Ścicieli	2008	2008	Olsztyn	Poland	35	[203]
RSM-181	PARASITOS GASTRINTESTINAIS EM Felis Catus Linnaeus, 1758 DE MOSSORÓ, RN	2023	2022	Mossoro	Brazil	72	[204]
RSM-182	Parasitos Gastrintestinais Em Fezes De Gatos Domiciliados No Município De Pelotas, RS, Brasil	2021	2018	Pelotas	Brazil	60	[205]
RSM-183	Parasitos Gastrintestinais Em Gatos Da Cidade De Porto Alegre, Rio Grande Do Sul	2017	2014	Porto Alegre	Brazil	339	[206]
RSM-184	Parasitos Gastrointestinais De Caninos E Felinos: Uma Questão De Saúde Pública	2021	2015		Brazil	78	[207]
RSM-185	Parásitos Intestinales En Perros Y Gatos Con Dueño De La Ciudad De Barranquilla, Colombia	2018	2014	Barranquilla	Colombia	45	[208]
RSM-186	Parásitos Zoonóticos Presentes En Gatos Domésticos (Felis Silvestris Catus) En Un Centro De Control Canino Y Felino, En Nuevo León; México	2021	2020	New Leon	Mexico	189	[209]
RSM-187	Parasitoses Gastrointestinais E Pulmonares Em Canídeos E Felídeos Da Região Oeste De Portugal Continental	2017	2017		Portugal	70	[210]
RSM-188	Parasitoses Pulmonares E Gastrointestinais Em Felinos Domésticos No Minho, Portugal	2016	2016	Viana de Castelo, Guimaraes, Ponte de Lima, Vila Verde, Braga, Amares	Portugal	68	[211]
RSM-189	Parasitosis Intestinales En Gatos De Querétaro	2023	2022	Queretaro	Mexico	180	[212]
RSM-190	Parasitosis Zoonóticas En Mascotas Caninas Y Felinas De Niños De Educación Primaria Del Cono Norte De Lima, Perú	2011	2011	Lima	Peru	49	[213]
RSM-191	Presence Of *Toxocara* Eggs On The Hair Of Dogs And Cats	2013	2010	Ankara	Turkey	30	[214]
RSM-192	Presence Of *Toxocara* Spp. In Domestic Cats In The State Of Mexico	2016	2016	Mexico City	Mexico	229	[215]
RSM-193	Prevalence And Associated Risk Factors Of Intestinal Parasites In Rural High-Mountain Communities Of The Valle Del Cauca—Colombia	2020	2020	Valle del Cauca	Colombia	7	[216]
RSM-194	Prevalence And Molecular Characterization Of *Toxocara cati* Infection In Feral Cats In Alexandria City, Northern Egypt	2021	2018	Alexandria	Egypt	100	[217]
RSM-195	Prevalence And Molecular Characterization Of Toxoplasmagondii And *Toxocara cati* Among Stray And Household Cats And Cat Owners In Tehran, Iran	2022	2017	Tehran	Iran	165	[218]
RSM-196	Prevalence And Public Health Relevance Of Enteric Parasites In Domestic Dogs And Cats In The Region Of Madrid (Spain) With An Emphasis On Giardia Duodenalis And Cryptosporidium Sp.	2023	2017	Madrid	Spain	35	[219]
RSM-197	Prevalence And Risk Factors Associated With Cat Parasites In Italy: A Multicenter Study	2021	2019		Italy	987	[220]
RSM-198	Prevalence And Risk Factors Associated With Endoparasitosis Of Dogs And Cats In Espírito Santo, Brazil	2016	2016	Espirito Santo	Brazil	160	[221]
RSM-199	Prevalence And Risk Factors For Patent *Toxocara* Infections In Cats And Cat Owners' Attitude Towards Deworming	2016	2010		Netherlands	670	[311]
RSM-200	Prevalence And Risk Factors Of Intestinal Parasites In Cats From China	2015	2013	Henan, Beijing	China	360	[222]
RSM-201	Prevalence Of Antibodies To Toxoplasma Gondii And Intestinal Parasites In Stray, Farm And Household Cats In Spain	2004	2004		Spain	382	[223]
RSM-202	Prevalence Of Cryptosporidian Infection In Cats In Turin And Analysis Of Risk Factors	2007	2007	Turin	Italy	200	[224]
RSM-203	Prevalence Of Endoparasites In Household Cat (Felis Catus) Populations From Transylvania (Romania) And Association With Risk Factors	2010	2007	Transylvania	Romania	414	[225]
RSM-204	Prevalence Of Endoparasites In Northern Mississippi Shelter Cats	2019	2017	Mississippi	USA	55	[226]

Table 1. *Cont.*

Code	Study Title	Publication	Study	Location	Country	N	Ref.
RSM-205	Prevalence Of Endoparasites In Stray And Fostered Dogs And Cats In Northern Germany	2012	2012	Lower Saxony	Germany	837	[227]
RSM-206	Prevalence Of Endoparasitic And Viral Infections In Client-Owned Cats In Metropolitan Bangkok, Thailand, And The Risk Factors Associated With Feline Hookworm Infections	2021	2014	Bangkok	Thailand	509	[228]
RSM-207	Prevalence Of Enteric Zoonotic Agents In Cats Less Than 1 Year Old In Central New York State	2001	1998	New York	USA	263	[229]
RSM-208	Prevalence Of Enteric Zoonotic Organisms In Cats	2000	1993	Colorado	USA	206	[230]
RSM-209	Prevalence Of Faecal-Borne Parasites In Colony Stray Cats In Northern Italy	2013	2008	Milan	Italy	139	[231]
RSM-210	Prevalence Of Fecal-Borne Parasites Detected By Centrifugal Flotation In Feline Samples From Two Shelters In Upstate New York	2011	2006	New York	USA	1629	[232]
RSM-211	Prevalence Of Fleas And Gastrointestinal Parasites In Free-Roaming Cats In Central Mexico	2013	2010		Mexico	358	[233]
RSM-212	Prevalence Of Gastro-Intestinal And Haemoparasitic Infections Among Domestic Cats Of Kerala	2023	2023	Kerala	India	122	[234]
RSM-213	Prevalence Of Gastrointestinal Helminth Parasites Of Zoonotic Significance In Dogs And Cats In Lower Northern Thailand	2017	2014		Thailand	180	[235]
RSM-214	Prevalencia De Helmintos Gastrointestinales En Gatos Admitidos En La Policlínica Veterinaria De La Universidad Del Zulia	2008	2004	Zulia	Venezuela	64	[236]
RSM-215	The Occurrence Of Endoparasites In Slovakian Household Dogs And Cats	2021	2018	Bratislava	Slovakian	50	[237]
RSM-216	The Parasite Fauna Of Stray Domestic Cats (Felis Catus) In Dubai, United Arab Emirates	2009	2004	Dubai	United Arab Emirates	240	[238]
RSM-217	The Prevalence Of Endoparasites Of Free Ranging Cats (Felis Catus) From Urban Habitats In Southern Poland	2020	2020	Kraków	Poland	81	[239]
RSM-218	Prevalence Of Gastrointestinal Parasites In Domestic Cats (Felis Catus) Diagnosed By Different Coproparasitological Techniques In The Municipality Of Seropédica, Rio De Janeiro	2023	2020	Rio de Janeiro	Brazil	237	[240]
RSM-219	Prevalencia De Parásitos Gastrointestinales En Gatos Domésticos (Felis Silvestris Catus Schreber, 1775) En La Habana, Cuba	2020	2014	La Havana	Cuba	356	[241]
RSM-220	Prevalence Of Helminth And Coccidian Parasites In Swedish Outdoor Cats And The First Report Of Aelurostrongylus Abstrusus In Sweden: A Coprological Investigation	2017	2017		Sweden	205	[242]
RSM-221	Prevalence Of Hookworm Infection And Strongyloidiasis In Cats And Potential Risk Factor Of Human Diseases	2018	2016	Thasala	Thailand	15	[243]
RSM-222	Prevalence Of Internal Helminthes In Stray Cats (Felis Catus) In Mosul City, Mosul-Iraq	2012	2008	Mosul	Iraq	55	[244]
RSM-223	Prevalence Of Intestinal Endoparasites With Zoonotic Potential In Domestic Cats From Botucatu, SP, Brazil	2017	2011	Botucatu	Brazil	1725	[245]
RSM-224	Prevalence Of Intestinal Nematodes Of Dogs And Cats In The Netherlands	1997	1993	Utrecht, Amersfoort	Netherlands	292	[246]
RSM-225	Prevalence Of Intestinal Parasites Detected In Routine Coproscopic Methods In Dogs And Cats From The Masovian Voivodeship In 2012–2015	2019	2012	Warsaw	Poland	5809	[247]
RSM-226	Prevalence Of Intestinal Parasites In Breeding Cattery Cats In Japan	2016	2013		Japan	342	[248]
RSM-227	Kedilerde Bağırsak Parazitlerinin Yaygınlığı Ve Halk Sağlığı Bakımından Önemi	2016	2015	Kırıkkale	Turkey	100	[249]
RSM-228	Prevalence Of Intestinal Parasites In Companion Animals In Ontario And Quebec, Canada, During The Winter Months	2008	2008		Canada	47	[250]
RSM-229	Prevalence Of Intestinal Parasites In Dogs And Cats From The Kvarner Region In Croatia	2023	2019	Kvarner	Croatia	64	[251]
RSM-230	Prevalence Of Intestinal Parasites In Dogs And Cats Under Veterinary Care In Porto Alegre, Rio Grande Do Sul, Brazil	2007	2002	Porto Alegre	Brazil	288	[252]

Table 1. *Cont.*

Code	Study Title	Publication	Study	Location	Country	N	Ref.
RSM-231	Prevalence Of Intestinal Parasites In Feral Cats In Some Urban Areas Of England	1981	1978		United Kingdom	92	[253]
RSM-232	Prevalence Of Intestinal Parasites In Pet Shop Kittens In Japan	2013	2011		Japan	555	[254]
RSM-233	Prevalence Of Intestinal Parasites In Private-Household Cats In Japan	2012	2008		Japan	942	[255]
RSM-234	Prevalence Of Intestinal Parasites In Shelter Dogs And Cats In The Metropolitan Area Of Barcelona (Spain)	2009	1999	Barcelona	Spain	50	[256]
RSM-235	Prevalence Of Intestinal Parasites, Risk Factors And Zoonotic Aspects In Dog And Cat Populations From Goiás, Brazil	2023	2020	Goiás	Brazil	55	[257]
RSM-236	Prevalence Of Major Digestive And Respiratory Helminths In Dogs And Cats In France: Results Of A Multicenter Study	2022	2017		France	425	[3]
RSM-237	Prevalence Of Protozoa And Gastrointestinal Helminthes In Stray Cats In Zanjan Province, North-West Of Iran	2009	2007	Zanjan	Iran	100	[258]
RSM-238	Prevalence Of Selected Bacterial And Parasitic Agents In Feces From Diarrheic And Healthy Control Cats From Northern California	2012	2007	California	USA	269	[259]
RSM-239	Prevalence Of Selected Zoonotic And Vector-Borne Agents In Dogs And Cats In Costa Rica	2011	2009	San Isidro de El General	Costa Rica	9	[260]
RSM-240	Prevalence Of Some Gastrointestinal Parasites In Cats In The Perth Area	1983	1978	Perth	Australia	752	[261]
RSM-241	Prevalence Of Species Of *Toxocara* In Dogs, Cats And Red Foxes From The Poznan Region, Poland	2001	1997	Poznan	Poland	105	[262]
RSM-242	Prevalence Of *Toxocara cati* And Other Intestinal Helminths In Stray Cats In Shiraz, Iran	2007	2005	Shiraz	Iran	114	[263]
RSM-243	Prevalence Of *Toxocara cati* In Pet Cats And Its Zoonotic Importance In Tabriz City, Iran	2020	2014	Tabriz	Iran	50	[264]
RSM-244	Prevalence Of *Toxocara* Infection In Domestic Dogs And Cats In Urban Environment	2018	2011		Russia	1146	[265]
RSM-245	Prevalence Of *Toxocara* Spp. And Other Parasites In Dogs And Cats In Halifax, Nova Scotia	1978	1971	Halifax	Canada	299	[266]
RSM-246	Prevalence Of Toxocariasis And Its Related Risk Factors In Humans, Dogs And Cats In Northeastern Iran: A Population-Based Study	2019	2017	Khorasan Razavi	Iran	236	[267]
RSM-247	Prevalence Of Toxoplasma Gondii And Other Gastrointestinal Parasites In Domestic Cats From Households In Thika Region, Kenya	2017	2015	Thika	Kenya	103	[268]
RSM-248	Prevalence Of Toxoplasma Gondii And Other Intestinal Parasites In Cats In Tokachi Sub-Prefecture, Japan	2018	2013	Tokachi	Japan	351	[269]
RSM-249	Prevalence Of Toxoplasma Gondii Antibodies And Intestinal Parasites In Stray Cats From Nigde, Turkey	2008	2003	Nigde	Turkey	72	[270]
RSM-250	Prevalence Of Zoonotic Parasites In Feral Cats Of Central Virginia, USA	2018	2016	Virginia	USA	192	[271]
RSM-251	Prevalence Survey Of Gastrointestinal And Respiratory Parasites Of Shelter Cats In Northeastern Georgia, USA	2019	2019	Georgia	USA	103	[272]
RSM-252	Prevalence, Co-Infection And Seasonality Of Fecal Enteropathogens From Diarrheic Cats In The Republic Of Korea (2016–2019): A Retrospective Study	2021	2016		Republic of Korea	2789	[273]
RSM-253	Prevalência De Helmintos Em Gatos (Felis Catus Domesticus) De Goiânia	1974	1973	Goiânia	Brazil	37	[274]
RSM-254	Prevalencia De Infección Por *Toxocara cati* Y Giardia Duodenalis En Gato Domestico	2018	2018	Lima	Peru	70	[275]
RSM-255	Prevalência De Parasitas Gastrointestinais E Cardiorrespiratórios Em Gatos Domésticos Na Área Metropolitana De Lisboa	2020	2020	Lisbon	Portugal	77	[276]
RSM-256	Prevalência De Parasitas Gastrointestinais Em Felinos No Concelho De Vila Nova De Gaia	2022	2021	Vila Nova de Gaia	Portugal	102	[277]
RSM-257	Prevalencia De Parásitos Gastrointestinales En Gatos Domésticos (Felis Catus) En La Parroquia La Matriz Del Cantón Latacunga.	2019	2019	Latacunga	Ecuador	100	[278]

Table 1. *Cont.*

Code	Study Title	Publication	Study	Location	Country	N	Ref.
RSM-258	Prevalencia De Parásitos Gastrointestinales En Muestras Coprológicas De Caninos Y Felinos Remitidas Al Laboratorio Ejelab, Risaralda. Abril 2017- Abril 2018	2018	2017	Risaralda	Colombia	85	[279]
RSM-259	Prevalência De Parasitos Intestinais Em Gatos Errantes Em Goiânia—Goiás: Ênfase No Diagnóstico De Toxoplasma Gondii Eavaliação Da Acurácia De Técnicas Parasitológicas	2015	2012	Goiânia	Brazil	154	[280]
RSM-260	Prevalencia De Parásitos Intestinales En Los Habitantes Y Sus Mascotas En Los Barrios Hospital, San Lorenzo, Amanecer Y San Antonio Del Municipio De Amatitlán	2014	2014	Amatitlán	Guatemala	15	[281]
RSM-261	Prevalencia De *Toxocara cati* En Felinos Domésticos (Felis Catus) En El Sector La Venecia Ii Del Distrito Metropolitano De Quito	2021	2021	Quito	Ecuador	100	[282]
RSM-262	Prevalencia De *Toxocara cati* En Gatos Domésticos En El Sector De Balerio Estacio, De La Ciudad De Guayaquil	2018	2018	Guayaquil	Ecuador	80	[283]
RSM-263	Principal Endoparasitoses Of Domestic Cats In Sardinia	2004	2000	Sardinia	Italy	183	[284]
RSM-264	Rastreio De Parasitas Gastrointestinais E Pulmonares Em Canídeos E Felídeos Da Região Autónoma Dos Açores—Ilhas De São Miguel E Terceira	2020	2019	Terceira, São Miguel	Portugal	115	[285]
RSM-265	Rastreio De Parasitas Gastrointestinais E Pulmonares Em Gatos De Gatis Nos Distritos De Lisboa E Setúbal, Portugal	2017	2015	Lisbon	Portugal	260	[286]
RSM-266	Recent Investigation On The Prevalence Of Gastrointestinal Nematodes In Cats From France And Germany	2003	1998		Germany	441	[287]
RSM-267	Results Of Parasitological Examinations Of Faecal Samples From Cats And Dogs In Germany Between 2003 And 2010	2011	2003	Freiburg	Germany	8560	[288]
RSM-268	Retrospective Survey Of Parasitism Identified In Feces Of Client-Owned Cats In North America From 2007 Through 2018	2020	2007		USA	2568	[289]
RSM-269	Risk Factors Associated With Cat Parasites In A Feline Medical Center	2021	2021	Mexico City	Mexico	528	[290]
RSM-270	Role Of Small Mammals In The Epidemiology Of Toxocariasis	1995	1995		Slovakian	116	[291]
RSM-271	The Prevalence Of Giardia And Other Intestinal Parasites In Children, Dogs And Cats From Aboriginal Communities In The Kimberley	1993	1993	Kimberley	Australia	33	[292]
RSM-272	The Prevalence Of Intestinal Helminths In Stray Cats In Central Scotland	1980	1980	Glasgow	United Kingdom	72	[293]
RSM-273	The Prevalence Of Intestinal Nematodes In Cats And Dogs From Lancashire, North-West England	2016	2016	Lancashire	United Kingdom	131	[294]
RSM-274	The Prevalence Of Intestinal Parasites In Dogs And Cats In Calgary, Alberta	2011	2008	Calgary	Canada	153	[295]
RSM-275	The Prevalence Of Intestinal Parasites Of Domestic Cats And Dogs In Vladivostok, Russia During 2014–2017	2018	2014	Vladivostok	Russia	135	[296]
RSM-276	The Prevalence Of Potentially Zoonotic Intestinal Parasites In Dogs And Cats In Moscow, Russia	2023	2018	Moscow	Russia	1350	[297]
RSM-277	The Prevalence Of *Toxocara cati* In Domestic Cats In Mexico City	2003	2003	Mexico City	Mexico	520	[298]
RSM-278	The Prevalence Of *Trichuris* Spp. Infection In Indoor And Outdoor Cats On St. Kitts	2015	2015		Saint Kitts and Nevis	41	[299]
RSM-279	*Toxocara* Canis And *Toxocara cati* In Stray Dogs And Cats In Bangkok, Thailand: Molecular Prevalence And Risk Factors	2022	2022	Bangkok	Thailand	500	[300]
RSM-280	*Toxocara cati* And Other Parasitic Enteropathogens: More Commonly Found In Owned Cats With Gastrointestinal Signs Than In Clinically Healthy Ones	2021	2021	Cluj-Napoca	Romania	137	[301]
RSM-281	*Toxocara cati* Infections In Stray Cats In Northern Iran	2007	2004	Mazandaran	Iran	100	[302]
RSM-282	*Toxocara* Infection In Dogs And Cats In Isfahan Province Of Iran In 2021	2023	2023	Isfahan	Iran	230	[303]
RSM-283	*Toxocara* Nematodes In Stray Cats From Shiraz, Southern Iran: Intensity Of Infection And Molecular Identification Of The Isolates	2013	2011	Shiraz	Iran	30	[304]

Table 1. *Cont.*

Code	Study Title	Publication	Study	Location	Country	N	Ref.
RSM-284	What Is The Role Of Swiss Domestic Cats In Environmental Contamination With Echinococcus Multilocularis Eggs?	2023	2022		Switzerland	146	[305]
RSM-285	*Cryptosporidium* Spp. In Dogs And Cats In Poland	2021	2016		Poland	101	[306]
RSM-286	Zoonotic And Other Gastrointestinal Parasites In Cats In Lumajang, East Java, Indonesia	2020	2018		Indonesia	120	[307]
RSM-287	Zoonotic Helminths Parasites In The Digestive Tract Of Feral Dogs And Cats In Guangxi, China	2015	2012	Guangxi	China	39	[308]
RSM-288	Zoonotic Parasites In Fecal Samples And Fur From Dogs And Cats In The Netherlands	2009	2007		Netherlands	63	[309]
RSM-289	Animal Toxocariasis In A Megalopolis Epidemic Aspects	2015	2015		Russia	44	[310]

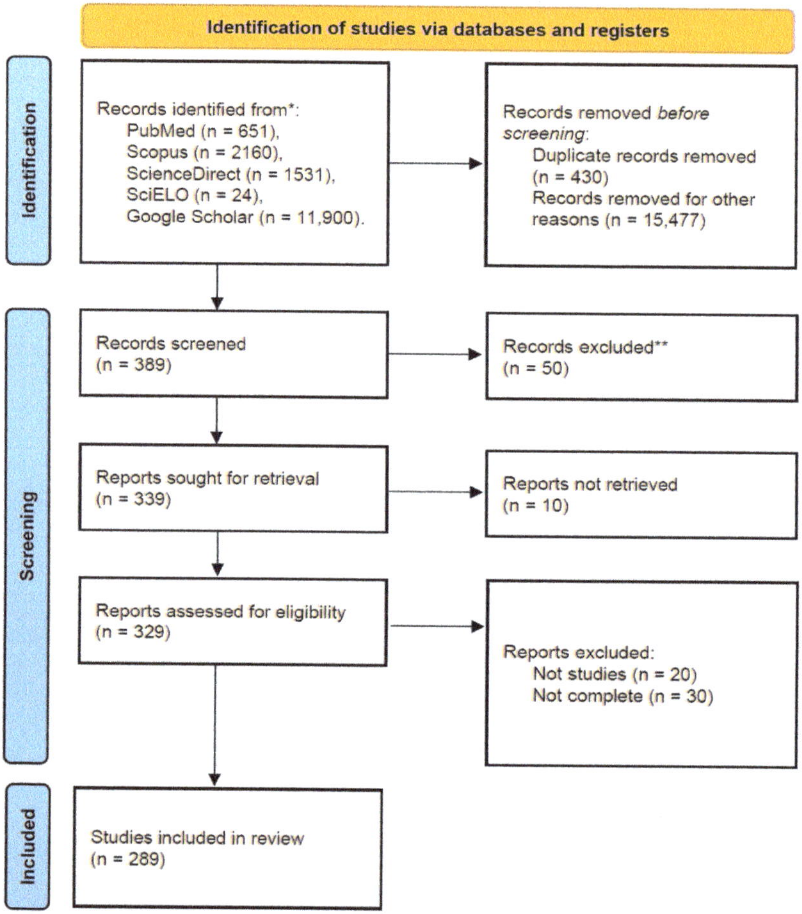

Figure 1. The 2020 PRISMA flow diagram. * All included databases, raw results. ** At an initial quality screening, including lack of inclusion criteria.

3.2. Characteristics of Included Studies

The characteristics of the included articles are summarised in Table 1. A total of 289 articles were included, in which 168,643 cats were evaluated, 92.6% by coproparasitological techniques, 5.4% by necropsy (histology), and 2.0% by PCR. The studies ranged from 1973 to 2023, but there were 30 (7.71%) in 2019 (Table 1). The studies were distributed as follows: Brazil (71 studies), the United States (23 studies), Italy (22 studies), Iran (21 studies), and Portugal (19 studies), among other 57 countries (Table 1). All faecal samples were evaluated using Direct Smear, Graham, Kinyou, Mini Parasep®Sf, Sporulation, Flotac, Bailinger, Mcmaster, Wisconsin, Centrifugal Flotation, Acid-Fast, Flotation (Faust), Charles, Meriflour, Mini-Flotac, Concentration Flotation, Sedimentation (Hoffman), Centrifugal Sedimentation, Baermann, Mifc (Merthiolate-Iodine-Formaldehyde-Concentration), Modified Telemann, Teuscher, Fulleborn, Fecal Smear (Ziehl-Neelsen), Sheater, Formalin Ether (Ritchie), Willis, and Gordon E. Whitlock techniques, among others, searching for *Toxocara* eggs, larvae, and adult parasites. PCR was also used from faecal samples to detect *Toxocara cati*.

3.3. Risk of Bias Assessment

In the risk of bias assessment, twenty studies were at high risk of bias, while the remaining 269 were at low risk of bias.

3.4. Prevalence of *Toxocara cati* in Cats Found Using Coproparasitological Methods

The global pooled prevalence of *Toxocara cati* in cats found using coproparasitological methods was 17.0% (95.0% CI: 16.2–17.8%), with high heterogeneity (I^2 = 97.981%, τ^2 = 0.004, Q^2 = 15652.788) (Figure 2). In the subgroup analysis by year (Figure 3), 1996 was the year with the highest reported pooled prevalence of *Toxocara cati* infection (90.6%; 95% CI 73.9–100.0%), followed by 1991 (60.0%; 95% CI 42.5–77.5%), and 2001 (36.2%; 95% CI 19.2–53.2%) (Figure 3). In the subgroup analysis according to country (Figure 4), Nepal had the highest prevalence of *Toxocara cati* infection (94.4%; 95% CI 89.7–99.2%), followed by the United Kingdom (90.9%; 95% CI 73.9–100.0%) and Bangladesh (76.9%; 95% CI 54.0–99.8%), among other countries (Figure 4). In the subgroup analysis according to continents or regions (Figure 5), Asia had the highest prevalence of *Toxocara cati* infection (27.9%; 95% CI 24.5–31.4%) (I^2 = 99.11%), followed by Africa (21.4%; 95% CI 7.1–35.6%) (I^2 = 93.03%) and North America (18.5%; 95% CI 15.2–21.9%) (I^2 = 98.1%) (Figure 5).

Considering the types of cats (Feral, Stray, Shelter, Domestic, and Breed), we found that the highest prevalence of *Toxocara cati* infection was in feral cats (42.6%; 95% CI 29.8–55.4%) (I^2 = 96.72%), followed by stray cats (29.9%; 95% CI 25.3–34.4%) (I^2 = 98.34%) and shelter cats (20.1%; 95% CI 16.1–24.1%) (I^2 = 97.79%) (Figure 6).

Regarding the coproparasitological methods, we found that the highest prevalence of *Toxocara cati* infection was obtained through a faecal direct smear (26.1%; 95% CI 22.7–29.5%) (I^2 = 98.77%), followed by flotation (Faust) (19.9%; 95% CI 18.4–21.4%) (I^2 = 98.46%) and centrifugal flotation (16.2%; 95% CI 12.3–20.1%) (I^2 = 96.71%) (Figure 7).

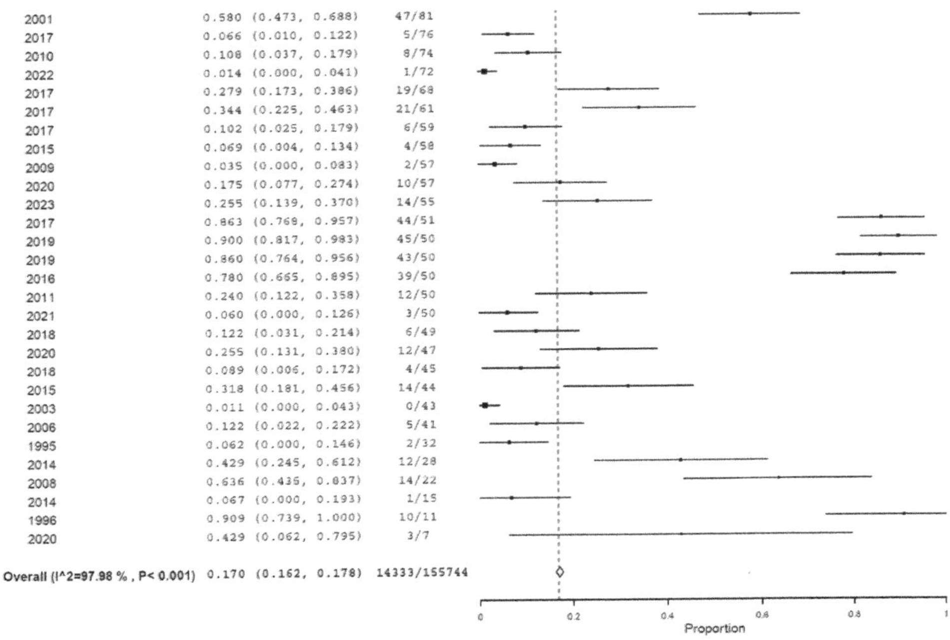

Figure 2. Prevalence of *Toxocara cati* in cats found using coproparasitological methods.

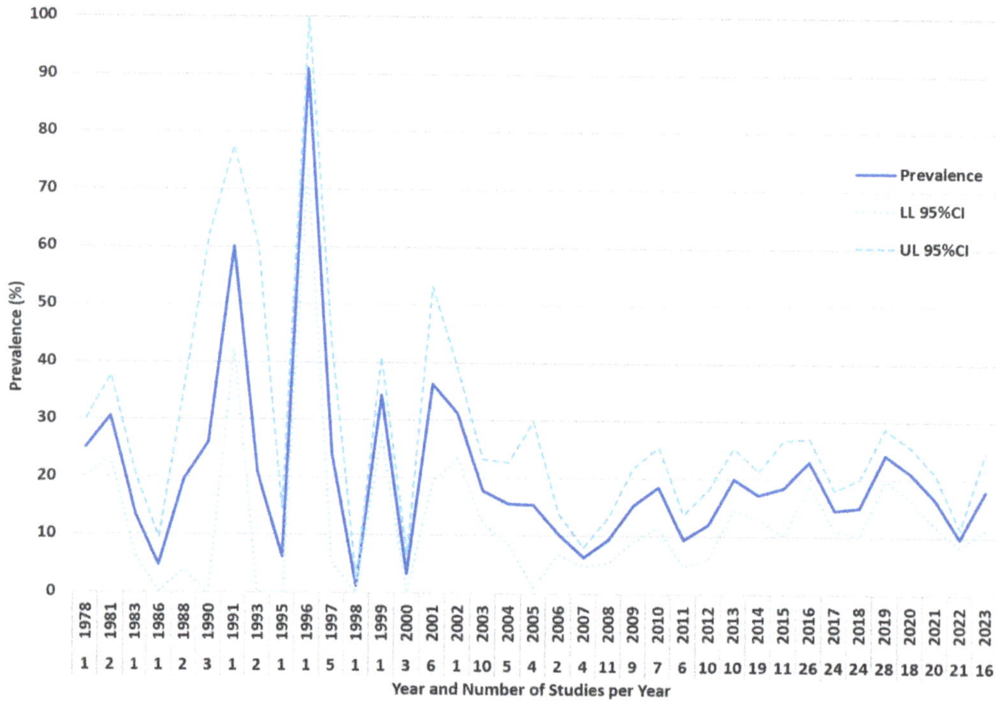

Figure 3. Prevalence of *Toxocara cati* in cats found using coproparasitological methods by years.

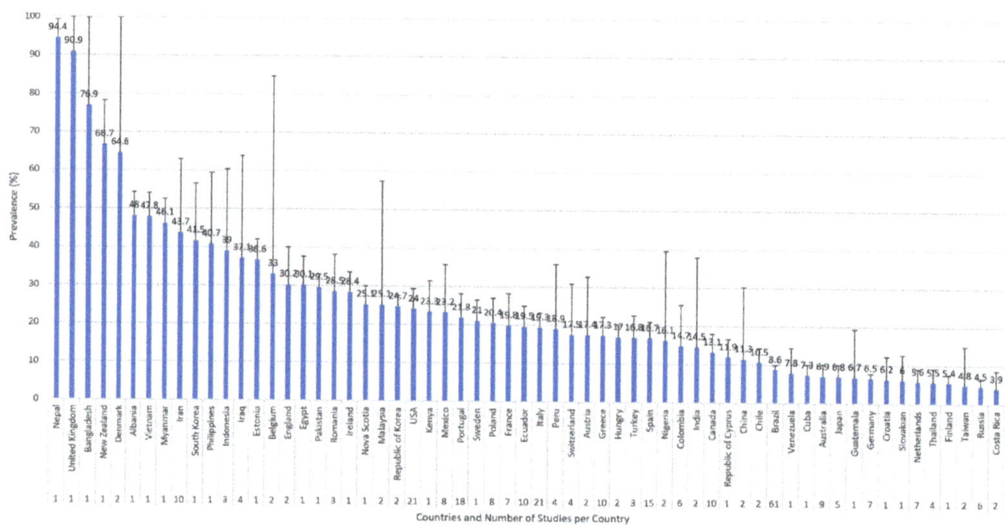

Figure 4. Prevalence of *Toxocara cati* in cats found using coproparasitological methods by countries. Error bars show the upper 95% CI value.

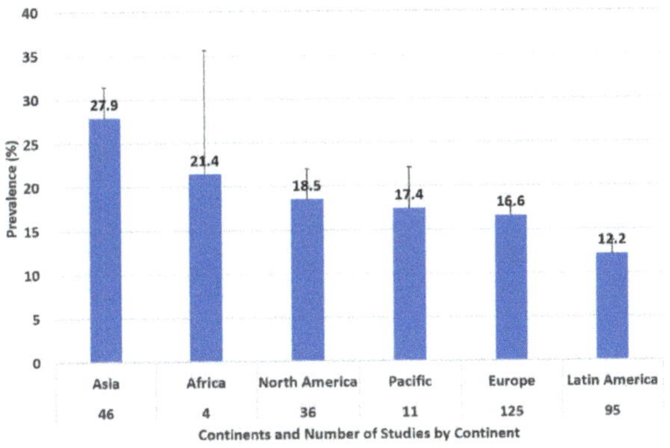

Figure 5. Prevalence of *Toxocara cati* in cats found using coproparasitological methods by continents. Error bars show the upper 95% CI value.

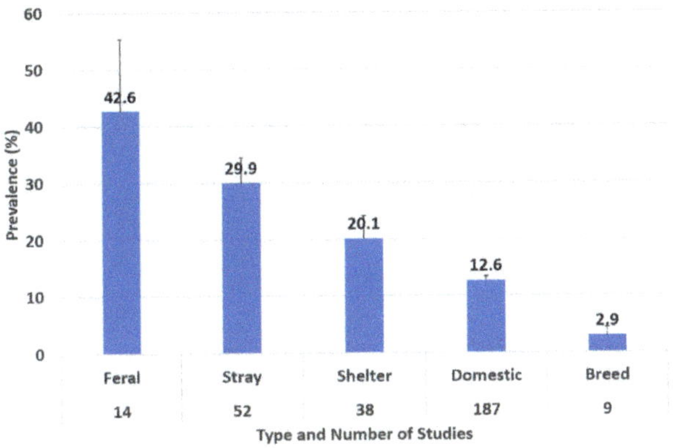

Figure 6. Prevalence of *Toxocara cati* in cats found using coproparasitological methods by cat type. Error bars show the upper 95% CI value.

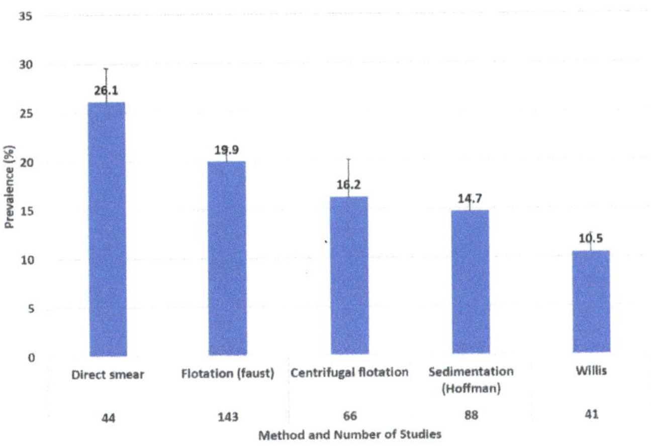

Figure 7. Prevalence of *Toxocara cati* in cats found using different coproparasitological methods. Error bars show the upper 95% CI value.

3.5. Prevalence of Toxocara Found Using Necropsy (Histology)

The pooled prevalence of toxocariasis found using the necropsy (histology) of gastrointestinal tissues was 30.0% (95.0% CI: 26.1–33.8%) with high heterogeneity (I^2 = 98.64%, τ^2 = 0.022, Q^2 = 4545.122) (Figure 8).

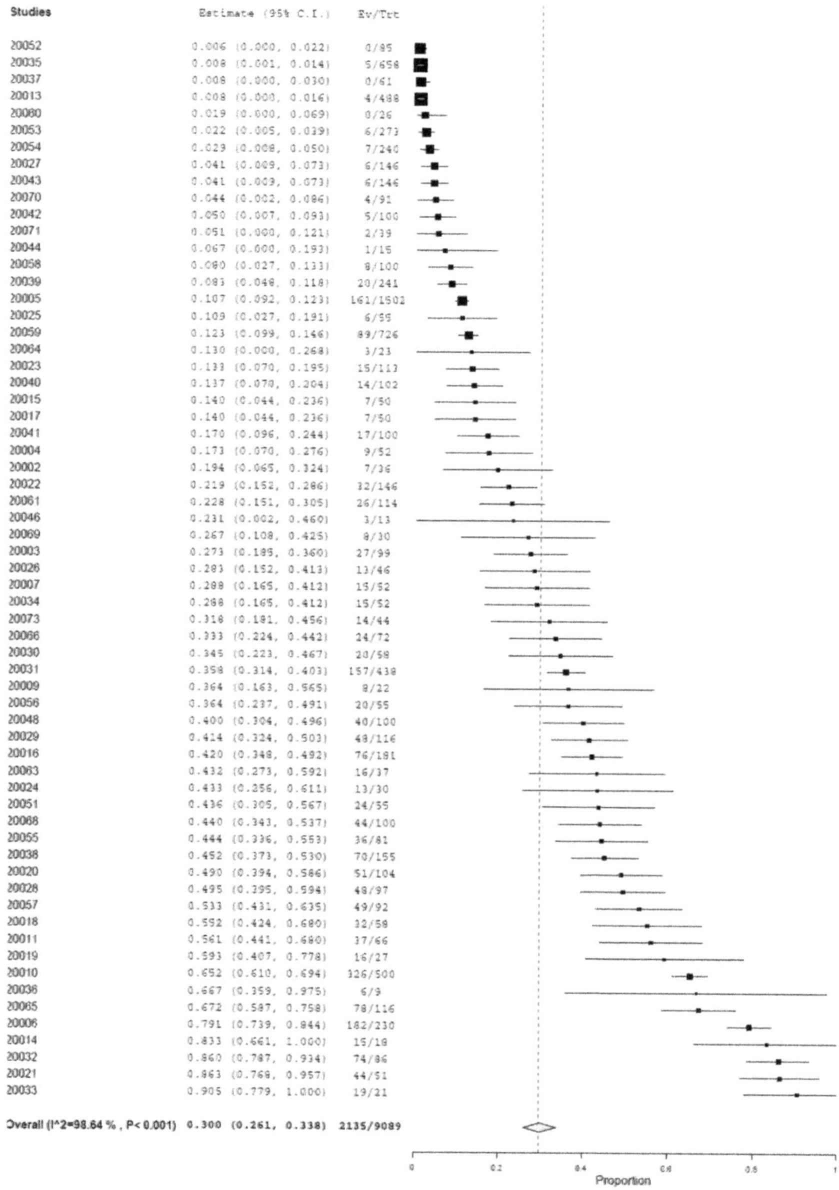

Figure 8. Prevalence of toxocariasis in cats found using necropsy (histology).

3.6. Prevalence of Toxocara catis Found Using PCR

The pooled prevalence of *Toxocara cati* infection found using PCR in four studies (N = 3454) was 4.9% (95.0% CI: 1.9–7.9%) with high heterogeneity (I^2 = 97.48%, τ^2 = 0.001, Q^2 = 119.225) (Figure 9).

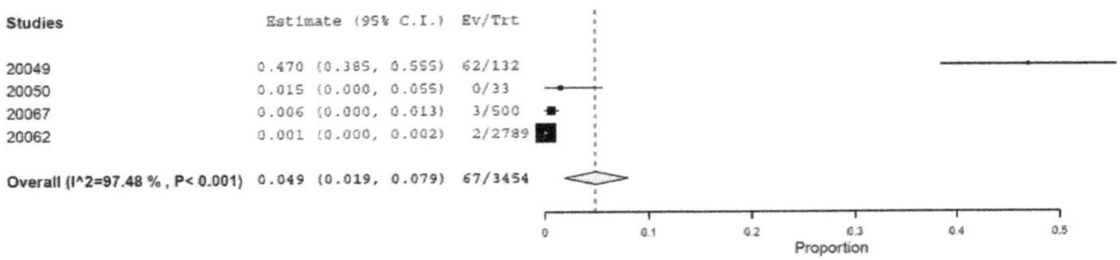

Figure 9. Prevalence of *Toxocara cati* in cats found using PCR.

4. Discussion

Toxocariasis, a helminth parasitic disease, is widespread, particularly in low- and middle-income nations. Despite its significant clinical implications, including the potential for fatal outcomes in humans and animals, mainly domestic ones, such as dogs and cats, many countries, particularly those with limited resources, do not actively monitor this condition [312]. There is a lack of epidemiological surveillance in many regions of the planet for toxocariasis in humans and animals.

This systematic review and meta-analysis, aimed at determining the pooled prevalence of *Toxocara cati* in cat populations worldwide using a comprehensive search approach across six databases, found a relevant prevalence. The findings underscore the considerable diversity in parasite prevalence across various countries and continents, as indicated by previous studies [20,313,314]. Through an extensive exploration of studies published between 1973 and 2023 across diverse geographic regions, this study facilitated the execution of a meta-analysis to ascertain the global prevalence of *T. cati*. This broad temporal and geographical scope enabled a robust synthesis of data, based on more than 150,000 animals, to provide insights into the prevalence patterns of this parasite on a global scale. As expected, *Toxocara cati*, compared with *T. canis*, is more neglected [315], making it difficult to understand that this pathogen affects other domestic and non-domestic animals and humans [316–319]. Toxocariasis in humans is also neglected, especially in developing countries [320,321]. Very few studies, and even more cases reported, can confirm *T. cati* infection specifically in humans, by serological or molecular methods, as compared with just toxocariasis or *Toxocara* spp infection in humans, mainly due to a lack of confirmation or a lack of specific tests to confirm species at diagnosis. It is also believed that there are no implications at all regarding the implicated *Toxocara* species [316,322]. Recent studies suggest that no proteins from *T. canis* and *T. cati* exist that could be used as a diagnostic tool to enable differential serodiagnostics of these species in humans. In addition, a heterogenic protein pattern between individual hosts has been found, which was most pronounced in *T. cati*-infected pigs [322].

Most of the studies on toxocariasis in animals have traditionally focused on dogs and *Toxocara canis* [12,323,324]. Comparatively, there have been a lack of studies on toxocariasis in cats, mainly due to *T. cati*. Cats are also relevant hosts of zoonotic diseases, specifically zoonotic parasites [325,326]. In this systematic review, it was observed that coproparasitological methods are still the predominant means of establishing toxocariasis in cats, showing a relevant prevalence that seems to be higher during specific years and places, probably, as shown before, influenced by seasonal, environmental, and even climatic factors [327–330]. For example, as expected, the country with the highest prevalence was Nepal, a country with a low Human Capital Index (0.5) and included in the group of lower-middle income economies (LMIE), according to the World Bank (https://datahelpdesk.worldbank.org/knowledgebase/articles/906519-world-bank-country-and-lending-groups) (accessed on 1 February 2024). Higher prevalences were also observed in other LMIE, such as Bangladesh, Vietnam, and Myanmar, all of them in Asia, which resulted in it being the continent with the highest prevalence. Culturally, there is a high level of contact with and apparent care of cats by humans in many Asian and Middle Eastern countries, such as Turkey (17%), Egypt (30%), and China (11%), among others. A recent study showed that cats are more popular than dogs in 91 countries, and dogs are more prevalent in 76 countries (https://www.budgetdirect.com.au/pet-insurance/guides/cats-vs-dogs-which-does-the-world-prefer.html) (accessed on 1 February 2024). However, the number of articles and the number of samples analysed per study for some countries would be insufficient to understand the relationships between prevalence and associated factors, despite the fact that the prevalence is weighted in the meta-analysis by the number of studies and the sample size.

The type of cat significantly influences the prevalence of toxocariasis in cats. Those living in wild, non-urban areas (feral) presented the highest prevalence (43%), while domestic cats (13%) and breed cats (3%) showed the lowest values. Other studies show that this is a risk factor for higher prevalences [326]. In general, unattended cats without proper veterinary control and assessment are at risk of exposure and infection.

The main coproparasitological methods vary slightly regarding the prevalence of toxocariasis, from 10.5% to 26.1%, with the faecal direct smear method associated with the highest prevalence.

Many studies assessed infection in dead animals, reporting a high prevalence, even higher than those studies assessing infection by coproparasitological methods. The prevalence of *Toxocara cati* at necropsy was 30%. In contrast, PCR prevalence was only 5%. Then, this was less sensitive than coproparasitological methods (17%). Again, the number of articles and the number of samples analysed per study by molecular methods such as PCR, would be insufficient to understand the differences in the sensitivity and specificity of methods, despite the fact that the prevalence is weighted in the meta-analysis by the number of studies and the sample size. To understand the sensitivity of PCR, specific studies of diagnostic test comparison should be performed, which was clearly outside the objectives of this systematic review, which focused on the prevalence of *T. cati* in cats.

Indeed, the histological diagnosis of *T. cati* can be limited, which makes a differential diagnosis with *T. canis* impossible. An infection due to *Toxocara* in a cat is not necessarily due to *T. cati*, as an infection due to *Toxocara* in a dog is not necessarily due to *T. canis*. Both species may infect other hosts, and in some, these may serve as paratenic hosts, just serving for infection without the reproduction and development of adult forms, as occurs in humans that are exclusively paratenic hosts [331–333]. More commercial tests and laboratories with standardised PCR for molecular diagnosis must be conducted. The molecular diagnosis of toxocariasis is only sometimes available for humans, where the primary tool is serological tests [312,333,334]. There is an urgent need for a molecular diagnosis of toxocariasis, with possibilities of sequencing and identifying species [335,336]. At the same time, better immunological tests are required, as ELISA and Western blot still need to be improved, mainly due to the antigen quality. Then, recombinant antigens-based tests are preferred and it is recommended that they are widely available [337–339].

Although dogs have been studied more, the present systematic review shows that infections due to *Toxocara cati* in cats may be even higher (17%) than those due to *T. canis* in dogs. A recent systematic review of *T. canis* in dogs found that the overall prevalence was 11.1% (95% CI, 10.6–11.7%) after studying more than 3 million dogs in 60 countries [12]. The authors concluded that young (<1 year of age), stray, rural, and male dogs had a significantly higher prevalence of infection than older, pet, urban, or female dogs [12]. Our results confirm the findings of a review from 2020, in which the prevalence of *Toxocara* infection in cats was 17.0% (16.1–17.8%), but there was a contrast regarding the continents, as this review found the highest prevalence in African countries (43.3%, 28.3–58). As mentioned earlier, we found this in Asia (28%). In Africa, we found 21.4 (7.1–35.6%). They found that the prevalence of *Toxocara* was higher in stray cats (28.6%, 25.1–32.1%) [11]. Our review found 29.9% in stray cats (25.3–34.4%) but higher results in feral cats (42.6%).

As indicated, cats may also serve as a source of human infections due to *Toxocara*. Cats play a crucial role globally as primary hosts for *Toxocara*, releasing eggs into the environment and thereby heightening public health concerns. Health authorities and cat caregivers must prioritise efforts toward preventing and managing this zoonotic disease in feline populations. This is especially crucial in regions with elevated risk factors and prevalence rates, necessitating heightened vigilance and proactive measures [11].

Cats, as with other species, may also be infected with another member of the family *Toxocaridae*, as is the case of *Toxascaris leonina*; nevertheless, fewer studies about it are available [340,341]. Regardless, studies and systematic reviews so far are lacking and needed [341].

This systematic review has certain limitations, including the fact that we were unable to assess the age or gender of cats, as this was not reported in most of the studies. This aspect could also be important in the prevalence and risk of *T. cati* infection, as has been suggested in *T. canis* [312].

5. Conclusions

The significance of toxocariasis in cats is its potential to infect humans, in addition to the damage that it may cause to felines. Humans can become accidental hosts by ingesting *Toxocara cati* eggs through contaminated soil, water, or food. Once ingested, the larvae can migrate to various tissues in the body, causing visceral larva migrans (VLM) or ocular larva migrans (OLM), which can result in serious health complications, including vision impairment, organ damage, and even neurological disorders. Preventive measures should be considered, given the high prevalence found in this systematic review and previous studies. The zoonotic aspect of toxocariasis in cats and dogs highlights the interconnectedness of animal and human health, including OneHealth, emphasising the importance of preventive measures such as deworming protocols for cats [311,342–344], proper hygiene practices, and public education on the risks associated with exposure to contaminated environments. By addressing toxocariasis in cats, feline health can be safeguarded, and the potential transmission of this parasitic infection to humans and other animals can also be minimised, promoting the well-being of both animals and humans. Finally, in farming animals, toxocariasis can lead to reduced productivity. Infected animals may exhibit decreased weight gain, reduced milk production (in dairy cattle), decreased fertility, and lower overall performance. This can directly impact farm income by reducing the quantity and quality of products produced.

Author Contributions: J.L.B.-A.: Conceptualisation, methodology, formal analysis, investigation, writing—original draft preparation, and writing—review and editing; A.C.E.-N.: Conceptualisation, formal analysis, investigation, writing—original draft preparation, and writing—review and editing; D.K.B.-A.: Conceptualisation, formal analysis, investigation, writing—original draft preparation, and writing—review and editing; A.J.R.-M.: Conceptualisation, investigation, writing—original draft preparation, and writing review and editing. All authors have read and agreed to the published version of the manuscript.

Funding: Universidad Continental, Huancayo, Peru, covered the APC in this article.

Institutional Review Board Statement: Not applicable.

Informed Consent Statement: Not applicable.

Data Availability Statement: Available upon reasonable request.

Acknowledgments: We would like to thank Olinda Delgado, from the Immunoparasitology Section of the Tropical Medicine Institute of the Universidad Central de Venezuela, in Caracas, DC, Venezuela, who has been a mentor to many of those interested in toxocariasis and *Toxocara* research in Latin America, guiding and inspiring on this zoonotic disease.

Conflicts of Interest: The authors declare no conflicts of interest.

References

1. Cross, A.R.; Baldwin, V.M.; Roy, S.; Essex-Lopresti, A.E.; Prior, J.L.; Harmer, N.J. Zoonoses under our noses. *Microbes Infect.* **2019**, *21*, 10–19. [CrossRef] [PubMed]
2. Rodriguez-Morales, A.J.; González-Leal, N.; Montes-Montoya, M.C.; Fernández-Espíndola, L.; Bonilla-Aldana, D.K.; Azeñas-Burgoa, J.M.; de Medina, J.C.D.; Rotela-Fisch, V.; Bermudez-Calderon, M.; Arteaga-Livias, K.; et al. Cutaneous Larva Migrans. *Curr. Trop. Med. Rep.* **2021**, *8*, 190–203. [CrossRef]
3. Bourgoin, G.; Callait-Cardinal, M.P.; Bouhsira, E.; Polack, B.; Bourdeau, P.; Roussel Ariza, C.; Carassou, L.; Lienard, E.; Drake, J. Prevalence of major digestive and respiratory helminths in dogs and cats in France: Results of a multicenter study. *Parasit. Vectors* **2022**, *15*, 314. [CrossRef] [PubMed]
4. Quintero-Cusguen, P.; Gutiérrez-Álvarez, A.M.; Patiño, D.R. Toxocariosis. *Acta Neurol. Colomb.* **2021**, *37*, 169–173. [CrossRef]
5. Holland, C.V. Knowledge gaps in the epidemiology of *Toxocara*: The enigma remains. *Parasitology* **2017**, *144*, 81–94. [CrossRef] [PubMed]
6. Bowman, D.D. Ascaris and *Toxocara* as foodborne and waterborne pathogens. *Res. Vet. Sci.* **2021**, *135*, 1–7. [CrossRef] [PubMed]
7. Healy, S.R.; Morgan, E.R.; Prada, J.M.; Betson, M. First report demonstrating the presence of *Toxocara* spp. eggs on vegetables grown in community gardens in Europe. *Food Waterborne Parasitol.* **2022**, *27*, e00158. [CrossRef]
8. Bakhshani, A.; Maleki, M.; Haghparast, A.; Parande Shirvan, S.; Borji, H. A survey on *Toxocara cati* eggs on the hair of stray cats: A potential risk factor for human toxocariasis in Northeastern Iran. *Comp. Immunol. Microbiol. Infect. Dis.* **2019**, *64*, 10–13. [CrossRef] [PubMed]
9. Benavides-Melo, C.J.; Vallejo-Timarán, D.A.; Astaiza-Martínez, J.M.; Bastidas-Coral, Y.S.; Portilla-Armero, J.A. Identificación de Huevos de *Toxocara* spp. en Zonas Verdes de Conjuntos Cerrados del Municipio de Pasto—Colombia. *Biosalud* **2017**, *16*, 44–52. [CrossRef]
10. Alvarado-Borja, V.; Valladares-Carranza, B.; Ortega-Santana, C.; Rivero-Pérez, N.; Bañuelos-Valenzuela, R.; Zaragoza-Bastida, A.; Delgadillo-Ruíz, L.; Velázquez-Ordoñez, V. Infección por *Toxocara canis* y su importancia en la salud animal y en la salud pública: Una revisión. *Salud Tecnol. Vet.* **2023**, *11*, 51–66. [CrossRef]
11. Rostami, A.; Sepidarkish, M.; Ma, G.; Wang, T.; Ebrahimi, M.; Fakhri, Y.; Mirjalali, H.; Hofmann, A.; Macpherson, C.N.L.; Hotez, P.J.; et al. Global prevalence of *Toxocara* infection in cats. *Adv. Parasitol.* **2020**, *109*, 615–639. [CrossRef] [PubMed]
12. Rostami, A.; Riahi, S.M.; Hofmann, A.; Ma, G.; Wang, T.; Behniafar, H.; Taghipour, A.; Fakhri, Y.; Spotin, A.; Chang, B.C.H.; et al. Global prevalence of *Toxocara* infection in dogs. *Adv. Parasitol.* **2020**, *109*, 561–583. [CrossRef] [PubMed]
13. Jimenez Castro, P.D.; Sapp, S.G. Role of cats in human toxocarosis. *Companion Anim.* **2020**, *26*, 6–14. [CrossRef] [PubMed]
14. Ma, G.; Holland, C.V.; Wang, T.; Hofmann, A.; Fan, C.K.; Maizels, R.M.; Hotez, P.J.; Gasser, R.B. Human toxocariasis. *Lancet Infect. Dis.* **2018**, *18*, e14–e24. [CrossRef] [PubMed]
15. Healy, S.R.; Morgan, E.R.; Prada, J.M.; Betson, M. Brain food: Rethinking food-borne toxocariasis. *Parasitology* **2022**, *149*, 1–9. [CrossRef] [PubMed]
16. Auer, H.; Walochnik, J. Toxocariasis and the clinical spectrum. *Adv. Parasitol.* **2020**, *109*, 111–130. [CrossRef]
17. Szwabe, K.; Blaszkowska, J. Stray dogs and cats as potential sources of soil contamination with zoonotic parasites. *Ann. Agric. Environ. Med.* **2017**, *24*, 39–43. [CrossRef] [PubMed]
18. Raissi, V.; Masoumi, M.T.; Ibrahim, A.; Etemadi, S.; Getso, M.; Jalali, P.; Babaei Pouya, N.; Zareie, M.; Ehsani Amraei, F.; Raiesi, O. Spatial analysis of *Toxocara* spp. eggs in soil as a potential for serious human infection. *Comp. Immunol. Microbiol. Infect. Dis.* **2021**, *75*, 101619. [CrossRef] [PubMed]
19. Fakhri, Y.; Gasser, R.B.; Rostami, A.; Fan, C.K.; Ghasemi, S.M.; Javanian, M.; Bayani, M.; Armoon, B.; Moradi, B. *Toxocara* eggs in public places worldwide—A systematic review and meta-analysis. *Environ. Pollut.* **2018**, *242*, 1467–1475. [CrossRef] [PubMed]
20. Bonilla-Aldana, D.K.; Morales-Garcia, L.V.; Ulloque Badaracco, J.R.; Mosquera-Rojas, M.D.; Alarcón-Braga, E.A.; Hernandez-Bustamante, E.A.; Al-Kassab-Córdova, A.; Benites-Zapata, V.A.; Rodriguez-Morales, A.J.; Delgado, O. Prevalence of *Toxocara* eggs in Latin American parks: A systematic review and meta-analysis. *Infez. Med.* **2023**, *31*, 329–349. [PubMed]
21. Öge, H.; Öge, S.; Özbakış, G.; Gürcan, S. Comparison of *Toxocara* eggs in hair and faecal samples from owned dogs and cats collected in Ankara, Turkey. *Vet. Parasitol.* **2014**, *206*, 227–231. [CrossRef] [PubMed]
22. Cárdenas Camacho, J.; Lesmes Infante, K.I.; Torres Tocasuche, M.C.; Alcantara-Neves, N.M.; Jaramillo-Hernández, D.A. Evaluación de técnicas coprodiagnósticas para *Toxocara canis*. *Rev. Investig. Vet. Perú* **2021**, *32*, e18861. [CrossRef]
23. Raulf, M.K.; Jordan, D.; Auer, H.; Warnecke, J.M.; Lepenies, B.; Strube, C. A new ELISA and western blot technique based on recombinant TES antigen and/or larval antigen for the detection of toxocariasis in humans. *Parasitology* **2021**, *148*, 333–340. [CrossRef] [PubMed]
24. Elsemore, D.A. Antigen detection: Insights into *Toxocara* and other ascarid infections in dogs and cats. *Adv. Parasitol.* **2020**, *109*, 545–559. [CrossRef] [PubMed]
25. Torkan, S.; Ghandehari-Alavijeh, M.R.; Khamesipour, F. Survey of the prevalence of *Toxocara cati* in stray cats in Isfahan city, Iran by PCR method. *Trop. Biomed.* **2017**, *34*, 550–555. [PubMed]
26. Zeng, X.; Zhang, Y.; Kwong, J.S.; Zhang, C.; Li, S.; Sun, F.; Niu, Y.; Du, L. The methodological quality assessment tools for preclinical and clinical studies, systematic review and meta-analysis, and clinical practice guideline: A systematic review. *J. Evid.-Based Med.* **2015**, *8*, 2–10. [CrossRef] [PubMed]

27. Yamamoto, N.; Kon, M.; Saito, T.; Maeno, N.; Koyama, M.; Sunaoshi, K.; Yamaguchi, M.; Morishima, Y.; Kawanaka, M. Prevalence of intestinal canine and feline parasites in Saitama Prefecture, Japan. *Kansenshogaku Zasshi* **2009**, *83*, 223–228. [CrossRef]
28. Echeverry, D.M.; Giraldo, M.I.; Castaño, J.C. Prevalence of intestinal helminths in cats in Quindío, Colombia. *Biomedica* **2012**, *32*, 430–436.
29. Raue, K.; Heuer, L.; Böhm, C.; Wolken, S.; Epe, C.; Strube, C. 10-year parasitological examination results (2003 to 2012) of faecal samples from horses, ruminants, pigs, dogs, cats, rabbits and hedgehogs. *Parasitol. Res.* **2017**, *116*, 3315–3330. [CrossRef] [PubMed]
30. Kalef, D.A.; Al-khayat, F. A Comparative Study of Some Intestinal Parasites in Fecal Samples of Domestic and Stray Cats in Baghdad, Iraq. *Comp. Parasitol. J.* **2022**, *89*, 31–35. [CrossRef]
31. Raab, O.; Greenwood, S.; Vanderstichel, R.; Gelens, H. A cross-sectional study of *Tritrichomonas foetus* infection in feral and shelter cats in Prince Edward Island, Canada. *Can. Vet. J.* **2016**, *57*, 265–270.
32. Hoopes, J.H.; Polley, L.; Wagner, B.; Jenkins, E.J. A retrospective investigation of feline gastrointestinal parasites in western Canada. *Can. Vet. J.* **2013**, *54*, 359–362. [PubMed]
33. Labarthe, N.; Serrão, M.L.; Ferreira, A.M.; Almeida, N.K.; Guerrero, J. A survey of gastrointestinal helminths in cats of the metropolitan region of Rio de Janeiro, Brazil. *Vet. Parasitol.* **2004**, *123*, 133–139. [CrossRef]
34. Pomroy, W.E. A survey of helminth parasites of cats from Saskatoon. *Can. Vet. J.* **1999**, *40*, 339–340. [PubMed]
35. Baker, M.K.; Lange, L.; Verster, A.; van der Plaat, S. A survey of helminths in domestic cats in the Pretoria area of Transvaal, Republic of South Africa. Part 1: The prevalence and comparison of burdens of helminths in adult and juvenile cats. *J. S. Afr. Vet. Assoc.* **1989**, *60*, 139–142.
36. Engbaek, K.; Madsen, H.; Larsen, S.O. A survey of helminths in stray cats from Copenhagen with ecological aspects. *Z. Parasitenkd.* **1984**, *70*, 87–94. [CrossRef] [PubMed]
37. Overgaauw, P.A.; Boersema, J.H. A survey of *Toxocara* infections in cat breeding colonies in The Netherlands. *Vet. Q.* **1998**, *20*, 9–11. [CrossRef] [PubMed]
38. Moskvina, T.V.; Zheleznova, L.V. A survey on endoparasites and ectoparasites in domestic dogs and cats in Vladivostok, Russia 2014. *Vet. Parasitol. Reg. Stud. Rep.* **2015**, *1–2*, 31–34. [CrossRef] [PubMed]
39. Borji, H.; Razmi, G.; Ahmadi, A.; Karami, H.; Yaghfoori, S.; Abedi, V. A survey on endoparasites and ectoparasites of stray cats from Mashhad (Iran) and association with risk factors. *J. Parasit. Dis.* **2011**, *35*, 202–206. [CrossRef] [PubMed]
40. Hajipour, N. A survey on the prevalence of *Toxocara cati*, *Toxocara canis* and *Toxascaris leonina* eggs in stray dogs and cats' faeces in Northwest of Iran: A potential risk for human health. *Trop. Biomed.* **2019**, *36*, 143–151. [PubMed]
41. Hajipour, N.; Imani Baran, A.; Yakhchali, M.; Banan Khojasteh, S.M.; Sheikhzade Hesari, F.; Esmaeilnejad, B.; Arjmand, J. A survey study on gastrointestinal parasites of stray cats in Azarshahr, (East Azerbaijan province, Iran). *J. Parasit. Dis.* **2016**, *40*, 1255–1260. [CrossRef] [PubMed]
42. Kostopoulou, D.; Claerebout, E.; Arvanitis, D.; Ligda, P.; Voutzourakis, N.; Casaert, S.; Sotiraki, S. Abundance, zoonotic potential and risk factors of intestinal parasitism amongst dog and cat populations: The scenario of Crete, Greece. *Parasit. Vectors* **2017**, *10*, 43. [CrossRef] [PubMed]
43. Langham, N.P.E.; Charleston, W.A.G. An investigation of the potential for spread of *Sarcocystis* spp. and other parasites by feral cats. *N. Z. J. Agric. Res.* **1990**, *33*, 429–435. [CrossRef]
44. Castro, O.; Valledor, S.; Crampet, A.; Casás, G. Aporte al conocimiento de los metazoos parásitos del gato doméstico en el Departamento de Montevideo, Uruguay. *Veterinaria* **2013**, *49*, 28–37.
45. Paula, M.; Oliveira, N.; Santos, J.; Lima, P.; Cardinot, C.; Rocha, T.; Barbosa, F.; Silva-Neto, A.; Franciscato, C. Avaliação das Endoparasitoses Intestinais que Acometem Cães e Gatos Mantidos em um Abrigo. *Ars. Vet.* **2021**, *37*, 273. [CrossRef]
46. Bergamo, M.; Alves, I.; Martins, M.; Lins, J.; Munhoz, P.; Albuquerque, A.C. Frequência de Parasitoses com Potencial Zoonótico em Cães e Gatos Naturalmente Infectados na Cidade de Maringá-PR. 2022, pp. 236–249. Available online: https://doi.org/10.37885/221010398 (accessed on 15 January 2024).
47. Vanparijs, O.F.J.; Thienpont, D.C. Canine and Feline Helminth and Protozoan Infections in Belgium. *J. Parasitol.* **1973**, *59*, 327–330. [CrossRef] [PubMed]
48. Araújo, A.M.D.D.; Caldas, M.M. Caracterização Da Ocorrência De Parasitas Gastrointestinais De Gatos Na Zona De Pesca Da Ilha De Faro. *Rev. Lusof. Ciencvered. Vet.* **2018**, *9*, 8–16. [CrossRef]
49. Alves, M.E.M. Caracterização Molecular de *Cryptosporidium* spp. e Ocorrência dos Principais Parasitas Gastrointestinais em Amostras Fecais de cães e Gatos Naturalmente Infectados. Master Thesis. 2018. Available online: http://repositorio.ufsm.br/handle/1/13839 (accessed on 15 January 2024).
50. Sátiva Cifuentes, S.N. Caracterización de la Infestación Parasitológica Gastrointestinal y Respiratoria en Gatos Ferales (Felis silvestris Catus) de la Ciudad de Córdoba, Comunidad Autónoma de Andalucía, España. Bachelor's Thesis, Universidad de Ciencias Aplicadas y Ambientales Facultad de Ciencias Agropecuarias Bogotá, Bogota, Colombia, 2022.
51. Remesar, S.; García-Dios, D.; Calabuig, N.; Prieto, A.; Díaz-Cao, J.M.; López-Lorenzo, G.; López, C.; Fernández, G.; Morrondo, P.; Panadero, R.; et al. Cardiorespiratory nematodes and co-infections with gastrointestinal parasites in new arrivals at dog and cat shelters in north-western Spain. *Transbound. Emerg. Dis.* **2022**, *69*, e3141–e3153. [CrossRef] [PubMed]
52. Dybing, N.A.; Jacobson, C.; Irwin, P.; Algar, D.; Adams, P.J. Challenging the dogma of the 'Island Syndrome': A study of helminth parasites of feral cats and black rats on Christmas Island. *Australas. J. Environ. Manag.* **2018**, *25*, 99–118. [CrossRef]

53. Knaus, M.; Rapti, D.; Shukullari, E.; Kusi, I.; Postoli, R.; Xhaxhiu, D.; Silaghi, C.; Hamel, D.; Visser, M.; Winter, R.; et al. Characterisation of ecto-and endoparasites in domestic cats from Tirana, Albania. *Parasitol. Res.* **2014**, *113*, 3361–3371. [CrossRef] [PubMed]
54. Kajero, O.T.; Janoušková, E.; Bakare, E.A.; Belizario, V.; Divina, B.; Alonte, A.J.; Manalo, S.M.; Paller, V.G.; Betson, M.; Prada, J.M. Co-infection of intestinal helminths in humans and animals in the Philippines. *Trans. R. Soc. Trop. Med. Hyg.* **2022**, *116*, 727–735. [CrossRef]
55. Lima, J.A.S. Comparação da Prevalência de Parasitos Entéricos em Gatos Errantes e Domiciliados em Goiânia-Goiás, Análise da Acurácia de Técnicas Parasitológicas e Avaliação da Copro-PCR para o Diagnóstico de *Toxoplasma gondii*. Master Thesis. 2016. Available online: https://repositorio.bc.ufg.br/tede/items/e822b543-00cf-4115-bf7a-fc75d1e64edd (accessed on 15 January 2024).
56. Shabbir, M.Z.; Rabbani, M.; Yaqub, T.; Ahmad, A.; Rehman, Z.u.; Umair, S. Comparative Clinical Epidemiology of Toxocariosis in Dogs and Cats. *Pak. J. Zool.* **2010**, *42*, 129–133.
57. Rashid, Z.M.; Aziz, S.A.; Ali, O.J.; Kakarash, N.K.; Marif, H.F.F. Coprological detection of Toxocariosis in domicile and stray dogs and cats in Sulaimani province, Iraq. *Iraqi J. Vet. Sci.* **2022**, *36*, 1047–1051. [CrossRef]
58. Mancianti, F.; Nardoni, S.; Ariti, G.; Parlanti, D.; Giuliani, G.; Papini, R.A. Cross-sectional survey of Toxoplasma gondii infection in colony cats from urban Florence (Italy). *J. Feline Med. Surg.* **2010**, *12*, 351–354. [CrossRef] [PubMed]
59. Veronesi, F.; Gazzonis, A.L.; Napoli, E.; Brianti, E.; Santoro, A.; Zanzani, S.A.; Olivieri, E.; Diaferia, M.; Giannetto, S.; Pennisi, M.G.; et al. Cross-sectional survey on Tritrichomonas foetus infection in Italian cats. *Vet. Parasitol. Reg. Stud. Rep.* **2016**, *6*, 14–19. [CrossRef] [PubMed]
60. Shukla, R.; Giraldo, P.; Kraliz, A.; Finnigan, M.; Sanchez, A.L. *Cryptosporidium* spp. and other zoonotic enteric parasites in a sample of domestic dogs and cats in the Niagara region of Ontario. *Can. Vet. J.* **2006**, *47*, 1179–1184. [PubMed]
61. Miró, G.; Rupérez, C.; Checa, R.; Gálvez, R.; Hernández, L.; García, M.; Canorea, I.; Marino, V.; Montoya, A. Current status of L. infantum infection in stray cats in the Madr. Reg. (Spain): Implic. Recent Outbreak Hum. Leishmaniosis? *Parasit. Vectors* **2014**, *7*, 112. [CrossRef] [PubMed]
62. Abu-Madi, M.A.; Pal, P.; Al-Thani, A.; Lewis, J.W. Descriptive epidemiology of intestinal helminth parasites from stray cat populations in Qatar. *J. Helminthol.* **2008**, *82*, 59–68. [CrossRef] [PubMed]
63. Tun, S.; Ithoi, I.; Mahmud, R.; Samsudin, N.I.; Kek Heng, C.; Ling, L.Y. Detection of Helminth Eggs and Identification of Hookworm Species in Stray Cats, Dogs and Soil from Klang Valley, Malaysia. *PLoS ONE* **2015**, *10*, e0142231. [CrossRef] [PubMed]
64. Cerda Bolados, C. Determinación de la Presencia de Enteroparásitos en Gatos Clínicamente Sanos en Cuatro Comunas de Santiago, Mediante los Métodos de Teuscher y Teleman. DVM Thesis, Universidad de Las Américas, Providencia, Chile, 2018.
65. Marin Baquerizo, A.N. Determinación de la Presencia de Helmintos Gastro Intestinales en Gatos en las Parroquias Urbano Marginales de la Ciudad de Babahoyo. Bachelor's Thesis, UTB, Babahoyo, Ecuador, 2023.
66. Gallegos Rosales, G.S. Determinación de Prevalencia de Parásitos Intestinales y Externos en Gatos Domésticos (*felis catus*) en Determinadas Zonas del Ecuador. Bachelor's Thesis, Universidad de las Américas, Quito, Ecuador, 2012.
67. Geng, J.; Elsemore, D.; Oudin, N.; Ketzis, J. Diagnosis of feline whipworm infection using a coproantigen ELISA and the prevalence in feral cats in southern Florida. *Vet. Parasitol. Reg. Stud. Rep.* **2018**, *14*, 181–186. [CrossRef] [PubMed]
68. Fernández Anchía, L. Diagnóstico de Parásitos Gastrointestinales en Caninos y Felinos: Estudio Retrospectivo en dos Laboratorios Veterinaries. DVM Thesis, Universidad Nacional, Heredia, Costa Rica, 2021.
69. Stalliviere, F.M.; Bellato, V.; Souza, A.P.d.; Sartor, A.A.; Moura, A.B.d.; Rosa, L.D. Ectoparasitos e helmintos intestinais em *Felis catus* domesticus, da cidade de Lages, SC, Brasil e aspectos sócioeconômicos e culturais das famílias dos proprietários dos animais. *Rev. Bras. Parasitol. Vet.* **2009**, *18*, 26–31. [CrossRef] [PubMed]
70. Tull, A.; Moks, E.; Saarma, U. Endoparasite prevalence and infection risk factors among cats in an animal shelter in Estonia. *Folia Parasitol.* **2021**, *68*, 010. [CrossRef] [PubMed]
71. Gates, M.C.; Nolan, T.J. Endoparasite prevalence and recurrence across different age groups of dogs and cats. *Vet. Parasitol.* **2009**, *166*, 153–158. [CrossRef] [PubMed]
72. Tamponi, C.; Varcasia, A.; Pinna, S.; Melis, E.; Melosu, V.; Zidda, A.; Sanna, G.; Pipia, A.P.; Zedda, M.T.; Pau, S.; et al. Endoparasites detected in faecal samples from dogs and cats referred for routine clinical visit in Sardinia, Italy. *Vet. Parasitol. Reg. Stud. Rep.* **2017**, *10*, 13–17. [CrossRef]
73. Morandi, B.; Greenwood, S.J.; Conboy, G.A.; Galuppi, R.; Poglayen, G.; VanLeeuwen, J.A. Endoparasites in dogs and cats diagnosed at the Veterinary Teaching Hospital (VTH) of the University of Prince Edward Island between 2000 and 2017. A large-scale retrospective study. *Prev. Vet. Med.* **2020**, *175*, 104878. [CrossRef]
74. Barutzki, D.; Schaper, R. Endoparasites in dogs and cats in Germany 1999–2002. *Parasitol. Res.* **2003**, *90* (Suppl. 3), S148–S150. [CrossRef]
75. Silva, W.I.; Lima, E.F.; Silva, J.O.; Alves, M.M.; Alves, C.L.P.; Silva, A.L.P.; Lima, J.A.; Feitosa, T.F.; Vilela, V.L.R. Endoparasites in domestic cats (*Felis catus*) in the semi-arid region of Northeast Brazil. *Rev. Bras. Parasitol. Vet.* **2023**, *32*, e012123. [CrossRef]
76. Knaus, M.; Kusi, I.; Rapti, D.; Xhaxhiu, D.; Winter, R.; Visser, M.; Rehbein, S. Endoparasites of cats from the Tirana area and the first report on Aelurostrongylus abstrusus (Railliet, 1898) in Albania. *Wien. Klin. Wochenschr.* **2011**, *123* (Suppl. S1), 31–35. [CrossRef] [PubMed]
77. Ramos, N.V.; Silva, M.L.E.; Barreto, M.S.; Barros, L.A.; Mendes-de-Almeida, F. Endoparasites of household and shelter cats in the city of Rio de Janeiro, Brazil. *Rev. Bras. Parasitol. Vet.* **2020**, *29*, e012819. [CrossRef]

78. Hoopes, J.; Hill, J.E.; Polley, L.; Fernando, C.; Wagner, B.; Schurer, J.; Jenkins, E. Enteric parasites of free-roaming, owned, and rural cats in prairie regions of Canada. *Can. Vet. J.* **2015**, *56*, 495–501. [PubMed]
79. Diaz, F.; Anccasi, M. Enteroparásitos en perros (*Canis familiaris*) y gatos (*Felis catus*) de la provincia de puno. *Rev. Investig. Altoandinas—J. High Andean Investig.* **2013**, *15*, 117–122. [CrossRef]
80. Pereira, N.D.V.; Souza, F.S.; Piranda, E.M.; Cançado, P.H.D.; Lisbôa, R.S. Enteroparasitos encontrados em cães e gatos atendidos em duas clínicas veterinárias na cidade de Manaus, AM. *Amaz. Sci.* **2012**, *1*, 8–17.
81. Rodríguez-Ponce, E.; González, J.F.; Conde de Felipe, M.; Hernández, J.N.; Raduan Jaber, J. Epidemiological survey of zoonotic helminths in feral cats in Gran Canaria island (Macaronesian archipelago-Spain). *Acta Parasitol.* **2016**, *61*, 443–450. [CrossRef] [PubMed]
82. Henry, P.; Huck-Gendre, C.; Franc, M.; Williams, T.L.; Bouhsira, E.; Lienard, E. Epidemiological Survey on Gastrointestinal and Pulmonary Parasites in Cats Around Toulouse (France). *Helminthologia* **2022**, *59*, 385–397. [CrossRef] [PubMed]
83. O'Lorcain, P. Epidemiology of *Toxocara* spp. in stray dogs and cats in Dublin, Ireland. *J. Helminthol.* **1994**, *68*, 331–336. [CrossRef]
84. García-Galán, A.; Muñoz Hernández, C.; Bernal Gómez, A.; Ortuño, M.; Risueño Iranzo, J.; Ortiz, J.; Goyena, E.; Berriatua, E. Coprological study of parasitosis in cats from the periurban area of Murcia city and their zoonotic implications. *An. Vet. Murcia* **2017**, *33*, 7–19.
85. Caiza Chicaiza, M.R. Estudio de la Prevalencia de Parásitos Gastrointestinales Zoonosicos en Perros y Gatos en el Barrio Carapungo de la Ciudad de Quito. DVM Thesis, Unidad Académica de Ciencias Agropecuarias y Recursos Naturales, Latacunga, Ecuador, 2010.
86. Díaz Valencia, Y.A.; Otalvaro Cruz, S.M.; Rodríguez García, M.L. Estudio Retrospectivo de Casos de Parasitosis Gastrointestinales Presentados en Caninos y Felinos en la Clínica Veterinaria Zooluciones Versátiles en la Ciudad de Bogotá. DVM Thesis, Universidad Antonio Nariño, Bogotá, Colombia, 2021.
87. Tavares, F.G. Estudo das Parasitoses Gastrintestinais de Cães e Gatos Domésticos no Município de São José dos Campos–SP. DVM Thesis, Trabalho de Conclusão de Curso-Universidade Vale do Paraíba, São Paulo, Brazil, 2005.
88. Nagamori, Y.; Payton, M.E.; Duncan-Decocq, R.; Johnson, E.M. Fecal survey of parasites in free-roaming cats in northcentral Oklahoma, United States. *Vet. Parasitol. Reg. Stud. Rep.* **2018**, *14*, 50–53. [CrossRef]
89. Symeonidou, I.; Gelasakis, A.I.; Arsenopoulos, K.; Angelou, A.; Beugnet, F.; Papadopoulos, E. Feline gastrointestinal parasitism in Greece: Emergent zoonotic species and associated risk factors. *Parasit. Vectors* **2018**, *11*, 227. [CrossRef] [PubMed]
90. Lin, D.S.; Lai, S.S.; Bowman, D.D.; Jacobson, R.H.; Barr, M.C.; Giovengo, S.L. Feline immunodeficiency virus, feline leukaemia virus, Toxoplasma gondii, and intestinal parasitic infections in Taiwanese cats. *Br. Vet. J.* **1990**, *146*, 468–475. [CrossRef]
91. Näreaho, A.; Puomio, J.; Saarinen, K.; Jokelainen, P.; Juselius, T.; Sukura, A. Feline intestinal parasites in Finland: Prevalence, risk factors and anthelmintic treatment practices. *J. Feline Med. Surg.* **2012**, *14*, 378–383. [CrossRef] [PubMed]
92. Bishop, G.T.; DeBess, E. Feline parasites and the emergence of feline lungworm in the Portland metropolitan area, Oregon, USA 2016–2017. *JFMS Open Rep.* **2021**, *7*, 20551169211053595. [CrossRef]
93. Karamon, J.; Sroka, J.; Dąbrowska, J.; Bilska-Zając, E.; Zdybel, J.; Kochanowski, M.; Różycki, M.; Cencek, T. First report of Echinococcus multilocularis in cats in Poland: A monitoring study in cats and dogs from a rural area and animal shelter in a highly endemic region. *Parasit. Vectors* **2019**, *12*, 313. [CrossRef]
94. Mundim, T.C.D.; Oliveira Júnior, S.D.; Rodrigues, D.C.; Cury, M.C. Freqüência de helmintos em gatos de Uberlândia, Minas Gerais. *Arq. Bras. Med. Veterinária Zootec.* **2004**, *56*, 562–563. [CrossRef]
95. Rodríguez-Vivas, R.I.; Cob-Galera, L.A.; Domínguez-Alpizar, J.L. Frecuencia de parásitos gastrointestinales en animales domésticos diagnosticados en Yucatán, México. *Rev. Bioméd.* **2001**, *12*, 19–25. [CrossRef]
96. Meza Mansilla, S.C. Frecuencia de Parásitos Gastrointestinales en Felinos Domésticos (*Felis catus*) en el Distrito de Jesús María-Lima. DVM Thesis, Universidad Peruana Cayetano Heredia, Lima, Peru, 2022.
97. Bittencourt, L.H.F.B.; Gross, M. Frequência de endoparasitas em gatos internados em quatro clínicas de Cascavel, Paraná. *Arq. Bras. Med. Vet. FAG* **2018**, *1*, 74–81.
98. Brener, B.; Mattos, D.; Lisboa, L.; Arashiro, E.; Millar, P.; Sudré, A.; Duque, V. Freqüência de enteroparasitas em amostras fecais de cães e gatos dos municípios do Rio de Janeiro e Niterói. *Rev. Bras. Ciência Vet.* **2005**, *12*, 102–105. [CrossRef]
99. Ferraz, A.; Dallmann, P.R.J.; Yonamine, G.R.; Segalla, C.C.; Moreira, T.F.B.; de Oliveira Nobre, M.; Nizoli, L.Q. Frequência de helmintos diagnosticados em cães e gatos no Laboratório de Doenças Parasitárias da Faculdade de Veterinária/UFPel. *Rev. Acad. Ciência Anim.* **2022**, *20*, e20003. [CrossRef]
100. Ferreira, F.P.; Dias, R.C.F.; Martins, T.A.; Constantino, C.; Pasquali, A.K.S.; Vidotto, O.; Freire, R.L.; Navarro, I.T. Frequência de parasitas gastrointestinais em cães e gatos do município de Londrina, PR, com enfoque em saúde pública. *Semin. Ciências Agrárias* **2013**, *34*, 3851–3858. [CrossRef]
101. Farias, N.; Christovão, M.; Stobbe, N. Freqüência de parasitas intestinais em cães (*Canis familiaris*) e gatos (*Felis catus* domestica) em Araçatuba, São Paulo. *Rev. Bras. Parasitol. Vet.* **1995**, *4*, 57–60.
102. Ferraz, A.; dos Santos Pires, B.; dos Santos, E.M.; Evaristo, T.A.; de Castro, T.A.; Dallmann, P.R.J.; de Oliveira Nobre, M.; Nizoli, L.Q. Frequência de parasitos gastrintestinais, presentes em fezes de cães e gatos, analisadas no laboratório de doenças parasitárias da UFPEL, durante o ano de 2017. *Sci. Anim. Health* **2019**, *7*, 41–53. [CrossRef]
103. Gennari, S.M.; Ferreira, J.I.G.d.S.; Pena, H.F.d.J.; Labruna, M.B.; Azevedo, S.d.S. Frequency of gastrointestinal parasites in cats seen at the University of São Paulo Veterinary Hospital, Brazil. *Rev. Bras. Parasitol. Vet.* **2016**, *25*, 423–428. [CrossRef] [PubMed]

104. Funada, M.R.; Pena, H.F.d.J.; Soares, R.M.; Amaku, M.; Gennari, S.M. Freqüência de parasitos gastrintestinais em cães e gatos atendidos em hospital-escola veterinário da cidade de São Paulo. *Arq. Bras. Med. Veterinária Zootec.* **2007**, *59*, 1338–1340. [CrossRef]
105. Calvete, C.; Lucientes, J.; Castillo, J.A.; Estrada, R.; Gracia, M.J.; Peribáñez, M.A.; Ferrer, M. Gastrointestinal helminth parasites in stray cats from the mid-Ebro Valley, Spain. *Vet. Parasitol.* **1998**, *75*, 235–240. [CrossRef] [PubMed]
106. Ridwan, Y.; Sudarnika, E.; Dewi, T.I.T.; Budiono, N.G. Gastrointestinal helminth parasites of pets: Retrospective study at the veterinary teaching hospital, IPB University, Bogor, Indonesia. *Vet. World* **2023**, *16*, 1043–1051. [CrossRef] [PubMed]
107. Borthakur, S.K.; Mukharjee, S.N. Gastrointestinal helminthes in stray cats (*Felis catus*) from Aizawl, Mizoram, India. *Southeast Asian J. Trop. Med. Public Health* **2011**, *42*, 255–258.
108. Darabi, E.; Kia, E.B.; Mohebali, M.; Mobedi, I.; Zahabiun, F.; Zarei, Z.; Khodabakhsh, M.; Khanaliha, K. Gastrointestinal Helminthic Parasites of Stray Cats (*Felis catus*) in Northwest Iran. *Iran. J. Parasitol.* **2021**, *16*, 418–425. [CrossRef]
109. Yakhchali, M.; Hajipour, N.; Malekzadeh-Viayeh, R.; Esmaeilnejad, B.; Nemati-Haravani, T.; Fathollahzadeh, M.; Jafari, R. Gastrointestinal Helminths and Ectoparasites in the Stray Cats (Felidae: *Felis catus*) of Ahar Municipality, Northwestern Iran. *Iran. J. Parasitol.* **2017**, *12*, 298–304. [PubMed]
110. Sheikh, M.M.; Tak, H.; Fazili, M. Gastrointestinal Helminths of Cat *Felis catus* in Kashmir. *Int. J. Sci. Technol. Res.* **2020**, *9*, 314–319.
111. Lee, S.H.; Ock, Y.; Choi, D.; Kwak, D. Gastrointestinal Parasite Infection in Cats in Daegu, Republic of Korea, and Efficacy of Treatment Using Topical Emodepside/Praziquantel Formulation. *Korean J. Parasitol.* **2019**, *57*, 243–248. [CrossRef] [PubMed]
112. Ubirajara Filho, C.R.C.; Santos, K.K.F.; Lima, T.A.R.F.; Alves, L.C.; Carvalho, G.A.; Ramos, R.A.N. Gastrointestinal parasites in dogs and cats in line with the One Health' approach. *Arq. Bras. Med. Veterinária Zootec.* **2022**, *74*, 43–50. [CrossRef]
113. Lima, V.F.S.; Ramos, R.A.N.; Lepold, R.; Borges, J.C.G.; Ferreira, C.D.; Rinaldi, L.; Cringoli, G.; Alves, L.C. Gastrointestinal parasites in feral cats and rodents from the Fernando de Noronha Archipelago, Brazil. *Rev. Bras. Parasitol. Vet.* **2017**, *26*, 521–524. [CrossRef] [PubMed]
114. Ngui, R.; Lee, S.C.; Yap, N.J.; Tan, T.K.; Aidil, R.M.; Chua, K.H.; Aziz, S.; Sulaiman, W.Y.; Ahmad, A.F.; Mahmud, R.; et al. Gastrointestinal parasites in rural dogs and cats in Selangor and Pahang states in Peninsular Malaysia. *Acta Parasitol.* **2014**, *59*, 737–744. [CrossRef] [PubMed]
115. Sauda, F.; Malandrucco, L.; De Liberato, C.; Perrucci, S. Gastrointestinal parasites in shelter cats of central Italy. *Vet. Parasitol. Reg. Stud. Rep.* **2019**, *18*, 100321. [CrossRef] [PubMed]
116. Pereira, P.F.; Barbosa, A.D.S.; Moura, A.P.P.; Vasconcellos, M.L.; Uchôa, C.M.A.; Bastos, O.M.P.; Amendoeira, M.R.R. Gastrointestinal parasites in stray and shelter cats in the municipality of Rio de Janeiro, Brazil. *Rev. Bras. Parasitol. Vet.* **2017**, *26*, 383–388. [CrossRef] [PubMed]
117. Monteiro, M.F.; Ramos, R.A.; Calado, A.M.; Lima, V.F.; Ramos, I.C.; Tenório, R.F.; Faustino, M.A.; Alves, L.C. Gastrointestinal parasites of cats in Brazil: Frequency and zoonotic risk. *Rev. Bras. Parasitol. Vet.* **2016**, *25*, 254–257. [CrossRef]
118. Takeuchi-Storm, N.; Mejer, H.; Al-Sabi, M.N.; Olsen, C.S.; Thamsborg, S.M.; Enemark, H.L. Gastrointestinal parasites of cats in Denmark assessed by necropsy and concentration McMaster technique. *Vet. Parasitol.* **2015**, *214*, 327–332. [CrossRef] [PubMed]
119. Rojekittikhun, W.; Chaisiri, K.; Mahittikorn, A.; Pubampen, S.; Sa-Nguankiat, S.; Kusolsuk, T.; Maipanich, W.; Udonsom, R.; Mori, H. Gastrointestinal parasites of dogs and cats in a refuge in Nakhon Nayok, Thailand. *Southeast Asian J. Trop. Med. Public Health* **2014**, *45*, 31–39. [PubMed]
120. McGlade, T.R.; Robertson, I.D.; Elliot, A.D.; Read, C.; Thompson, R.C. Gastrointestinal parasites of domestic cats in Perth, Western Australia. *Vet. Parasitol.* **2003**, *117*, 251–262. [CrossRef] [PubMed]
121. Adams, P.J.; Elliot, A.D.; Algar, D.; Brazell, R.I. Gastrointestinal parasites of feral cats from Christmas Island. *Aust. Vet. J.* **2008**, *86*, 60–63. [CrossRef] [PubMed]
122. Ryan, G.E. Gastro-intestinal parasites of feral cats in New South Wales. *Aust. Vet. J.* **1976**, *52*, 224–227. [CrossRef] [PubMed]
123. Mohsen, A.; Hossein, H. Gastrointestinal parasites of stray cats in Kashan, Iran. *Trop. Biomed.* **2009**, *26*, 16–22. [PubMed]
124. Abbas, I.; Al-Araby, M.; Elmishmishy, B.; El-Alfy, E.S. Gastrointestinal parasites of cats in Egypt: High prevalence high zoonotic risk. *BMC Vet. Res.* **2022**, *18*, 420. [CrossRef] [PubMed]
125. López-Arias, Á.; Villar, D.; López-Osorio, S.; Calle-Vélez, D.; Chaparro-Gutiérrez, J.J. Giardia is the most prevalent parasitic infection in dogs and cats with diarrhea in the city of Medellín, Colombia. *Vet. Parasitol. Reg. Stud. Rep.* **2019**, *18*, 100335. [CrossRef] [PubMed]
126. Vanparijs, O.; Hermans, L.; van der Flaes, L. Helminth and protozoan parasites in dogs and cats in Belgium. *Vet. Parasitol.* **1991**, *38*, 67–73. [CrossRef] [PubMed]
127. Lefkaditis, M.; Pastiu, A.; Rodi-Buriel, A.; Sossidou, A.; Panorias, A.; Eleftheriadis, T.; Vasile, C.; Mihalca, A. Helminth burden in stray cats from Thessaloniki, Greece. *Helminthologia* **2014**, *51*, 73–76. [CrossRef]
128. Coman, B.J.; Jones, E.H.; Driesen, M.A. Helminth parasites and arthropods of feral cats. *Aust. Vet. J.* **1981**, *57*, 324–327. [CrossRef] [PubMed]
129. Scholz, T.; Uhlírová, M.; Ditrich, O. Helminth parasites of cats from the Vientiane province, Laos, as indicators of the occurrence of causative agents of human parasitoses. *Parasite* **2003**, *10*, 343–350. [CrossRef] [PubMed]
130. Cowper, S.G. Helminth parasites of dogs and cats and toxoplasmosis antibodies in cats in Swansea, South Wales. *Ann. Trop. Med. Parasitol.* **1978**, *72*, 455–459. [CrossRef] [PubMed]

131. Cheryl, R.; Dennis, J.R. Helminth Parasites of the House Cat, *Felis catus*, in Connecticut, U.S.A. *Comp. Parasitol.* **2003**, *70*, 115–119. [CrossRef]
132. Ramos, D.G.d.S. Helmintofauna de Gatos (*Felis silvestres catus*, LINNAEUS, 1758) da Região Metropolitana de Cuiabá. 2012. Available online: https://www.ufmt.br/ppgvet/dissertacao/helmintofauna-de-gatos-felis-silvestres-catus-linnaeus-1758-da-regiao-metropolitana-de-cuiaba/ (accessed on 1 December 2023).
133. De Quadros, R.M.; Trevisani, N.; de Moura, A.B.; Ramos, C.J.R. Helmintofauna parasitária em gatos errantes de Lages, Santa Catarina, Brasil. *Rev. Bras. Hig. Sanidade Anim. RBHSA* **2021**, *15*, 10.
134. Little, S.; Adolph, C.; Downie, K.; Snider, T.; Reichard, M. High Prevalence of Covert Infection With Gastrointestinal Helminths in Cats. *J. Am. Anim. Hosp. Assoc.* **2015**, *51*, 359–364. [CrossRef] [PubMed]
135. Millán, J.; Casanova, J.C. High prevalence of helminth parasites in feral cats in Majorca Island (Spain). *Parasitol. Res.* **2009**, *106*, 183–188. [CrossRef] [PubMed]
136. Montoya, A.; García, M.; Gálvez, R.; Checa, R.; Marino, V.; Sarquis, J.; Barrera, J.P.; Rupérez, C.; Caballero, L.; Chicharro, C.; et al. Implications of zoonotic and vector-borne parasites to free-roaming cats in central Spain. *Vet. Parasitol.* **2018**, *251*, 125–130. [CrossRef]
137. Ketzis, J.; Bork-Larsen, H.; Bustria, J.; Conan, A. Importation of cats and risk of parasite spread: A Caribbean perspective and case study from St Kitts. *Parasit. Vectors* **2020**, *13*, 488. [CrossRef] [PubMed]
138. Barros Núñez, M. Incidencia de Parásitos Gastrointestinales en Gatos en la Ciudad de Guayaquil. DVM Thesis, Universidad de Guayaquil, Facultad de Medicina Veterinaria y Zootecnia, Guayaquil, Ecuador, 2013.
139. Torres, A.; Aisela, F. Incidencia de Parásitos en Gatos (felis silvestris catus) en el Centro de Bienestar Animal Tecámac Municipio de Tecámac. DVM Thesis, Universidad Autónoma Metropolitana, Mexico city, Mexico, 2020.
140. Castillo Pilamunga, V.M. Incidencia de *Toxocara cati* en felinos domésticos de la parroquia Veracruz, cantón Pastaza, provincia de Pastaza. *Univ. Cienc.* **2024**, *13*, 138–147.
141. Sobral, M.C.G.D.O. Infecções por Parasitos Gastrintestinais em Gatos Domésticos de Araguaína, Tocantins. Master's Thesis, Universidade Federal do Tocantins, Palmas, Brazil, 2017.
142. Sohn, W.M.; Chai, J.Y. Infection status with helminthes in feral cats purchased from a market in Busan, Republic of Korea. *Korean J. Parasitol.* **2005**, *43*, 93–100. [CrossRef] [PubMed]
143. Côrtes, V.D.A.; Paim, G.V.; Alencar Filho, R.A.d. Infestação por ancilostomídeos e toxocarídeos em cães e gatos apreendidos em vias públicas, São Paulo (Brasil). *Rev. Saúde Pública* **1988**, *22*, 341–343. [CrossRef] [PubMed]
144. Nath, T.C.; Eom, K.S.; Choe, S.; Islam, S.; Sabuj, S.S.; Saha, E.; Tuhin, R.H.; Ndosi, B.A.; Kang, Y.; Kim, S.; et al. Insights to helminth infections in food and companion animals in Bangladesh: Occurrence and risk profiling. *Parasite Epidemiol. Control* **2022**, *17*, e00245. [CrossRef] [PubMed]
145. Gregory, G.G.; Munday, B.L. Internal parasites of feral cats from the Tasmanian Midlands and King Island. *Aust. Vet. J.* **1976**, *52*, 317–320. [CrossRef]
146. Riggio, F.; Mannella, R.; Ariti, G.; Perrucci, S. Intestinal and lung parasites in owned dogs and cats from central Italy. *Vet. Parasitol.* **2012**, *193*, 78–84. [CrossRef] [PubMed]
147. Umeche, N.; Ima, A.E. Intestinal helminthic infections of cats in Calabar, Nigeria. *Folia Parasitol.* **1988**, *35*, 165–168.
148. Okaeme, A.N. Intestinal helminths of cats in the Kainji Lake area, Nigeria. *Vet. Res. Commun.* **1986**, *10*, 237–240. [CrossRef] [PubMed]
149. Abu-Madi, M.A.; Behnke, J.M.; Prabhaker, K.S.; Al-Ibrahim, R.; Lewis, J.W. Intestinal helminths of feral cat populations from urban and suburban districts of Qatar. *Vet. Parasitol.* **2010**, *168*, 284–292. [CrossRef] [PubMed]
150. Blasco, X.; Manteca, X.; López-Béjar, M.; Carbajal, A.; Castellà, J.; Ortuño, A. Intestinal Parasites and Fecal Cortisol Metabolites in Multi-Unowned-Cat Environments: The Impact of Housing Conditions. *Animals* **2021**, *11*, 1300. [CrossRef] [PubMed]
151. Zottler, E.M.; Bieri, M.; Basso, W.; Schnyder, M. Intestinal parasites and lungworms in stray, shelter and privately owned cats of Switzerland. *Parasitol. Int.* **2019**, *69*, 75–81. [CrossRef] [PubMed]
152. Arruda, I.F.; Ramos, R.C.F.; Barbosa, A.D.S.; Abboud, L.C.S.; Dos Reis, I.C.; Millar, P.R.; Amendoeira, M.R.R. Intestinal parasites and risk factors in dogs and cats from Rio de Janeiro, Brazil. *Vet. Parasitol. Reg. Stud. Rep.* **2021**, *24*, 100552. [CrossRef] [PubMed]
153. Ferreira, F.S.; Pereira-Baltasar, P.; Parreira, R.; Padre, L.; Vilhena, M.; Távora Tavira, L.; Atouguia, J.; Centeno-Lima, S. Intestinal parasites in dogs and cats from the district of Évora, Portugal. *Vet. Parasitol.* **2011**, *179*, 242–245. [CrossRef] [PubMed]
154. López, D.J.; Abarca, V.K.; Paredes, M.P.; Inzunza, T.E. Parásitos intestinales en caninos y felinos con cuadros digestivos en Santiago, Chile: Consideraciones en Salud Pública. *Rev. Méd. Chile* **2006**, *134*, 193–200. [CrossRef]
155. Huh, S.; Sohn, W.M.; Chai, J.Y. Intestinal parasites of cats purchased in Seoul. *Korean J. Parasitol.* **1993**, *31*, 371–373. [CrossRef]
156. Zanzani, S.A.; Gazzonis, A.L.; Scarpa, P.; Berrilli, F.; Manfredi, M.T. Intestinal parasites of owned dogs and cats from metropolitan and micropolitan areas: Prevalence, zoonotic risks, and pet owner awareness in northern Italy. *BioMed Res. Int.* **2014**, *2014*, 696508. [CrossRef] [PubMed]
157. Reginaldo, G.M.d.S. Estudo de Endoparasitos e Ectoparasitos em Gatos Domésticos de área Urbana. 2023. Available online: https://bdtd.ibict.br/vufind/Record/UNSP_9b30ad67d01ee02e2198d0e9f7e66920 (accessed on 15 January 2024).
158. Kurnosova, O.P.; Arisov, M.V.; Odoyevskaya, I.M. Intestinal Parasites of Pets and Other House-kept Animals in Moscow. *Helminthologia* **2019**, *56*, 108–117. [CrossRef]

159. Blasco, X.; Salas, A.; Manuelian, C.; Torre, C.; Ortuño, A. Intestinal parasitic infection in multi-cat shelters in Catalonia. *Isr. J. Vet. Med.* **2017**, *72*, 16–21.
160. Schuster, R.K.; Kaufmann, A.; Hering, S. Investigations on the endoparasite fauna of the domestic cat in eastern Brandenburg [Untersuchungen zur endoparasitenfauna der hauskatze in Ostbrandenburg]. *Berl. Munch. Tierarztl. Wochenschr.* **1997**, *110*, 48–50.
161. Vrhovec, M.G.; Alnassan, A.A.; Pantchev, N.; Bauer, C. Is there any change in the prevalence of intestinal or cardiopulmonary parasite infections in companion animals (dogs and cats) in Germany between 2004–2006 and 2015–2017? An assessment of the impact of the first ESCCAP guidelines. *Vet. Parasitol.* **2022**, *312*, 109836. [CrossRef] [PubMed]
162. Anh, N.T.; Thuy, D.T.; Hoan, D.H.; Hop, N.T.; Dung, D.T. Levels of *Toxocara* infections in dogs and cats from urban Vietnam together with associated risk factors for transmission. *J. Helminthol.* **2016**, *90*, 508–510. [CrossRef] [PubMed]
163. Mohd Zain, S.N.; Sahimin, N.; Pal, P.; Lewis, J.W. Macroparasite communities in stray cat populations from urban cities in Peninsular Malaysia. *Vet. Parasitol.* **2013**, *196*, 469–477. [CrossRef] [PubMed]
164. Alves, M.E.M.; Martins, F.D.C.; Bräunig, P.; Pivoto, F.L.; Sangioni, L.A.; Vogel, F.S.F. Molecular detection of *Cryptosporidium* spp. and the occurrence of intestinal parasites in fecal samples of naturally infected dogs and cats. *Parasitol. Res.* **2018**, *117*, 3033–3038. [CrossRef] [PubMed]
165. Azimian, H.; Shokrani, H.; Fallahi, S. Molecular evaluation of *Toxocara* species in stray cats using loop-mediated isothermal amplification (lamp) technique as a rapid, sensitive and simple screening assay. *Vet. Med. Sci.* **2021**, *7*, 647–653. [CrossRef] [PubMed]
166. Palmer, C.S.; Thompson, R.C.A.; Traub, R.J.; Rees, R.; Robertson, I.D. National study of the gastrointestinal parasites of dogs and cats in Australia. *Vet. Parasitol.* **2008**, *151*, 181–190. [CrossRef]
167. Hansen, A.; Skarbye, L.; Vinther, L.; Willesen, J.; Pipper, C.; Olsen, C.; Mejer, H. Occurrence and clinical significance of Aelurostrongylus abstrusus and other endoparasites in Danish cats. *Vet. Parasitol.* **2016**, *234*, 31–39. [CrossRef] [PubMed]
168. Diakou, A.; Sofroniou, D.; Di Cesare, A.; Kokkinos, P.; Traversa, D. Occurrence and zoonotic potential of endoparasites in cats of Cyprus and a new distribution area for *Troglostrongylus brevior*. *Parasitol. Res.* **2017**, *116*, 3429–3435. [CrossRef] [PubMed]
169. Traversa, D.; Morelli, S.; Cassini, R.; Crisi, P.E.; Russi, I.; Grillotti, E.; Manzocchi, S.; Simonato, G.; Beraldo, P.; Viglietti, A.; et al. Occurrence of canine and feline extra-intestinal nematodes in key endemic regions of Italy. *Acta Trop.* **2019**, *193*, 227–235. [CrossRef] [PubMed]
170. Pivoto, F.L.; Lopes, L.F.D.; Vogel, F.S.F.; Botton, S.d.A.; Sangioni, L.A. Ocorrência de parasitos gastrointestinais e fatores de risco de parasitismo em gatos domésticos urbanos de Santa Maria, RS, Brasil. *Ciênc. Rural* **2013**, *43*, 1453–1458. [CrossRef]
171. Nutter, F.B.; Dubey, J.P.; Levine, J.F.; Breitschwerdt, E.B.; Ford, R.B.; Stoskopf, M.K. Seroprevalences of antibodies against *Bartonella henselae* and *Toxoplasma gondii* and fecal shedding of *Cryptosporidium* spp., *Giardia* spp., and *Toxocara cati* in feral and pet domestic cats. *J. Am. Vet. Med. Assoc.* **2004**, *225*, 1394–1398. [CrossRef]
172. Beugnet, F.; Guillot, J.; Polack, B.; Chermette, R. Enquête sur le parasitisme digestif des chiens et des chats de particuliers de la région parisienne. *Rev. Méd. Vét.* **2000**, *151*, 443–446.
173. Ramos, D.G.; Scheremeta, R.G.; Oliveira, A.C.; Sinkoc, A.L.; Pacheco Rde, C. Survey of helminth parasites of cats from the metropolitan area of Cuiabá, Mato Grosso, Brazil. *Rev. Bras. Parasitol. Vet.* **2013**, *22*, 201–206. [CrossRef] [PubMed]
174. Coelho, W.M.D.; Amarante, A.F.T.d.; Soutello, R.V.G.d.; Meireles, M.V.; Bresciani, K.D.S. Ocorrência de parasitos gastrintestinais em amostras fecais de felinos no município de Andradina, São Paulo. *Rev. Bras. Parasitol. Vet.* **2009**, *18*, 46–49. [CrossRef] [PubMed]
175. Ragozo, A.M.A.; Muradian, V.; Silva, J.C.R.; Caravieri, R.; Amajoner, V.R.; Magnabosco, C.; Gennari, S.M. Ocorrência de parasitos gastrintestinais em fezes de gatos das cidades de São Paulo e Guarulhos. *Braz. J. Vet. Res. Anim. Sci.* **2002**, *39*, 244–246. [CrossRef]
176. Gennari, S.M.; Kasai, N.; Pena HF, D.J.; Cortez, A. Ocorrência de protozoários e helmintos em amostras de fezes de cães e gatos da cidade de São Paulo. *Braz. J. Vet. Res. Anim. Sci.* **1999**, *36*, 87–91. [CrossRef]
177. Arruda, I.F.; Ramos, R.C.F.; da Silva Barbosa, A.; Villar-Echarte, G.; Augusto, A.M.; Troccoli, F.; Amendoeira, M.R.R. Occurrence of Toxoplasma gondii and other gastrointestinal parasites in free-roaming cats from the Rio de Janeiro zoo. *Braz. J. Vet. Med.* **2023**, *45*, e006023. [CrossRef] [PubMed]
178. Leite, L.C. Ocorrência de endoparasitas com potencial zoonótico de transmissão em fezes de gatos (*Felis catus* domesticus Linnaeus, 1758) domiciliados na área urbana e região metropolitana de Castro–Paraná-Brasil The occurrence of endoparasites with zoonotic. *Ambiência* **2012**, *8*, 923–930. [CrossRef]
179. Gavioli, F.A.; Borsa, A.; Diogo, J.E.; de Lara Pinto, A.Z.; Azevedo, L.S.; Sousa, V.R.F. Ocorrência de endoparasitos em gatos de Cuiabá, Mato Grosso, Brasil. *Arch. Vet. Sci.* **2011**, *16*, 25–30. [CrossRef]
180. Marques, S.M.T.; Menetrier, L.D.C.; Meyer, J. Ocorrência de Nematódeos e Protozoários em Gatos com Tutores da Cidade de Porto Alegre, RS, Brasil. *Rev. Agrár. Acad.* **2020**, *3*, 89–99. [CrossRef]
181. Silva, C.M.d.A. Ocorrência de Parasitas Gastrintestinais de Gatos (*Felis catus*) Domiciliados nos Municípios de Patos-PB e Parelhas-RN. 2017. Available online: http://dspace.sti.ufcg.edu.br:8080/jspui/handle/riufcg/24347?mode=full (accessed on 1 January 2024).
182. Tesserolli, G.L.; Fayzano, L.; Agottani, J.V.B. Ocorrência de parasitas gastrintestinais em fezes de cães e gatos, Curitiba-PR. *Rev. Acad. Ciênc. Anim.* **2005**, *3*, 31–34. [CrossRef]
183. Federman, H.B.; de Holanda, J.C.; Evangelista, A. Ocorrência de Parasitos em Gatos (*Felis catus* Domesticu5) e Pombos (Cohimba livia) Procedentes de Algumas Localidades de Minas Gerais. *Rev. Patol. Trop./J. Trop. Pathol.* **1973**, *2*, 207–215.

184. Torrico, K.J.; Santos, K.R.; Martins, T.; Silva, P.E.; Flavio, M.; Takahira, R.K.; Lopes, R.S. Ocorrência de parasitas gastrintestinais em cães e gatos na rotina do laboratório de enfermidades parasitárias da FMVZ/Unesp-Botucatu, SP. *Rev. Bras. Parasitol. Vet. Braz. J. Vet. Parasitol.* **2008**, *17*, 182–183.
185. Garcia-Campos, A.; Power, C.; O'Shaughnessy, J.; Browne, C.; Lawlor, A.; McCarthy, G.; O'Neill, E.; de Waal, T. One-year parasitological screening of stray dogs and cats in County Dublin, Ireland. *Parasitology* **2018**, *146*, 746–752. [CrossRef]
186. Dos Santos, S.I.H. Parasitas Respiratórios, Gastrointestinais e Auriculares em Gatos de Colónia, na Casa dos Animais de Lisboa. DVM Thesis, Universidade de Lisboa, Lisboa, Portugal, 2020.
187. Waap, H.; Gomes, J.; Nunes, T. Parasite communities in stray cat populations from Lisbon, Portugal. *J. Helminthol.* **2014**, *88*, 389–395. [CrossRef] [PubMed]
188. Villeneuve, A.; Polley, L.; Jenkins, E.; Schurer, J.; Gilleard, J.; Kutz, S.; Conboy, G.; Benoit, D.; Seewald, W.; Gagné, F. Parasite prevalence in fecal samples from shelter dogs and cats across the Canadian provinces. *Parasit. Vectors* **2015**, *8*, 281. [CrossRef] [PubMed]
189. Yamaguchi, N.; Macdonald, D.W.; Passanisi, W.C.; Harbour, D.A.; Hopper, C.D. Parasite prevalence in free-ranging farm cats, Felis silvestris catus. *Epidemiol. Infect.* **1996**, *116*, 217–223. [CrossRef] [PubMed]
190. Wyrosdick, H.M.; Chapman, A.; Martinez, J.; Schaefer, J.J. Parasite prevalence survey in shelter cats in Citrus County, Florida. *Vet. Parasitol. Reg. Stud. Rep.* **2017**, *10*, 20–24. [CrossRef] [PubMed]
191. Fromont, E.; Morvilliers, L.; Artois, M.; Pontier, D. Parasite richness and abundance in insular and mainland feral cats: Insularity or density? *Parasitology* **2001**, *123*, 143–151. [CrossRef] [PubMed]
192. Joachim, A.; Auersperg, V.; Drüe, J.; Wiedermann, S.; Hinney, B.; Spergser, J. Parasites and zoonotic bacteria in the feces of cats and dogs from animal shelters in Carinthia, Austria. *Res. Vet. Sci.* **2023**, *164*, 105022. [CrossRef] [PubMed]
193. Beugnet, F.; Bourdeau, P.; Chalvet-Monfray, K.; Cozma, V.; Farkas, R.; Guillot, J.; Halos, L.; Joachim, A.; Losson, B.; Miró, G.; et al. Parasites of domestic owned cats in Europe: Co-infestations and risk factors. *Parasit. Vectors* **2014**, *7*, 291. [CrossRef] [PubMed]
194. Milstein, T.C.; Goldsmid, J.M. Parasites of feral cats from southern Tasmania and their potential significance. *Aust. Vet. J.* **1997**, *75*, 218–219. [CrossRef] [PubMed]
195. Krecek, R.C.; Moura, L.; Lucas, H.; Kelly, P. Parasites of stray cats (*Felis domesticus* L., 1758) on St. Kitts, West Indies. *Vet. Parasitol.* **2010**, *172*, 147–149. [CrossRef] [PubMed]
196. Capári, B.; Hamel, D.; Visser, M.; Winter, R.; Pfister, K.; Rehbein, S. Parasitic infections of domestic cats, *Felis catus*, in western Hungary. *Vet. Parasitol.* **2013**, *192*, 33–42. [CrossRef] [PubMed]
197. Barrientos, M.; Uchôa, C.; Alonso, C. Exame parasitológico de fezes de gatos (*Felis catus* domesticus) domiciliados e errantes da Região Metropolitana do Rio de Janeiro, Brasil. *Rev. Soc. Bras. Med. Trop.* **2003**, *36*, 331–334. [CrossRef]
198. Duarte, A.; Castro, I.; Pereira da Fonseca, I.M.; Almeida, V.; Madeira de Carvalho, L.M.; Meireles, J.; Fazendeiro, M.I.; Tavares, L.; Vaz, Y. Survey of infectious and parasitic diseases in stray cats at the Lisbon Metropolitan Area, Portugal. *J. Feline Med. Surg.* **2010**, *12*, 441–446. [CrossRef]
199. Adhikari, R.B.; Dhakal, M.A.; Ale, P.B.; Regmi, G.R.; Ghimire, T.R. Survey on the prevalence of intestinal parasites in domestic cats (*Felis catus* Linnaeus, 1758) in central Nepal. *Vet. Med. Sci.* **2023**, *9*, 559–571. [CrossRef]
200. Lins, S.B.H. Parasitos de Interesse Zoonótico em Felinos (*Felis catus* domesticus), Campo Grande, Mato Grosso do Sul. 2016. Available online: https://bdtd.ibict.br/vufind/Record/UFMS_35b9946e6aa98af60e0b8551ff3e1323 (accessed on 20 January 2024).
201. Souza-Dantas, L.M.d.; Bastos, O.P.M.; Brener, B.; Salomão, M.; Guerrero, J.; Labarthe, N.V. Técnica de centrífugo-flutuação com sulfato de zinco no diagnóstico de helmintos gastrintestinais de gatos domésticos. *Ciênc. Rural* **2007**, *37*, 904–906. [CrossRef]
202. Soe, B.K.; Hlaing, K.S.; Naing, T.W.; Thaw, Z.H.; Myint, A.W. The first study on the prevalence of gastrointestinal parasites in owned and sheltered cats in Yangon, Myanmar. *Vet. World* **2023**, *16*, 414–420. [CrossRef] [PubMed]
203. Iller, M.; Sokół, R. Ocena zależności pasożytów wewnętrznych psów i kotów od przygotowania hodowlano-weterynaryjnego właścicieli. *Wiadomości Parazytol.* **2008**, *54*, 245–247.
204. Rolim, C.; Aguiar, A.; Lima, L.; Souza, M.; Duarte, V.; Dantas, I.; Batista, A.; Guimarães, R.; Medeiros, F.; Coelho, W.; et al. Parasitos Gastrintestinais em *Felis catus* Linnaeus, 1758 de Mossoró, RN. *Multiplicidade Ciênc. Saúde* **2023**, *4*, 60–74. [CrossRef]
205. Ferraz, A.; dos Santos Pires, B.; Barwaldt, E.T.; dos Santos, E.M.; de Lima, C.M.; de Castro, T.A.; de Oliveira Nobre, M.; Nizoli, L.Q. Parasitos gastrintestinais em fezes de gatos domiciliados no município de Pelotas, RS, Brasil. *Vet. Notícias* **2021**, *27*, 52–67. [CrossRef]
206. Marques, S.M.T.; Oliveira, M.R.F.d.; Mattos, M.J.T.d. Parasitos gastrintestinais em gatos da cidade de Porto Alegre, Rio Grande do Sul. *Pubvet. Londrina* **2017**, *11*, 1132–1137. [CrossRef]
207. Ribeiro Berenguer, L.K.A.; Cavalcanti de Andrade Gomes, C.F.; Florêncio dos Santos, J.; Bianque Oliveira, J. Parasitos Gastrointestinais de Caninos e Felinos: UMA Questão de Saúde Pública. *Arch. Vet. Sci.* **2021**, *26*, 90–104. [CrossRef]
208. Sarmiento-Rubiano, L.A.; Delgado, L.; Ruiz, J.P.; Sarmiento, M.C.; Becerra, J. Parásitos intestinales en perros y gatos con dueño de la ciudad de Barranquilla, Colombia. *Rev. Investig. Vet. Perú* **2018**, *29*, 1403–1410. [CrossRef]
209. Aguilar Pérez, F.C. Parásitos Zoonóticos Presentes en Gatos Domésticos (Felis Silvestris Catus) en un Centro de Control Canino y Felino, en Nuevo León; México. DVM Thesis, Universidad Autónoma de Nuevo León, San Nicolás de los Garza, Mexico, 2021.
210. De Sousa Melo, A.C.M. Parasitoses Gastrointestinais E Pulmonares Em Canídeos E Felídeos Da Região Oeste De Portugal Continental. DVM Thesis, Universidade de Lisboa, Lisboa, Portugal, 2017.

211. Matos, B.d.M. Parasitoses pulmonares e gastrointestinais em felinos domésticos no Minho, Portugal. DVM Thesis, Universidade de Lisboa, Faculdade de Medicina Veterinária, Lisboa, Portugal, 2016.
212. Ponce, A.L.M. Parasitosis Intestinales en Gatos de Querétaro. 2023. Available online: https://ri-ng.uaq.mx/handle/123456789/4354 (accessed on 15 January 2024).
213. Noé Mocetti, N.; Ulloa, F.; Peña, P.; Santos, D.; Fernández, C.; Anchante, H.; Terashima, A.; Chávez, A.; Falcón, N. Parasitosis zoonóticas en mascotas caninas y felinas de niños de educación primaria del cono norte de Lima, Perú. Una Salud. *Rev. Sapuvet Salud Pública* **2011**, *2*, 15–24.
214. Öge, S.; Oge, H.; Gönenç, B.; Özbakış-Beceriklisoy, G.; Yildiz, C. Presence of *Toxocara* eggs on the hair of dogs and cats. *Ank. Univ. Vet. Fak. Derg.* **2013**, *60*, 171–176. [CrossRef]
215. Rodríguez Gallegos, L.M.; Romero Núñez, C.; Bautista Gómez, L.G.; Martínez Castañeda, J.S.; Heredia Cardenas, R. Presence of *Toxocara* spp. in Domestic Cats in the State of Mexico. *Acta Sci. Vet.* **2016**, *44*, 5. [CrossRef]
216. Peña-Quistial, M.G.; Benavides-Montaño, J.A.; Duque, N.J.R.; Benavides-Montaño, G.A. Prevalence and associated risk factors of Intestinal parasites in rural high-mountain communities of the Valle del Cauca-Colombia. *PLoS Negl. Trop. Dis.* **2020**, *14*, e0008734. [CrossRef]
217. El-Seify, M.A.; Marey, N.M.; Satour, N.; Elhawary, N.M.; Sultan, K. Prevalence and Molecular Characterization of *Toxocara cati* Infection in Feral Cats in Alexandria City, Northern Egypt. *Iran. J. Parasitol.* **2021**, *16*, 270–278. [CrossRef] [PubMed]
218. Karimi, P.; Shafaghi-Sisi, S.; Meamar, A.R.; Nasiri, G.; Razmjou, E. Prevalence and Molecular Characterization of Toxoplasma gondii and *Toxocara cati* Among Stray and Household Cats and Cat Owners in Tehran, Iran. *Front. Vet. Sci.* **2022**, *9*, 927185. [CrossRef] [PubMed]
219. Mateo, M.; Montoya, A.; Bailo, B.; Köster, P.C.; Dashti, A.; Hernández-Castro, C.; Saugar, J.M.; Matas, P.; Xiao, L.; Carmena, D. Prevalence and public health relevance of enteric parasites in domestic dogs and cats in the region of Madrid (Spain) with an emphasis on *Giardia duodenalis* and *Cryptosporidium* sp. *Vet. Med. Sci.* **2023**, *9*, 2542–2558. [CrossRef]
220. Genchi, M.; Vismarra, A.; Zanet, S.; Morelli, S.; Galuppi, R.; Cringoli, G.; Lia, R.; Diaferia, M.; Frangipane di Regalbono, A.; Venegoni, G.; et al. Prevalence and risk factors associated with cat parasites in Italy: A multicenter study. *Parasit. Vectors* **2021**, *14*, 475. [CrossRef] [PubMed]
221. Campos, D.R.; Oliveira, L.C.; de Siqueira, D.F.; Perin, L.R.; Campos, N.C.; Aptekmann, K.P.; Martins, I.V. Prevalence and risk factors associated with endoparasitosis of dogs and cats in Espírito Santo, Brazil. *Acta Parasitol.* **2016**, *61*, 544–548. [CrossRef] [PubMed]
222. Yang, Y.; Liang, H. Prevalence and Risk Factors of Intestinal Parasites in Cats from China. *BioMed Res. Int.* **2015**, *2015*, 967238. [CrossRef] [PubMed]
223. Miró, G.; Montoya, A.; Jiménez, S.; Frisuelos, C.; Mateo, M.; Fuentes, I. Prevalence of antibodies to *Toxoplasma gondii* and intestinal parasites in stray, farm and household cats in Spain. *Vet. Parasitol.* **2004**, *126*, 249–255. [CrossRef] [PubMed]
224. Rambozzi, L.; Menzano, A.; Mannelli, A.; Romano, S.; Isaia, M.C. Prevalence of cryptosporidian infection in cats in Turin and analysis of risk factors. *J. Feline Med. Surg.* **2007**, *9*, 392–396. [CrossRef]
225. Mircean, V.; Titilincu, A.; Vasile, C. Prevalence of endoparasites in household cat (*Felis catus*) populations from Transylvania (Romania) and association with risk factors. *Vet. Parasitol.* **2010**, *171*, 163–166. [CrossRef] [PubMed]
226. Loftin, C.M.; Donnett, U.B.; Schneider, L.G.; Varela-Stokes, A.S. Prevalence of endoparasites in northern Mississippi shelter cats. *Vet. Parasitol. Reg. Stud. Rep.* **2019**, *18*, 100322. [CrossRef] [PubMed]
227. Becker, A.C.; Rohen, M.; Epe, C.; Schnieder, T. Prevalence of endoparasites in stray and fostered dogs and cats in Northern Germany. *Parasitol. Res.* **2012**, *111*, 849–857. [CrossRef] [PubMed]
228. Jitsamai, W.; Khrutkham, N.; Hunprasit, V.; Chandrashekar, R.; Bowman, D.; Sukhumavasi, W. Prevalence of endoparasitic and viral infections in client-owned cats in metropolitan Bangkok, Thailand, and the risk factors associated with feline hookworm infections. *Vet. Parasitol. Reg. Stud. Rep.* **2021**, *25*, 100584. [CrossRef] [PubMed]
229. Spain, C.V.; Scarlett, J.M.; Wade, S.E.; McDonough, P. Prevalence of enteric zoonotic agents in cats less than 1 year old in central New York State. *J. Vet. Intern. Med.* **2001**, *15*, 33–38. [CrossRef]
230. Hill, S.L.; Cheney, J.M.; Taton-Allen, G.F.; Reif, J.S.; Bruns, C.; Lappin, M.R. Prevalence of enteric zoonotic organisms in cats. *J. Am. Vet. Med. Assoc.* **2000**, *216*, 687–692. [CrossRef]
231. Spada, E.; Proverbio, D.; Della Pepa, A.; Domenichini, G.; Bagnagatti De Giorgi, G.; Traldi, G.; Ferro, E. Prevalence of faecal-borne parasites in colony stray cats in northern Italy. *J. Feline Med. Surg.* **2013**, *15*, 672–677. [CrossRef]
232. Lucio-Forster, A.; Bowman, D.D. Prevalence of fecal-borne parasites detected by centrifugal flotation in feline samples from two shelters in upstate New York. *J. Feline Med. Surg.* **2011**, *13*, 300–303. [CrossRef] [PubMed]
233. Cantó, G.J.; Guerrero, R.I.; Olvera-Ramírez, A.M.; Milián, F.; Mosqueda, J.; Aguilar-Tipacamú, G. Prevalence of fleas and gastrointestinal parasites in free-roaming cats in central Mexico. *PLoS ONE* **2013**, *8*, e60744. [CrossRef] [PubMed]
234. Vincy, P.; Tresamol, P.V. Prevalence of gastro-intestinal and haemoparasitic infections among domestic cats of Kerala. *J. Parasit. Dis.* **2023**, *47*, 562–565. [CrossRef] [PubMed]
235. Pumidonming, W.; Salman, D.; Gronsang, D.; Abdelbaset, A.E.; Sangkaeo, K.; Kawazu, S.I.; Igarashi, M. Prevalence of gastrointestinal helminth parasites of zoonotic significance in dogs and cats in lower Northern Thailand. *J. Vet. Med. Sci.* **2017**, *78*, 1779–1784. [CrossRef] [PubMed]

236. Ramírez-Barrios, R.A.; Fernández, G.; Valera, Z.; Acosta, G.; Parra, O.; Barboza, G.A. Prevalencia de helmintos gastrointestinales en gatos admitidos en la policlínica veterinaria de la Universidad del Zulia. *Rev. Científica* **2008**, *18*, 374–380.
237. Šmigová, J.; Papajová, I.; Šoltys, J.; Pipiková, J.; Šmiga, Ľ.; Šnábel, V.; Takáčová, J.; Takáč, L. The occurence of endoparasites in Slovakian household dogs and cats. *Vet. Res. Commun.* **2021**, *45*, 243–249. [CrossRef] [PubMed]
238. Schuster, R.K.; Thomas, K.; Sivakumar, S.; O'Donovan, D. The parasite fauna of stray domestic cats (*Felis catus*) in Dubai, United Arab Emirates. *Parasitol. Res.* **2009**, *105*, 125–134. [CrossRef] [PubMed]
239. Wierzbowska, I.A.; Kornaś, S.; Piontek, A.M.; Rola, K. The Prevalence of Endoparasites of Free Ranging Cats (*Felis catus*) from Urban Habitats in Southern Poland. *Animals* **2020**, *10*, 748. [CrossRef] [PubMed]
240. Silva, Y.H.D.; Campos, D.R.; Lima, G.A.C.; Quintal, J.P.; Guimarães, B.G.; Rêgo, G.; Avelar, B.R.; Intrieri, J.M.; Correia, T.R.; Scott, F.B. Prevalence of gastrointestinal parasites in domestic cats (*Felis catus*) diagnosed by different coproparasitological techniques in the municipality of Seropédica, Rio de Janeiro. *Rev. Bras. Parasitol. Vet.* **2023**, *32*, e006223. [CrossRef] [PubMed]
241. Lemus-García, M.; Fimia-Duarte, R.; Iannacone, J.; Suarez-Fernández, Y.E. Prevalencia de parásitos gastrointestinales en gatos domésticos (Felis silvestris catus schreber, 1775) en La Habana, Cuba. *Paid. XXI* **2020**, *10*, 443–457. [CrossRef]
242. Grandi, G.; Comin, A.; Ibrahim, O.; Schaper, R.; Forshell, U.; Lind, E. Prevalence of helminth and coccidian parasites in Swedish outdoor cats and the first report of *Aelurostrongylus abstrusus* in Sweden: A coprological investigation. *Acta Vet. Scand.* **2017**, *59*, 19. [CrossRef]
243. Sedionoto, B.; Anamnart, W. Prevalence of hookworm infection and strongyloidiasis in cats and potential risk factor of human diseases. *E3S Web Conf.* **2018**, *31*, 06002. [CrossRef]
244. Al-Obaidi, Q. Prevalence of Internal Helminthes in Stray Cats (*Felis catus*) in Mosul City, Mosul-Iraq. *J. Anim. Vet. Adv.* **2012**, *11*, 2732–2736. [CrossRef]
245. De Souza, F.B.; Nakiri, I.M.; Lourenço, N.O.; da Silva, G.G.; Paschoalini, D.R.; Guimarães-Okamoto, P.T.C.; Melchert, A. Prevalence of Intestinal Endoparasites With Zoonotic Potential in Domestic Cats From Botucatu, SP, Brazil. *Top. Companion Anim. Med.* **2017**, *32*, 114–117. [CrossRef] [PubMed]
246. Overgaauw, P. Prevalence of intestinal nematodes of dogs and cats in The Netherlands. *Vet. Q.* **1997**, *19*, 14–17. [CrossRef] [PubMed]
247. Bartosik, J.; Łojek, J.; PuchaŁA, M.; Karabowicz, J.; Górski, P.; Długosz, E.; Zygner, W. Prevalence of intestinal parasites detected in routine coproscopic methods in dogs and cats from the Masovian voivodeship in 2012–2015. *Med. Weter.* **2019**, *75*, 293–297. [CrossRef]
248. Ito, Y.; Itoh, N.; Kimura, Y.; Kanai, K. Prevalence of intestinal parasites in breeding cattery cats in Japan. *J. Feline Med. Surg.* **2016**, *18*, 834–837. [CrossRef] [PubMed]
249. Korkmaz, U.F.; Gökpınar, S.; Yıldız, K. Prevalence of Intestinal Parasites in Cats and Their Importance in Terms of Public Health. *Turk. Parazitol. Derg.* **2016**, *40*, 194–198. [CrossRef] [PubMed]
250. Blagburn, B.; Schenker, R.; Gagne, F.; Drake, J. Prevalence of Intestinal Parasites in Companion Animals in Ontario and Quebec, Canada, during the Winter Months. *Vet. Ther. Res. Appl. Vet. Med.* **2008**, *9*, 169–175.
251. Faraguna, S.; Vlahek, I.; Miočić, K.; Andreanszky, T.; Pećin, M. Prevalence of Intestinal Parasites in Dogs and Cats from the Kvarner Region in Croatia. *Acta Vet.* **2023**, *73*, 41–54. [CrossRef]
252. Lorenzini, G.; Tasca, T.; De Carli, G.A. Prevalence of intestinal parasites in dogs and cats under veterinary care in Porto Alegre, Rio Grande do Sul, Brazil. *Braz. J. Vet. Res. Anim. Sci.* **2007**, *44*, 137–145. [CrossRef]
253. Nichol, S.; Ball, S.J.; Snow, K.R. Prevalence of intestinal parasites in feral cats in some urban areas of England. *Vet. Parasitol.* **1981**, *9*, 107–110. [CrossRef] [PubMed]
254. Itoh, N.; Ito, Y.; Kato, A.; Kanai, K.; Chikazawa, S.; Hori, Y.; Hoshi, F.; Higuchi, S. Prevalence of intestinal parasites in pet shop kittens in Japan. *J. Feline Med. Surg.* **2013**, *15*, 908–910. [CrossRef]
255. Itoh, N.; Ikegami, H.; Takagi, M.; Ito, Y.; Kanai, K.; Chikazawa, S.; Hori, Y.; Hoshi, F.; Higuchi, S. Prevalence of intestinal parasites in private-household cats in Japan. *J. Feline Med. Surg.* **2012**, *14*, 436–439. [CrossRef] [PubMed]
256. Gracenea, M.; Gómez, M.; Torres, J. Prevalence of intestinal parasites in shelter dogs and cats in the metropolitan area of Barcelona (Spain). *Acta Parasitol.* **2009**, *54*, 73–77. [CrossRef]
257. Souza, J.B.B.; Silva, Z.M.A.; Alves-Ribeiro, B.S.; Moraes, I.S.; Alves-Sobrinho, A.V.; Saturnino, K.C.; Ferraz, H.T.; Machado, M.R.F.; Braga, Í.A.; Ramos, D.G.S. Prevalence of Intestinal Parasites, Risk Factors and Zoonotic Aspects in Dog and Cat Populations from Goiás, Brazil. *Vet. Sci.* **2023**, *10*, 492. [CrossRef] [PubMed]
258. Esmaeilzadeh, M.; Shamsfard, M.; Kazemi, A.; Khalafi, S.A.; Altome, S.A. Prevalence of Protozoa and Gastrointestinal Helminthes in Stray Cats in Zanjan Province, North-West of Iran. *Iran. J. Parasitol.* **2009**, *4*, 71–75.
259. Queen, E.V.; Marks, S.L.; Farver, T.B. Prevalence of selected bacterial and parasitic agents in feces from diarrheic and healthy control cats from Northern California. *J. Vet. Intern. Med.* **2012**, *26*, 54–60. [CrossRef] [PubMed]
260. Scorza, A.V.; Duncan, C.; Miles, L.; Lappin, M.R. Prevalence of selected zoonotic and vector-borne agents in dogs and cats in Costa Rica. *Vet. Parasitol.* **2011**, *183*, 178–183. [CrossRef] [PubMed]
261. Shaw, J.; Dunsmore, J.; Jakob-Hoff, R. Prevalence of some gastrointestinal parasites in cats in the Perth area. *Aust. Vet. J.* **1983**, *60*, 151–152. [CrossRef] [PubMed]
262. Luty, T. Prevalence of species of *Toxocara* in dogs, cats and red foxes from the Poznan region, Poland. *J. Helminthol.* **2001**, *75*, 153–156.

263. Zibaei, M.; Sadjjadi, S.M.; Sarkari, B. Prevalence of *Toxocara cati* and other intestinal helminths in stray cats in Shiraz, Iran. *Trop. Biomed.* **2007**, *24*, 39–43.
264. Shokri, A.; Asl, A.; Garedaghi, Y. Prevalence of *Toxocara cati* in pet cats and it's zoonotic importance in Tabriz city, Iran. *J. Zoonotic Dis.* **2020**, *4*, 62–67. [CrossRef]
265. Kurnosova, O.; Odoyevskaya, I.; Petkova, S.; Dilcheva, V. Prevalence of *Toxocara* infection in domestic dogs and cats in Urban environment. *Bull. Russ. State Med. Univ.* **2018**, *7*, 89–92. [CrossRef]
266. Malloy, W.F.; Embil, J.A. Prevalence of *Toxocara* spp. and other parasites in dogs and cats in Halifax, Nova Scotia. *Can. J. Comp. Med.* **1978**, *42*, 29–31. [PubMed]
267. Rezaiemanesh, M.R.; Afzalaghaee, M.; Hamidi, S.; Eshaghzadeh, A.; Paydar, M.; Hejazi, S.H. Prevalence of toxocariasis and its related risk factors in humans, dogs and cats in northeastern Iran: A population-based study. *Trans. R. Soc. Trop. Med. Hyg.* **2019**, *113*, 399–409. [CrossRef] [PubMed]
268. Nyambura Njuguna, A.; Kagira, J.M.; Muturi Karanja, S.; Ngotho, M.; Mutharia, L.; Wangari Maina, N. Prevalence of Toxoplasma gondii and Other Gastrointestinal Parasites in Domestic Cats from Households in Thika Region, Kenya. *BioMed Res. Int.* **2017**, *2017*, 7615810. [CrossRef] [PubMed]
269. Salman, D.; Pumidonming, W.; Oohashi, E.; Igarashi, M. Prevalence of Toxoplasma gondii and other intestinal parasites in cats in Tokachi sub-prefecture, Japan. *J. Vet. Med. Sci.* **2018**, *80*, 960–967. [CrossRef] [PubMed]
270. Karatepe, B.; Babür, C.; Karatepe, M.; Kiliç, S.; Dündar, B. Prevalence of Toxoplasma gondii antibodies and intestinal parasites in stray cats from Nigde, Turkey. *Ital. J. Anim. Sci.* **2008**, *7*, 113–118. [CrossRef]
271. Taetzsch, S.J.; Gruszynski, K.R.; Bertke, A.S.; Dubey, J.P.; Monti, K.A.; Zajac, A.M.; Lindsay, D.S. Prevalence of zoonotic parasites in feral cats of Central Virginia, USA. *Zoonoses Public Health* **2018**, *65*, 728–735. [CrossRef] [PubMed]
272. Hoggard, K.R.; Jarriel, D.M.; Bevelock, T.J.; Verocai, G.G. Prevalence survey of gastrointestinal and respiratory parasites of shelter cats in northeastern Georgia, USA. *Vet. Parasitol. Reg. Stud. Rep.* **2019**, *16*, 100270. [CrossRef] [PubMed]
273. Oh, Y.I.; Seo, K.W.; Kim, D.H.; Cheon, D.S. Prevalence, co-infection and seasonality of fecal enteropathogens from diarrheic cats in the Republic of Korea (2016–2019): A retrospective study. *BMC Vet. Res.* **2021**, *17*, 367. [CrossRef] [PubMed]
274. Campos, D.B.; Garibaldi, I.M.; Carneiro, J.R. Prevalência de helmintos em gatos (*Felis catus* domesticus) de Goiânia. *Rev. Patol. Trop./J. Trop. Pathol.* **1974**, *3*, 355–359.
275. Dueñas Peralta, R.A. Prevalencia de Infección Por Toxocara cati y Giardia Duodenalis en Gato Doméstico. 2018. Available online: https://repositorio.uap.edu.pe/handle/20.500.12990/2613 (accessed on 1 December 2023).
276. De, M.I.T.G.P. Prevalência De Parasitas Gastrointestinais e cardiorrespiratórios Em Gatos domésticos Na Área Metropolitana De Lisboa. DVM Thesis, Universidade de Lisboa, Lisboa, Portugal, 2020.
277. Monteiro, J.F.J. Prevalência de Parasitas Gastrointestinais em Felinos no Concelho de Vila Nova de Gaia. 2022. Available online: http://hdl.handle.net/20.500.11960/3104 (accessed on 15 January 2024).
278. Briones Silva, K.I. *Prevalencia de Parásitos Gastrointestinales en Gatos Domésticos (Felis catus) en la Parroquia La Matriz del Cantón Latacunga*; Universidad Técnica de Cotopaxi (UTC): Latacunga, Ecuador, 2019; Available online: http://repositorio.utc.edu.ec/handle/27000/6233 (accessed on 15 January 2024).
279. Pineda Cardona, J.M.; Roldan Duarte, O.M. Prevalencia de Parásitos Gastrointestinales en Muestras Coprológicas de Caninos y Felinos Remitidas al Laboratorio Ejelab, Risaralda. 2018. Available online: https://hdl.handle.net/11059/10126 (accessed on 15 January 2024).
280. Rezende, H.H.A. Prevalência de Parasitos Intestinais em Gatos Errantes em Goiânia–Goiás: Ênfase no Diagnóstico de Toxoplasma Gondii e Avaliação da Acurácia de Técnicas Parasitológicas. 2015. Available online: https://repositorio.bc.ufg.br/tede/items/5db3b9bc-c29c-418b-8124-01a4ca093893/full (accessed on 15 January 2024).
281. Fuentes Soto, E.A. Prevalencia de Parásitos Intestinales en Los Habitantes y Sus Mascotas en Los Barrios Hospital, San Lorenzo, Amanecer y San Antonio del Municipio de Amatitlán. 2014. Available online: http://www.repositorio.usac.edu.gt/id/eprint/1595 (accessed on 15 January 2024).
282. Gaguancela Padilla, M.B. Prevalencia de *Toxocara cati* en Felinos Domésticos (*Felis catus*) en el Sector la Venecia II del Distrito Metropolitano de Quito. DVM Thesis, Universidad Técnica de Cotopaxi (UTC), Latacunga, Ecuador, 2021.
283. Tapia Anrango, J.D. Prevalencia de *Toxocara cati* en Gatos Domésticos en el Sector de Balerio Estacio, de la Ciudad de Guayaquil. DVM Thesis, Facultad de Medicina Veterinaria y Zootecnia, Universidad de Guayaquil, Guayaquil, Ecuador, 2018.
284. Porqueddu, M.; Scala, A.; Tilocca, V. Principal endoparasitoses of domestic cats in Sardinia. *Vet. Res. Commun.* **2004**, *28* (Suppl. S1), 311–313. [CrossRef] [PubMed]
285. Teixeira, R.P.C. Rastreio de Parasitas Gastrointestinais e Pulmonares em Canídeos e Felídeos da Região Autónoma dos Açores, Ilhas de São Miguel e Terceira. DVM Thesis, Universidade de Lisboa, Lisboa, Portugal, 2020.
286. Carvalho, I.T. Rastreio de Parasitas Gastrointestinais e Pulmonares em Gatos de Gatis nos Distritos de Lisboa e Setúbal, Portugal. DVM Thesis, Universidade de Lisboa, Lisboa, Portugal, 2017.
287. Coati, N.; Hellmann, K.; Mencke, N.; Epe, C. Recent investigation on the prevalence of gastrointestinal nematodes in cats from France and Germany. *Parasitol. Res.* **2003**, *90* (Suppl. 3), S146–S147. [CrossRef] [PubMed]
288. Barutzki, D.; Schaper, R. Results of parasitological examinations of faecal samples from cats and dogs in Germany between 2003 and 2010. *Parasitol. Res.* **2011**, *109* (Suppl. 1), S45–S60. [CrossRef] [PubMed]

289. Nagamori, Y.; Payton, M.E.; Looper, E.; Apple, H.; Johnson, E.M. Retrospective survey of parasitism identified in feces of client-owned cats in North America from 2007 through 2018. *Vet. Parasitol.* **2020**, *277*, 109008. [CrossRef] [PubMed]
290. Iturbe Cossío, T.L.; Montes Luna, A.D.; Ruiz Mejia, M.; Flores Ortega, A.; Heredia Cárdenas, R.; Romero Núñez, C. Risk factors associated with cat parasites in a feline medical center. *JFMS Open Rep.* **2021**, *7*, 20551169211033183. [CrossRef] [PubMed]
291. Dubinský, P.; Havasiová-Reiterová, K.; Petko, B.; Hovorka, I.; Tomasovicová, O. Role of small mammals in the epidemiology of toxocariasis. *Parasitology* **1995**, *110 Pt 2*, 187–193. [CrossRef] [PubMed]
292. Meloni, B.P.; Thompson, R.C.; Hopkins, R.M.; Reynoldson, J.A.; Gracey, M. The prevalence of Giardia and other intestinal parasites in children, dogs and cats from aboriginal communities in the Kimberley. *Med. J. Aust.* **1993**, *158*, 157–159. [CrossRef] [PubMed]
293. McColm, A.A.; Hutchison, W.M. The prevalence of intestinal helminths in stray cats in central Scotland. *J. Helminthol.* **1980**, *54*, 255–257. [CrossRef] [PubMed]
294. Wright, I.; Stafford, K.; Coles, G. The prevalence of intestinal nematodes in cats and dogs from Lancashire, north-west England. *J. Small Anim. Pract.* **2016**, *57*, 393–395. [CrossRef] [PubMed]
295. Joffe, D.; Van Niekerk, D.; Gagné, F.; Gilleard, J.; Kutz, S.; Lobingier, R. The prevalence of intestinal parasites in dogs and cats in Calgary, Alberta. *Can. Vet. J.* **2011**, *52*, 1323–1328.
296. Moskvina, T.; Atopkin, D. The prevalence of intestinal parasites of domestic cats and dogs in Vladivostok, Russia during 2014–2017. *Zool. Ecol.* **2018**, *28*, 180–184. [CrossRef]
297. Kurnosova, O.P.; Panova, O.A.; Arisov, M.V. The prevalence of potentially zoonotic intestinal parasites in dogs and cats in Moscow, Russia. *Helminthologia* **2023**, *60*, 44–51. [CrossRef] [PubMed]
298. Martínez-Barbabosa, I.; Vázquez Tsuji, O.; Cabello, R.R.; Cárdenas, E.M.; Chasin, O.A. The prevalence of *Toxocara cati* in domestic cats in Mexico City. *Vet. Parasitol.* **2003**, *114*, 43–49. [CrossRef] [PubMed]
299. Ketzis, J.K.; Shell, L.; Chinault, S.; Pemberton, C.; Pereira, M.M. The prevalence of Trichuris spp. infection in indoor and outdoor cats on St. *Kitts. J. Infect. Dev. Ctries.* **2015**, *9*, 111–113. [CrossRef] [PubMed]
300. Phoosangwalthong, P.; Luong, N.; Wongwigkan, J.; Kamyingkird, K.; Phasuk, J.; Pattanatanang, K.; Thammasonthijarern, N.; Kengradomkij, C.; Chimnoi, W.; Odermatt, P.; et al. *Toxocara canis* and *Toxocara cati* in Stray Dogs and Cats in Bangkok, Thailand: Molecular Prevalence and Risk Factors. *Parasitologia* **2022**, *2*, 88–94. [CrossRef]
301. Ursache, A.L.; Györke, A.; Mircean, V.; Dumitrache, M.O.; Codea, A.R.; Cozma, V. *Toxocara cati* and Other Parasitic Enteropathogens: More Commonly Found in Owned Cats with Gastrointestinal Signs Than in Clinically Healthy Ones. *Pathogens* **2021**, *10*, 198. [CrossRef] [PubMed]
302. Sharif, M.; Nasrolahei, M.; Ziapour, S.P.; Gholami, S.; Ziaei, H.; Daryani, A.; Khalilian, A. *Toxocara cati* infections in stray cats in northern Iran. *J. Helminthol.* **2007**, *81*, 63–66. [CrossRef] [PubMed]
303. Pourshahbazi, G.; Khanahmad, H.; Khadivi, R.; Yousefi, H.A.; Mobarakeh, S.; Boldaji, F.H.; Darani, H.Y. *Toxocara* Infection in Dogs and Cats in Isfahan Province of Iran in 2021. *Adv. Biomed. Res.* **2023**, *12*, 201. [CrossRef]
304. Mikaeili, F.; Mirhendi, H.; Hosseini, M.; Asgari, Q.; Kia, E.B. *Toxocara* nematodes in stray cats from shiraz, southern iran: Intensity of infection and molecular identification of the isolates. *Iran. J. Parasitol.* **2013**, *8*, 593–600. [PubMed]
305. Furtado Jost, R.; Müller, N.; Marreros, N.; Moré, G.; Antoine, L.; Basso, W.; Frey, C.F. What is the role of Swiss domestic cats in environmental contamination with Echinococcus multilocularis eggs? *Parasit. Vectors* **2023**, *16*, 353. [CrossRef] [PubMed]
306. Piekara-Stępińska, A.; Piekarska, J.; Gorczykowski, M. *Cryptosporidium* spp. in dogs and cats in Poland. *Ann. Agric. Environ. Med.* **2021**, *28*, 345–347. [CrossRef] [PubMed]
307. Rabbani, I.A.; Mareta, F.J.; Kusnoto; Hastutiek, P.; Lastuti, N.D.R.; Mufasirin; Suharsono; Sardjana, I.K.W.; Sukmanadi, M.; Suwanti, L.T. Zoonotic and other gastrointestinal parasites in cats in Lumajang, East Java, Indonesia. *Infect. Dis. Rep.* **2020**, *12*, 8747. [CrossRef] [PubMed]
308. Fang, F.; Li, J.; Huang, T.; Guillot, J.; Huang, W. Zoonotic helminths parasites in the digestive tract of feral dogs and cats in Guangxi, China. *BMC Vet. Res.* **2015**, *11*, 211. [CrossRef] [PubMed]
309. Overgaauw, P.A.; van Zutphen, L.; Hoek, D.; Yaya, F.O.; Roelfsema, J.; Pinelli, E.; van Knapen, F.; Kortbeek, L.M. Zoonotic parasites in fecal samples and fur from dogs and cats in The Netherlands. *Vet. Parasitol.* **2009**, *163*, 115–122. [CrossRef] [PubMed]
310. Panova, O.A.; Glamazdin, I.G.; Spiridonov, S.E. Animal Toxocariasis in a Megalopolis: Epidemic Aspects. *Med. Parazitol.* **2015**, *3*, 39–41.
311. Nijsse, R.; Ploeger, H.W.; Wagenaar, J.A.; Mughini-Gras, L. Prevalence and risk factors for patent *Toxocara* infections in cats and cat owners' attitude towards deworming. *Parasitol. Res.* **2016**, *115*, 4519–4525. [CrossRef] [PubMed]
312. Delgado, O.; Rodríguez-Morales, A.J. Aspectos clínico-epidemiológicos de la toxocariasis: Una enfermedad desatendida en Venezuela y América Latina. *Boletín Malariol. Salud Ambient.* **2009**, *49*, 1–33.
313. Henke, K.; Ntovas, S.; Xourgia, E.; Exadaktylos, A.K.; Klukowska-Rötzler, J.; Ziaka, M. Who Let the Dogs Out? Unmasking the Neglected: A Semi-Systematic Review on the Enduring Impact of Toxocariasis, a Prevalent Zoonotic Infection. *Int. J. Environ. Res. Public Health* **2023**, *20*, 6972. [CrossRef]
314. Bhangale, G.N.; Narladkar, B.W.; Tayde, R.S. Systematic review and meta-analysis on the risk potential of zoonotic toxocariasis from soil contamination of public places in India. *Vet. Parasitol. Reg. Stud. Rep.* **2021**, *24*, 100560. [CrossRef] [PubMed]
315. Fisher, M. *Toxocara cati*: An underestimated zoonotic agent. *Trends Parasitol.* **2003**, *19*, 167–170. [CrossRef] [PubMed]

316. Fukae, J.; Kawanabe, T.; Akao, N.; Kado, M.; Tokoro, M.; Yokoyama, K.; Hattori, N. Longitudinal myelitis caused by visceral larva migrans associated with *Toxocara cati* infection: Case report. *Clin. Neurol. Neurosurg.* **2012**, *114*, 1091–1094. [CrossRef] [PubMed]
317. Sakai, R.; Kawashima, H.; Shibui, H.; Kamata, K.; Kambara, C.; Matsuoka, H. *Toxocara cati*-induced ocular Toxocariasis. *Arch. Ophthalmol.* **1998**, *116*, 1686–1687. [PubMed]
318. Eberhard, M.L.; Alfano, E. Adult *Toxocara cati* infections in U.S. Child. Report Four Cases. *Am. J. Trop. Med. Hyg.* **1998**, *59*, 404–406. [CrossRef] [PubMed]
319. Rodan, K.S.; Buckley, J.J. Infection with adult *Toxocara cati*. *Br. Med. J.* **1969**, *2*, 188. [CrossRef] [PubMed]
320. Rodriguez-Morales, A.J.; Bonilla-Aldana, D.K.; Gallego-Valencia, V.; Gómez-DeLaRosa, S.H.; López-Echeverri, C.; Peña-Verjan, N.M.; Vargas-Díaz, K.; Ramírez, A.; Diaz-Henao, W.; Murillo-García, D.R.; et al. Toxocariasis in Colombia: More Than Neglected. *Curr. Trop. Med. Rep.* **2020**, *7*, 17–24. [CrossRef]
321. Bolivar-Mejia, A.; Alarcón-Olave, C.; Calvo-Betancourt, L.S.; Paniz-Mondolfi, A.; Delgado, O.; Rodriguez-Morales, A.J. Toxocariasis in the Americas: Burden and Disease Control. *Curr. Trop. Med. Rep.* **2014**, *1*, 62–68. [CrossRef]
322. Poulsen, C.S.; Skov, S.; Yoshida, A.; Skallerup, P.; Maruyama, H.; Thamsborg, S.M.; Nejsum, P. Differential serodiagnostics of *Toxocara canis* and *Toxocara cati*—Is it possible? *Parasite Immunol.* **2015**, *37*, 204–207. [CrossRef] [PubMed]
323. Holland, C.V. A walk on the wild side: A review of the epidemiology of *Toxocara canis* and *Toxocara cati* in wild hosts. *Int. J. Parasitol. Parasites Wildl.* **2023**, *22*, 216–228. [CrossRef]
324. Ketzis, J.K.; Lucio-Forster, A. *Toxocara canis* and *Toxocara cati* in domestic dogs and cats in the United States, Mexico, Central America and the Caribbean: A review. *Adv. Parasitol.* **2020**, *109*, 655–714. [CrossRef] [PubMed]
325. Del Pozo, A.A.; Angulo-Cruzado, M.; Amenero-Vega, R.; Álvarez-Lulichac, A.; Fernández-Cosavalente, H.; Barboza-Meca, J.; Rodriguez-Morales, A.J. Hepatosplenic abscesses in an immunocompetent child with cat-scratch disease from Peru. *Ann. Clin. Microbiol. Antimicrob.* **2019**, *18*, 23. [CrossRef] [PubMed]
326. Ramírez-Ocampo, S.; Cotte-Alzate, J.D.; Escobedo, Á.A.; Rodríguez-Morales, A.J. Prevalence of zoonotic and non-zoonotic genotypes of Giardia intestinalis in cats: A systematic review and meta-analysis. *Infez. Med.* **2017**, *25*, 326–338. [PubMed]
327. Mherzi, N.; Lamchouri, F.; Toufik, H. Assessment of the effects of seasonal changes, urban discharges and leachates on the parasitological and bacteriological qualities of soil and water from Oued Larbaâ (North-eastern, Morocco). *Environ. Monit. Assess.* **2021**, *193*, 628. [CrossRef] [PubMed]
328. An-Nori, A.; El Fels, L.; Ezzariai, A.; El Hayani, B.; El Mejahed, K.; El Gharous, M.; Hafidi, M. Effectiveness of helminth egg reduction by solar drying and liming of sewage sludge. *Environ. Sci. Pollut. Res. Int.* **2021**, *28*, 14080–14091. [CrossRef] [PubMed]
329. Lukashev, A.N.; Ruzina, M.N.; Akhmadishina, L.V. *Toxocara* prevalence in dogs, cats and the environment in Russia. *Adv. Parasitol.* **2020**, *109*, 801–817. [CrossRef] [PubMed]
330. Bradbury, R.S.; Panicker, I.S. *Toxocara* seroprevalence in Canada-Climate, environment and culture. *Adv. Parasitol.* **2020**, *109*, 291–316. [CrossRef] [PubMed]
331. Martínez-Pulgarín, D.F.; Muñoz-Urbano, M.; Gomez-Suta, L.D.; Delgado, O.M.; Rodriguez-Morales, A.J. Ocular toxocariasis: New diagnostic and therapeutic perspectives. *Recent. Pat. Antiinfect. Drug Discov.* **2015**, *10*, 35–41. [CrossRef]
332. Bolívar-Mejía, A.; Rodríguez-Morales, A.J.; Paniz-Mondolfi, A.E.; Delgado, O. Cardiovascular manifestations of human toxocariasis. *Arch. Cardiol. Mex.* **2013**, *83*, 120–129. [CrossRef] [PubMed]
333. Delgado, O.M.; Rosas-Bustamante, J.; Ortegoza, J.; Duarte, E.; Coraspe, V.; Rivas, M.; Silva, S.; Rodríguez-Morales, A.J. Acute cases of toxocariasis classified by IGG antibodies avidity in Venezuela. *J. Egypt. Soc. Parasitol.* **2011**, *41*, 611–614. [PubMed]
334. Van De, N.; Trung, N.V.; Chai, J.Y. Molecular diagnosis of an ocular toxocariasis patient in Vietnam. *Korean J. Parasitol.* **2013**, *51*, 563–567. [CrossRef] [PubMed]
335. Xu, Y.; Zheng, W.B.; Li, H.Y.; Cai, L.; Zou, Y.; Xie, S.C.; Zhu, X.Q.; Elsheikha, H.M. RNA sequencing reveals dynamic expression of spleen lncRNAs and mRNAs in Beagle dogs infected by *Toxocara canis*. *Parasit. Vectors* **2022**, *15*, 279. [CrossRef] [PubMed]
336. Meng, X.; Xie, Y.; Gu, X.; Zheng, Y.; Liu, Y.; Li, Y.; Wang, L.; Zhou, X.; Zuo, Z.; Yang, G. Sequencing and analysis of the complete mitochondrial genome of dog roundworm *Toxocara canis* (Nematoda: Toxocaridae) from USA. *Mitochondrial DNA B Resour.* **2019**, *4*, 2999–3001. [CrossRef] [PubMed]
337. Mesa-Arango, J.A.; Olave-Velandia, A.M.; García-Montoya, G.M.; Isaza-Agudelo, J.P.; Jiménez-Ruiz, A.; Alzate, J.F. Evaluation of new *Toxocara canis* chimeric antigens as an alternative to conventional TES-Ag for anti-*Toxocara* antibodies detection. *Heliyon* **2022**, *8*, e11144. [CrossRef] [PubMed]
338. Yunus, M.H.; Tan Farrizam, S.N.; Abdul Karim, I.Z.; Noordin, R. A Lateral Flow Rapid Test for Human Toxocariasis Developed Using Three *Toxocara canis* Recombinant Antigens. *Am. J. Trop. Med. Hyg.* **2018**, *98*, 32–38. [CrossRef] [PubMed]
339. Varghese, A.; Raina, O.K.; Chandra, D.; Mirdha, B.R.; Kelawala, N.H.; Solanki, J.B.; Kumar, N.; Ravindran, R.; Arun, A.; Rialch, A.; et al. Sero-detection of *Toxocara canis* infection in human with T.canis recombinant arginine kinase, cathepsin L-1 and TES-26 antigens. *Acta Parasitol.* **2017**, *62*, 775–778. [CrossRef] [PubMed]
340. Eslahi, A.V.; Badri, M.; Khorshidi, A.; Majidiani, H.; Hooshmand, E.; Hosseini, H.; Taghipour, A.; Foroutan, M.; Pestehchian, N.; Firoozeh, F.; et al. Prevalence of Toxocara and Toxascaris infection among human and animals in Iran with meta-analysis approach. *BMC Infect. Dis.* **2020**, *20*, 20. [CrossRef] [PubMed]
341. Rostami, A.; Riahi, S.M.; Fallah Omrani, V.; Wang, T.; Hofmann, A.; Mirzapour, A.; Foroutan, M.; Fakhri, Y.; Macpherson, C.N.L. Robin B Gasser Global Prevalence Estimates of *Toxascaris leonina* Infection in Dogs and Cats. *Pathogens* **2020**, *9*, 503. [CrossRef] [PubMed]

342. Rodriguez-Morales, A.J.; Martinez-Pulgarin, D.F.; Muñoz-Urbano, M.; Gómez-Suta, D.; Sánchez-Duque, J.A.; Machado-Alba, J.E. Bibliometric Assessment of the Global Scientific Production of Nitazoxanide. *Cureus* **2017**, *9*, e1204. [CrossRef] [PubMed]
343. Failoc-Rojas, V.E.; Silva-Díaz, H.; Maguiña, J.L.; Rodriguez-Morales, A.J.; Díaz-Velez, C.; Apolaya-Segura, M.; Valladares-Garrido, M.J. Evidence-based indications for ivermectin in parasitic diseases: An integrated approach to context and challenges in Peru. *Parasite Epidemiol. Control* **2023**, *23*, e00320. [CrossRef] [PubMed]
344. Delgado, O.M.; Fernandez, G.; Silva, S.; Ramirez, O.; Romero, J.; Rodriguez-Morales, A.J. Preliminary evidence of nitazoxanide activity on *Toxocara canis* in a mouse model. *Int. J. Antimicrob. Agents* **2008**, *31*, 182–184. [CrossRef] [PubMed]

MDPI AG
Grosspeteranlage 5
4052 Basel
Switzerland
Tel.: +41 61 683 77 34

Animals Editorial Office
E-mail: animals@mdpi.com
www.mdpi.com/journal/animals

www.ingramcontent.com/pod-product-compliance
Lightning Source LLC
Chambersburg PA
CBHW061608210926
51850CB00009B/2258
9783725846115